DATE DUE

The Library Store   #47-0107

ACS SYMPOSIUM SERIES **586**

Immunoanalysis of Agrochemicals

Emerging Technologies

Judd O. Nelson, EDITOR
University of Maryland

Alexander E. Karu, EDITOR
University of California—Berkeley

Rosie B. Wong, EDITOR
American Cyanamid Agricultural Research Division

Developed from a symposium sponsored
by the Division of Agrochemicals
at the 207th National Meeting
of the American Chemical Society,
San Diego, California,
March 13–17, 1994

American Chemical Society, Washington, DC 1995

Library of Congress Cataloging-in-Publication Data

Immunoanalysis of agrochemicals: emerging technologies / Judd O. Nelson, Alexander E. Karu, Rosie B. Wong, editors.

    p.   cm.—(ACS symposium series, ISSN 0097–6156; 586)

"Developed from a symposium sponsored by the Division of Agrochemicals at the 207th National Meeting of the American Chemical Society, San Diego, California, March 13–17, 1994."

Includes bibliographical references and indexes.

ISBN 0–8412–3149–4

    1. Agricultural chemicals—Analysis—Congresses. 2. Immunoassy—Congresses. I. Nelson, Judd O. II. Karu, Alexander E. III. Wong, Rosie B. IV. American Chemical Society. Division of Agrochemicals. V. American Chemical Society. Meeting (207th: 1994: San Diego, Calif.) VI. Series.

RA1270.A4I46   1995
615.9′07—dc20   95–5968
                                                              CIP

This book is printed on acid-free, recycled paper.

Foreword

THE ACS SYMPOSIUM SERIES was first published in 1974 to provide a mechanism for publishing symposia quickly in book form. The purpose of this series is to publish comprehensive books developed from symposia, which are usually "snapshots in time" of the current research being done on a topic, plus some review material on the topic. For this reason, it is necessary that the papers be published as quickly as possible.

Before a symposium-based book is put under contract, the proposed table of contents is reviewed for appropriateness to the topic and for comprehensiveness of the collection. Some papers are excluded at this point, and others are added to round out the scope of the volume. In addition, a draft of each paper is peer-reviewed prior to final acceptance or rejection. This anonymous review process is supervised by the organizer(s) of the symposium, who become the editor(s) of the book. The authors then revise their papers according to the recommendations of both the reviewers and the editors, prepare camera-ready copy, and submit the final papers to the editors, who check that all necessary revisions have been made.

As a rule, only original research papers and original review papers are included in the volumes. Verbatim reproductions of previously published papers are not accepted.

M. Joan Comstock
Series Editor

Contents

Preface

NEW CONCEPTS AND TECHNOLOGICAL ADVANCES that will influence the design, versatility, and reliability of the next generation of assays for small toxic molecules were the primary focuses of the symposium upon which this book is based. Some symposium participants and authors not previously involved in environmental immunoanalysis are now discovering new ways of accomplishing goals in this field. Gaining new perspectives on issues that affect the use and acceptance of immunoassays and related methods was also our objective.

More new immunoassays are being reported worldwide every year. A few years ago, most studies using immunoassay were undertaken to validate a particular test. Researchers now are using immunoassay as a primary data-gathering method, especially when immunoassays can cost-effectively process numbers of samples that would be prohibitive with instrumental analysis. Immunoassays and related methods are also being put to new uses such as monitoring of manufacturing and remediation processes, and new pesticide discovery.

Technologies for small-molecule recognition, incorporating knowledge from molecular biology, physics, and chemistry, are advancing rapidly. Dramatic advances defining antibody structure have been augmented by powerful molecular biological methods that allow antibody genes to be cloned and expressed in bacteria. Synthetic combinatorial antibody libraries with diversity vastly greater than the mammalian repertoire offer the possibility of obtaining antibodies that would be difficult or impossible to derive by conventional immunization. With molecular modeling and in vitro mutagenesis it is now possible to engineer new properties into antibodies, enzymes, receptors, ion-channel subunits, and small recognition peptides. One of the potentially most significant advances is the demonstration that certain organic polymers can retain an "imprint" of a small molecule and specifically bind that compound in a detection method very similar to immunoassays.

New synthesis schemes and computational tools are contributing to the design of better haptens and competitor molecules. Quantitative structure–activity parameters, including properties such as electrostatic potential of small analytes, are being correlated with recognition by binding molecules. Although these techniques were first used to develop improved antibodies and immunoassays, they apply to other molecular recognition systems as well. Combinatorial chemistry enables diverse repertoires of antibodies and other recognition proteins to be screened for binding to large arrays of ligands and ligand mimics (mimotopes). This strategy has implications for the discovery of new pesticides, inhibitors, and drugs, as well as for antibody characterization and assay development.

Some of the distinctions between antibody-based and instrumental analytical methods are disappearing. Concepts from physics are paving the way for development of miniaturized multianalyte assays, automated instrument-based immunomethods, and a variety of sensor formats. Flow injection, fluorescence polarization, and assay techniques using liposomes and magnetic particles have increased throughput and made immunoassays more versatile. Notable advances in sensor technology include the theory and implementation of miniaturized multiantibody, multianalyte arrays and development of reusable sensors for repeated measurements. Carefully designed immunoaffinity methods will reduce the cost, complexity, and scale of residue recovery and sample cleanup and will increase reliability and sample throughput.

One session of the meeting was devoted to identifying ways to speed and simplify the evaluation of immunoassay methods to foster acceptance by regulatory agencies. The chapters in the last section of this volume present new industry and regulatory agency perspectives on appropriate roles for immunoassay and criteria for acceptance. These chapters include proposed quality standards for kit manufacture and a set of guidelines for the validation and use of immunoassays as stand-alone procedures or in conjunction with instrumental methods.

Throughout the symposium it was evident that considerable distance exists between the technologies that are being developed and those that are presently being validated and approved for regulatory purposes. The increase in practical use and validation of antibody-based and antibody-like small-molecule detection methods is encouraging. As more experience is gained with the present generation of assays, newer methods are likely to be accepted faster. Advanced techniques and formats will raise new validation and quality-assurance issues. However, they may also be more versatile and reproducible, and will eliminate problems inherent in some of the present assays. Our hope is that this collection of papers provides an overview of relevant state-of-the-art research, a glimpse of future directions, and a stimulus for more efficient validation of the current methods.

Acknowledgments

We thank the chapter authors and especially the peer reviewers for their thoroughness and cooperation, the Agrochemicals Division for financial support, Anne Wilson of ACS Books for shepherding us through the editorial process on schedule, and all of those who participated in making the symposium a source of new ideas and constructive solutions.

JUDD O. NELSON
University of Maryland
College Park, MD 20742

ALEXANDER E. KARU
University of California
Berkeley, CA 94720

ROSIE B. WONG
American Cyanamid Agricultural Research Division
Princeton, NJ 08543

November 15, 1994

Chapter 1

Impact of Emerging Technologies on Immunochemical Methods for Environmental Analysis

Bruce D. Hammock and Shirley J. Gee

Departments of Entomology and Environmental Toxicology,
University of California, Davis, CA 95616

There have been more reports concerning the application of
immunochemical technologies for the evaluation of environmental
contamination, food contamination, and the monitoring of
biomarkers of human exposure to environmental chemicals in the last
2 years than in the preceding twenty years. During this time classical
approaches have been employed to avoid confusion in process. With
the acceptance of immunochemical technology more innovative
concepts are now being applied for application to the environmental
field. More than simply screening of environmental samples, it is
likely that immunochemistry will be among the many hyphenated
technologies in the analytical field. Second, recent advances in
immunochemistry should be examined and advantages exploited to
solve environmental problems. As a true interdisciplinary field
immunochemistry incorporates advances in molecular modeling,
synthetic chemistry, antibody production, biosensor development,
data analysis and other areas. Technological development must be in
the context of regulatory and consumer acceptance, thus it is critical
that we maintain a dialog between developers and users of
immunoassays regarding the capabilities of the analytical methods
and the criteria for their acceptance by regulatory agencies.

In 1971 Ercegovich (1) collected a handful of papers on the use of immunoassay in
agriculture and environmental chemistry and discussed the possible application of
this technology. This study was followed a few years later by a more detailed
evaluation of the potential of immunochemistry by Hammock and Mumma (2). In
many ways both of these chapters were prophetic in that immunoassays now have
many applications in the environmental field. In the early days of environmental
immunoassay we largely transferred technologies from medicine and other fields to

0097–6156/95/0586–0001$12.00/0

our own. This trend continues as new developments in medical technologies are applied to agricultural and environmental chemistry. However, new technologies and ideas increasingly are being pioneered in the demanding environmental field, particularly as these tools reach the hands of classical analytical chemists. There is always a frustration among fundamental scientists to see how slowly a new technology is reduced to practice. The analytical community is justifiably skeptical of any new technology, not just immunoassay. Possibly the hesitancy of the analytical community to embrace new technology is good, since it is important to be able to compare analytical data among laboratories through out the world and to compare data generated through time. The focus of the immunoassay field for the last 15 years on the ELISA (enzyme linked immunosorbent assay) format has helped to introduce the technology by avoiding a complexity of terms and approaches and allowing the user to have a generally uniform set of equipment. It is important to quickly implement improved technologies, but not so quickly that confidence is lost in the validity of immunochemistry as an accurate and precise tool for economical trace analysis. The interaction of scientists and regulators from a variety of backgrounds as documented by this volume is the best approach to achieving this goal.

It is reassuring that the same technology outlined in classical texts like those of Williams and Chase (3) and Langone and Van Vunakis (4-7) can still be used to develop and perform successful immunoassays. However, many of the modern technologies discussed here can be applied immediately to make assay development easier and the resulting analytical tools more powerful. This text illustrates that the development of the next generation of immunoassays will be interdisciplinary as it draws on sophisticated technologies from many fields. Advances in each of these many fields often will impact several aspects of assay development. In this text edited by Judd Nelson, Alex Karu and Rosie Wong, we have an exciting glimpse of the many fronts along which this technology will evolve.

Binding Proteins

All immunoassays are competitive binding assays based on the law of mass action. The critical part of any immunoassay is the binding protein. It is important for those entering the field to realize that polyclonal antibody technology based on classical immunization protocols provide the antibodies for most of the commercial clinical and environmental assays. Nevertheless, a variety of other technologies promise to make the immunoassays of the future far superior to the ones that exist today.

Polyclonal Antibodies. Seldom are conferences held on the technology of producing and using polyclonal antibodies yet this is a technology that continues to develop. In environmental chemistry scientists are taking advantage of some of the subtle approaches to make the generation of a truly superior polyclonal antibody more routine. Many of the reagents involved in the production and use of polyclonal antibodies that previously were prepared by the experimentalist, now are commercially available, and the repertoire of such reagents continues to increase. Although rabbits and goats remain the mainstay for production of polyclonal antibodies, other species are being used with greater frequency when they offer special advantages. Bovine systems can produce the quantity of antibodies needed for immunoaffinity chromatography in a cost effective manner. Most antibodies are stable molecules, but many new approaches are being used to improve stability and to adapt them for use with solvents and complex sample matrices. As inexpensive and straight forward reagents that form the basis for of many highly specific and sensitive assays, polyclonal antibodies are unlikely to be replaced. Classical production of antibodies may even be integrated into more sophisticated recombinant technologies. As cloning procedures for antibodies from commonly used laboratory animals become available, polyclonal serum based assays could be developed first and then later the spleen of the immunized animal for cloning efforts. Finally, in polyclonal antisera several antibodies bind the analyte with different affinities and specificities. The individual antibodies differ in their susceptibility to interference by materials in the sample matrix. This is a potential advantage for analyte detection in difference matrices. In some cases the multiplicity of recognition systems of a polyclonal serum can actually give the analyst greater confidence in the resulting answer. It is likely that we will see this advantage of the polyclonal serum increasingly mimicked with more sophisticated reagents in which single binding proteins are either mixed or used in an array to provide multiple recognition sites for a single analyte.

In spite of many advantages there are numerous real and perceived limitations of polyclonal antibodies that are driving many new technologies discussed in this text. The ensemble of different affinities and specificities of antibodies in a serum changes with each boost and usually is different in each immunized animal. Commercially this problem is solved by using serum pools that are carefully characterized. Since the cost of antibody production is small relative to validation and characterization it is increasingly attractive to have a constant supply of identical antisera. Especially for biosensor applications, antibodies with clearly defined association and dissociation constants are desirable as are constant supplies of antibody fragments. Polyclonal antibodies raised against proteins usually react with many sites. This may prevent detection of subtle differences that are important for identifying a particular pathogen or recombinant protein. A variety of technologies including protein modeling now allow the preparation or isolation of peptides of interest from a complex protein. These can then be coupled to a protein or other polymer and used to raise polyclonal antibodies that are highly specific for a single epitope on a target molecule.

Monoclonal Antibodies. This exciting technology developed by Köhler and Milstein (*8*) has a well-established use in environmental chemistry. As environmental immunoassays are becoming more sophisticated, the high initial cost of developing monoclonal antibodies becomes less important compared to the subsequent expense of validating an assay and the comfort of having a conceptually immortal cell line that can provide an unlimited supply of uniform antibodies. Numerous laboratories now have the capability to prepare both mono and polyclonal antibodies for small molecules. Monoclonal antibodies are proving invaluable in the diagnosis of agricultural pathogens. The increasing proportion of monoclonal antibodies in environmental chemistry and their use as the basis for commercial kits for the detection of environmental chemicals is in part, a testament to the maturity of the field. Many superior monoclonal antibody based immunoassays were reported at the 1994 American Chemical Society Meeting.

Not every monoclonal antibody in a panel will yield a superior assay. Most laboratories immunize with several strains of mice, evaluate the polyclonal antibodies, and then screen for a collection of antibodies of the desired affinity and physical properties. For instance it may be desirable to have a very high affinity antibody for an assay, a low affinity antibody for an immunoaffinity column, an antibody with high off rate for a biosensor application, and an antibody resistant to solvent and matrix effects for a field assay. A careful screening strategy will give the investigator the monoclonal antibody that is desired. The development of a library of monoclonal antibodies for use in pattern recognition and other paradigms offers a very real hope for the development of pattern recognition systems colloquially referred to as environmental tasters. As agriculture moves toward the use of biotechnology in crop protection, the analysis of biopolymers is of increasing importance. The ability to select a monoclonal antibody for a single epitope is very attractive for the identification of pathogens and evaluation of protein products of biotechnology. Also in some of the recombinant approaches discussed below, the monoclonal antibody can provide a valuable starting point for the development of recombinant proteins. There are numerous other applications where monoclonal technology will prove essential for maximal success including the development of some biosensor formats.

Recombinant Antibodies. The excitement surrounding recombinant antibody technology rivals that of monoclonal technology over a decade ago. The potential rewards from the effort to obtain recombinant antibodies are great. The ability to produce antibodies at a fraction of the cost of the production of poly- or monoclonal systems is very attractive, especially when antibodies may be expressed in sufficient quantities to even aid in the accumulation of a specific chemical in the environment. Neither poly- nor monoclonal technologies are effective enough to produce commercial quantities of antibodies for low cost affinity chromatography and concentration systems. Utilizing the powerful tools of molecular biology, the ability to make chimeric systems so that a transduction system can be covalently associated with an antibody or to add or remove specific sites for use in biosensor development is a clear goal for the future. Advances in modeling should make it possible to modify the characteristics of the binding site in a rational way to adjust the sensitivity and specificity of the resulting assay. Many scientists in the field, however, did not anticipate the complexity of cloning, assembling and expressing antibody molecules. Considering the difficulty of the task, exceptional progress is being made although the technology is far from routine.

When considering the development of recombinant binding molecules, it is first important to decide if an antibody molecule or simply a binding protein is needed. There are very valid reasons for selecting the difficult goal of cloning a real antibody molecule. Among these reasons is its possible use in human or veterinary therapy, use of the sophisticated technology surrounding the antibody structure, integration of immune maturation into a cloning protocol, and retaining properties of a well-characterized monoclonal antibody in the recombinant system. However, the same methods used to manipulate and select antibody genes can also be applied to derive recombinant enzymes and other binding proteins. Figure 1 presents a decision tree that may be useful when examining recombinant proteins for use in competitive binding formats.

If recombinant antibodies are the target, there are also multiple decision points. Will the investigator use the affinity maturation system of the animal to obtain the antibody clone or use an *in vitro* system to produce the variability needed to select a truly superior antibody? The former route is attractive since cloning from a monoclonal cell line or the spleen of an immunized animal stacks the odds in favor of finding a clone with a sufficiently high affinity for use in environmental chemistry. In addition, cloning from an existing monoclonal cell line provides great power in the ability to model the antibody for which that gene codes. There are several systems now from which the heavy and light chain genes from a monoclonal cell line can be cloned. As discussed below under modeling and as illustrated in this text, this procedure alone is a powerful tool for the investigator. However, there are numerous problems including the fact that commonly used cloning procedures such as polymerase chain reaction (PCR) coupled with the apparent toxicity of antibody components in many systems often lead to numerous errors. Thus at least a limited screening system may be necessary even when cloning antibody genes from a monoclonal cell line.

The other alternative being pursued by several laboratories involves bypassing the animal entirely. In these approaches one or more of the hyper-variable regions is mutated in a semi-random way and then binding molecules are selected. The key to the technology is a panning procedure analogous to the way one would pan for gold. By having part of the antibody molecule on the surface of a phage or other organism, the organisms that have the desired binding properties can be selected. The same rules that we use in the design of haptens and in subsequent screening procedures to select good polyclonal sera or monoclonal cell lines apply here. However, in concept sequential selections with the same or even different haptens can be used to develop truly superior binding molecules. Technologies to mutate antibodies in a random or directed process or shuffle chains between panning operations may also improve the characteristics of the resulting recombinant proteins. One may be able to select binding proteins that would be impossible to generate in the whole animal. The same methods that are now being applied to humanize antibodies (loop grafting and changing of surface determinants) may also be applied to improve the performance or change properties (thermal and solvent

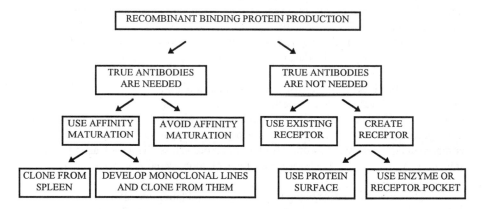

Figure 1.   Decision tree for recombinant binding protein production.

stability, handle recognition, etc.) of antibodies to be used in environmental analysis. The above approaches can also be used in combination. Panning procedures may be used to improve a cloned antibody from an animal or monoclonal cell line. Alternatively, site-directed mutagenesis or PCR procedures could be applied to change the properties of each.

At this time, the next logical step -- expression -- is difficult. In the recombinant field, expression is the actual production of the desired protein by a heterologous system. Work is currently moving along multiple fronts in the development of cloning and expression systems for antibodies. Not included in the decision tree is the multiplicity of possibilities for expression including expression of whole antibodies, single chain antibodies, or other partial fragments as well as another series of decisions regarding the possible need to screen the expression system and whether to use prokaryotic or eukaryotic systems. Different expression systems offer separate advantages. For large scale screening the phage systems in *E. coli* are very attractive. Humanized antibodies offer promise in the sequestration or detoxification of agricultural chemicals in exposed individuals. For the production of humanized antibodies or antibodies with many of the same characteristics of antibodies produced *in vivo*, eukaryotic expression systems offer many advantages. Here again there are many choices. The high integrity and high levels of production with the baculovirus system may be off set against the ease of scale up in a yeast or fungal system versus the use of plants to bioaccumulate a toxic waste on site.

Antibody-less Immunoassays. Several papers presented in this symposium demonstrate detection methods that use receptor molecules, enzymes, oligopeptides and even nonbiological molecules, based on the same principles as immunoassays.

Receptor Molecules. The most obvious alternative source of binding molecules are receptor proteins. For decades assays have been developed based on natural receptor molecules rather than antibodies. The affinity of these receptors may be superior to antibodies, and their specificity may even parallel a biological process. However, their use has been sporadic due to a limited number of receptors, difficulty in obtaining sufficient amounts of these usually rare proteins, and often due to the physical properties of the receptors that did not lend themselves to an assay format. These problems are being overcome by advances in molecular biology. Lipophilic receptors can be made soluble by removal of transmembrane domains or the addition of glycosylation sites. PCR and other techniques are leading to the isolation of whole families of receptors while expression systems allow many of these receptors to be produced more cheaply and in larger amounts than antibodies. Once a receptor is cloned and modeled, its binding site can be modified by site-directed mutagenesis to tailor its recognition to analytes other than the natural ligand.

Receptor molecules can be used directly in a format resembling a normal immunoassay. However, there is a difference between a receptor and an antibody. In the cell, the receptor is part of a biological transduction system that translates the binding response into a signal. This property will allow the development of the ultimate biosensors where an organism can be constructed which is sensitive to a specific agent due to the presence of a receptor molecule. The receptor system can be coupled to a molecular system providing an electrical response, cell proliferation or death, a color or luminescent event or various other readouts. Initially these systems will seem to be throwbacks to the old days of bioassays, and clearly they do have some disadvantages. However, they combine many of the advantages of a defined physical assay with a biological transduction system. They can be used in formats that would resemble an ELISA and of course can be more closely related to a biological effect. In fact on the continuum between what we would consider a classical physical and a classical biological based assay we would find immunoassays, isolated receptor assays, and cell based receptor assays closer to the classical physical assay while receptor based cell assays would be closer to bioassays.

Artificial Binding Proteins. As discussed under recombinant antibodies, several laboratories are taking the approach of generating a recombinatorial library by preparing a group of mutations in only one or several hyper-variable regions such as complementary determining region 3 (CDR3) of the heavy chain. It becomes obvious that in these systems the advantages of immune maturation are not used. In this situation one must pose the serious question -- is the antibody molecule only used for a framework? If the answer is yes then many small proteins are more stable and more easily handled by phage display systems. Probably the evolution of the humoral immune system was driven by the need to recognize foreign proteins. Thus the valleys on the surface of an antibody combining site are rather shallow. If a protein is wanted to mimic this situation then a simple globular protein may be sufficient. However if binding regions for a small molecule are needed, an enzyme or other protein with a deep active site may in fact be a better framework molecule. In concept mutants of the enzyme could be developed that would simultaneously recognize multiple faces of a single molecule resulting in a very high affinity. Thus, at the same time that we use hapten chemistry to drive the production of catalytic antibodies, we may see enzymes and other proteins modified to become binding proteins.

Nonbiological Polymers. An even greater leap may come as we develop nonprotein polymers that selectively bind certain molecules. It seems unlikely that the exceptionally high affinity seen with some of the above systems will be obtained in the near future. However, even very limited selectivity can prove to be a useful tool. The great power of reversed phase high performance liquid chromatography (HPLC) with very primitive hydrophobic ligands can give remarkable separations. These antibody-like polymers could be considered as highly selective stationary phases for chromatography, and they could bring another tremendous leap in the resolving power of a variety of concentration or chromatographic techniques.

Future materials research will likely lead to high affinity binding molecules of nonbiological origin.

Chelation Systems. As discussed in this text this method has been successfully applied to chelate systems for metal detection. Technically the metal assay reported in this book is not an immunoassay since antibodies are not used although it has the appearance of an ELISA type assay. Similar nonbiological and biological chelation systems can be applied to the detection of many metals. Similarly one can monitor DNA and RNA fragments in an ELISA type format by PCR amplification. It is likely that this convenient format introduced by ELISA will be used in many analytical applications extending beyond the use of antibodies.

Hapten Synthesis

The development of immunoassays to small molecules (haptens) involves the attachment of the small molecule (or a mimic of it) to a protein to raise antibodies and then the development of some system for detecting the antibody-analyte complex or free molecules. This can involve high specific activity radiosynthesis (classical radioimmunoassay) or the covalent attachment of the hapten to a reporter group that can range from an enzyme (ELISA) through a fluorophore. In general as the target analyte gets smaller, the care required in developing haptens increases. Large haptens generally have more than one reactive group. One must use selective reactions or these groups must be selectively blocked during hapten synthesis, and then de-protected to provide the best analyte mimic. New techniques are being developed to solve this problem. Strategies that direct the antibodies toward portions of the large molecule distal from the site of attachment alleviate problems of handle recognition.
     As the size of the target analyte decreases the chemistry of attachment to a protein for immunization has a greater contribution to the size and electronic properties of the molecule. Handle recognition can then occur where the hapten attached to the protein or reporter is recognized better than the analyte. This situation results in poor assay sensitivity and often undesirable specificity. This problem of hapten handle recognition often was ignored and key functional groups for recognition were blocked in the coupling process resulting in inferior assays. Now there is an appreciation that careful design of haptens, more than any other single factor, results in assays that are of superior quality. The concept that heterology is needed in the chemistry of linking the hapten to the protein, the handle between the hapten and the protein, and/or the position of attachment for small molecule immunoassays is now widely accepted. Several chapters in this volume provide excellent guidelines for the design of haptens and lead references to a literature rich in approaches for selective chemistry, including the development of heterobifunctional agents. An entire discipline of bioconjugate chemistry is developing, much of which is applicable to the design of haptens for immunoassay.
     As computer models of chemical structures become more accurate and user

friendly, we will see an increase in rational design of a series of haptens based on computer modeling, computer assisted design of synthetic pathways and the application of modern techniques in synthetic chemistry. These techniques among others will involve the use of solid phase chemistry, bioenzymatic synthesis, and sophisticated group selective chemistry.

Proteins such as bovine serum albumin, keyhole limpet hemocyanin and a small number of related proteins are currently used as carrier proteins for immunization. Immunochemists are now taking advantage of immunological and chemical properties of a large repertoire of commercially available proteins to facilitate exposure of key functional groups or improve immune response. Highly cross-linked proteins keep a defined structure for the carrier protein during immunization while a heavily glycosylated protein can provide both novel attachment chemistry and a hydrophilic and very antigenic surface for the protein. It is likely that for some haptens we will also see engineering of the protein to aid in the attachment of difficult functionalities. Haptens attached to defined polymers generally yield poor results, however, peptides attached to such polymers can yield monospecific polyclonal antibodies with high efficiency. This observation offers the possibility of synthesizing a hapten peptide conjugate either by classical chemistry or solid phase procedures followed by attachment of the resulting peptide to a polymer for immunization or for biosensor applications.

In the design of assays there is a general principle that the reporter hapten must bind to the antibody with less affinity than the analyte. However, this observation is determined intuitively. Scientists are now making more efforts to define the parameters of immunoassay so that the assays can be optimized and adapted to different uses based on mathematics rather than intuition. Chemistry has a key role since the basis to such calculations is the knowledge of how many haptens are covalently linked to an antigen or reporter molecule. For some molecules radioactivity or spectral properties such as UV or NMR can be used. However, these techniques are not general. The development of highly sophisticated mass spectral technologies including time of flight systems, fast atom bombardment and laser desorption technologies, and particularly electrospray offer systems that can determine hapten loading exactly. Such technologies will make the optimization and development of immunoassays less intuitive and more quantitative.

Structural Modeling of Antibodies, Analytes, and Haptens

Computer modeling will play an increasing role in the rational design of haptens, interpretation of the interaction of antibodies and analytes, and ultimately in the molecular engineering of antibody molecules. As the cost of computer systems drops, the power of processing and graphics systems increases, and modeling systems become both more reliable and user friendly, their use will increase dramatically. A major change over the last several years is that these trends have placed sophisticated modeling tools in the hands of synthetic chemists and immunologists. Rather than having to work through an expert in the computer system, the practicing scientist can use computers as tools to aid in creativity.

Computer simulations are all vague approximations of reality. However, computing power is making both empirically based and mechanistically based modeling systems more powerful. Immunologists long have used steric properties in hapten design. With the development of very powerful systems for evaluating the free energy parameters of substituents electronic and resonance features will increasingly be integrated into the design of optimal haptens. Systems for the visualization, energy minimization, and comparison of molecules are increasingly important in sophisticated hapten design as are expert systems to aid in planning the synthesis of haptens.

The modeling of proteins is far more complex and relies largely on empirically based systems. These systems also are becoming easier to use and more powerful. It is critical that these tools are used as an interactive tool for creativity rather than being viewed as giving a picture of how an antibody actually appears. Still more difficult but even more useful are programs that allow docking of analytes and haptens with computer generated antibody images. Properly constructed computer models of antibody combining sites are formulated from the high degree of sequence and conformational similarity in solved crystal structures. When used to approximate the relationship of antigen and hapten in the combining site, computer models are valuable for antibody engineering. The models provide insights to the critical binding interactions between antibody and analyte. The reality of these binding interactions can be tested by generating antibodies with specific amino acids altered and observing the changes in analyte and hapten recognition.

Our collective confidence in these models will increase and their value thus will expand rapidly as programs become more sophisticated and more reality checks are made. Such tests of models can be based on physical method such as circular dichroism, X-ray, and NMR analysis, site-directed mutagenesis, and competitive binding studies. In the short term these models will provide great insight into explaining assay specificity and designing optimal haptens. In the longer term models coupled with recombinant technologies will allow the tailored design of binding molecules for specific purposes.

Assay Formats

For the last 20 years the environmental field has simply adapted medical immunochemical technology. Yet in many respects environmental chemists have led the way in the implementation of nonisotopic immunoassays. Rosalyn Yalow points out that radioimmunoassay (RIA) has many advantages and the technology is well entrenched in clinical diagnostics. However, the added barrier that radioactive licenses and acquisition of detection equipment presented to an environmental chemistry laboratory led Ralph Mumma to encourage the ELISA format for immunoassay. At this time the ELISA format was widely considered to be a qualitative test of limited sensitivity. Scientists in environmental chemistry were among the first to demonstrate that ELISAs could be a highly reproducible format equaling or surpassing RIA in sensitivity. We also gained a format that could be adapted in many ways and could in fact yield field-based assays. There is no doubt that the ELISA format has introduced immunoassay into the environmental field and that it will continue to be a dominant technology. However, even the greatest advocates of the technology admit that it has too many steps and that many improvements can be made in assay design. Each competing technology has its advocates. However, it is important to recognize that most antibodies can be formatted in many ways. Once the immunochemical reagent has been obtained, the investigator is not limited to only one technology for its use. It is certain that we will see a variety of technologies in the future each with special applications in environmental chemistry.

The holy grail in immunodiagnostics is the biosensor where an antibody or other biological molecule acts as a receptor intimately tied to a physical system that acts as a transducer of the signal. With enzyme-based biosensors, 20 years passed between the demonstration of the biosensor concept and a commonly available analytical device. It should be realized that antibody based biosensors are far more complex than enzyme based biosensors since they have no associated catalytic event to aid in transduction. Also they do not release their ligand quickly, leading to slow response. Thus it is likely to be a long wait before we have antibody based biosensors in which a small probe can give a continuous readout of an analyte at low concentration. Antibody-based biosensors require advances in several technologies, including controlled delivery of the competing tracer, stabilization of the antibody (or binding protein), and better means of transducing the binding event into a signal. Antibody engineering methods may solve some of the problems of stability and reversibility of binding.

Many small improvements in an ELISA-like system will be introduced on the way to biosensors. Many of these technologies will synergize to yield superior assays. We undoubtedly will see many new amplification systems that will improve sensitivity, precision and/or result in fewer steps. Instrumentation with expanded wavelength ranges in the absorbance, fluorescence and luminescence modes allow the analyst to select from an increasing repertoire of endpoints. This is certain to lead to improved substrates and enzymes for the classical ELISA format. There is a great deal of interest in preparing recombinant antibodies. Of equal importance is the development of recombinant reporter enzymes that will have improved shelf life, stability in matrices and solvents, sites for attachment of haptens and biosensor components, and greater sensitivity. Fluorescent dyes attached to components of immunoassay systems or even polymers will play an increasing role as dye wavelengths and signal processing reduce the background fluorescence that has plagued many systems. One of the first pesticide immunoassays was based on fluorescence polarization, but sensitivity problems have plagued these and other homogenous systems. These dyes and other technologies will lead to fluorescence polarization and/or time resolved systems of greater sensitivity that will yield homogenous or nonseparation immunoassays. Running multiple parallel assays offers many advantages, but there are situations where the speed offered by a variety of flow injection based systems will gain in popularity for continuous monitoring. Autosamplers, in common use in analytical laboratories, will reduce the labor involved in these approaches. Another direction for immunoassays will be the development of more user-friendly field portable systems. Liposome based systems offer many advantages for such rapid methods. Findings along many fronts will aid in the development of on line immunoassays essential for wide spread use of the combined technologies described below.

The high sample through put of immunoassays offers a dramatic advantage over many classical procedures in that they are adaptable to robotic systems. Many classical procedures must be completely redesigned in order to use robotics systems efficiently. The early standardization on the 96 well format makes immunoassays and the solid phase concentration and clean up systems commonly used with them easy to adapt to a robotic system. It could be that the ELISA format will aid in the design of classical analytical procedures more amenable to the rapidly evolving laboratory robotic systems.

Many immunoassays will become much smaller. The ability to scale the size of an immunoassay using engineering principles will be facilitated as we develop more defined reagents and optimized assays based on physical parameters rather than by using trial and error coupled with intuition as largely is the case now. There are numerous drivers for this trend toward miniaturization. A reduction in the space dedicated to laboratory-based assays and improved mobility for field-based assays will be key factors. There also are other drivers including smaller sample size, faster assays, the ability to put many assays on a small card or probe, an improvement in signal to noise, and certainly an improvement in sensitivity. In solution, immunoassay sensitivity ultimately is limited by the affinity of the antibody regardless of the amplification system. Once this limit is reached, one must make smaller assays in order to obtain more sensitive assays. By analogy the quadrupole mass spectrometer revolutionized the analytical field by providing the first relatively inexpensive instrument that yielded high quality data. The early mass spectrometers did not provide good data, but when those data were collected and averaged, spectra were generated that rivaled the quality of the data obtained on very expensive magnetic sector instruments. With multiple miniaturized immunoassays that can be computer averaged, we could increase both the sensitivity and reliability of the signal. In addition, the use of multiple antibodies as discussed below will facilitate our confidence in the identification of the compound as well as its quantitation. Still other advantages of miniaturized assays are discussed in this text. As discussed below miniaturization coupled with chemometrics will facilitate multianalyte immunoassay by ELISA.

Sample Preparation and Tracking

In the past decade, with immunoassays one either attempted direct assays of simple matrices or used a pre-existing clean up system. Most of these clean up systems were designed for gas liquid chromatography (GLC) or HPLC applications, and it is rather simple to branch off at some point and exchange solvents into one compatible with immunoassay formats. This procedure has the advantage of using existing technologies and facilitating direct comparison of immunoassay with other technologies on the same sample extract. However, it is encouraging to see the development of concentration and clean up systems designed specifically to present the analyte to an immunoassay. These innovative technologies, including supercritical fluid extraction, a variety of solid phase extraction systems, the use of water-miscible solvents and even the judicious use of water based systems, are making immunoassays more user friendly and more powerful. They are reducing sample size and the use of solvents which present health hazards was well as disposal costs. Miniaturization in ELISA technologies is actually leading a trend in analytical chemistry toward miniaturized equipment that will include GLC, HPLC, and capillary electrophoresis. We should also see solid phase extraction systems and related sample handling systems integrated into a 96 well format to speed sample handling and processing with robotic systems. The standard format used with many ELISAs aids in the tracking of samples using X/Y coordinates. Such a spatial procedure for tracking both sequential and parallel samples could be applied advantageously to classical analysis. The high volume of samples that can be processed with ELISA undoubtedly will drive application of other sample tracking systems already used clinically such as bar codes and other recognition systems that could be also applied to other analytical technologies.

Hyphenated Technologies

Immunoassays often are thought of as technologies that compete with other physical assays. Certainly immunoassays can yield valuable qualitative and quantitative data as stand alone tools, but it is likely that we will see the assays coupled with other technologies for many applications. For instance immunoassays are excellent detector systems for HPLC. Standard HPLC systems allow very large samples of even crude matrices to be run where the HPLC system acts to concentrate as well as to purify the sample. The combined technologies allow the specificity of immunoassays to detect compounds when ultraviolet (UV) or other detector systems would lack the selectivity to detect a sample peak above a large background. A prior HPLC step can concentrate the sample as well as remove materials that could interfere with an immunoassay yielding improved sensitivity. Finally the combined technologies provide greatly improved confidence in the identification of the resulting compound since it is unlikely that a material interfering in an immunoassay would have the same retention time as an analyte. Microbore columns can reduce the size of peak-containing fractions from 10 to 100 fold. This technology seems made for immunochemical detection since these fractions can be detected directly in 96 well plates for analysis by ELISA. The solvent from microbore systems will not usually interfere with most immunoassays. The speed, small sample size and water miscible solvents used in capillary electrophoresis and the volatile solvents used in supercritical fluid chromatography both lend themselves well to immunoassay. Antibodies can be added directly to solvents in capillary electrophoresis to look for mobility shifts of an analyte as a confirmatory test. First we will see the development of off-line systems, but if these combined technologies prove useful, on-line systems will be sure to follow. As formats with fewer steps are developed integration of these on line detectors will become more simple.

Antibodies not only will appear as detector systems for chromatographs, but also can be used to concentrate and clean up samples for subsequent analysis by immunoassay, mass spectroscopy (MS) or other technologies. Antibodies that detect a compound and its metabolites or a class of compounds will be especially valuable in this regard. Several workers have shown that antibodies also can behave as highly selective reverse phase HPLC columns to separate even very closely related compounds since the immunoaffinity column is eluted with a gradient such as methanol or acetonitrile. Immunoassays have gained acceptance largely because they can reduce the major cost of environmental analysis which is sample processing. Clean up and processing also is the slowest, most complex and most error-prone phase of the assay process. For further advances to be made it will be critical to further improve sample handling. Immunoaffinity systems and the aqueous reaction system of immunoassays are certain to help reduce the problems associated with sample handling.

Acceptance and Quality Control

Having worked for over 20 years in environmental immunoassay it is frustrating to see how slowly the technology has been accepted in spite of clear demonstrations that it provides high quality data at a low price. However, in the last few years the rate of acceptance of the technology has been increasing dramatically. There are many factors involved in this positive process. Certainly a key factor is an industry association, the Analytical Environmental Immunochemical Consortium, that is setting a high standard of professionalism for those involved in this technology. The manufacturers of kits have also set a high standard for themselves that hopefully will be encouraged by regulatory agencies. The development of a process through the AOAC Research Institute to obtain independent evaluation of the performance of a kit will have an increasing impact. As we see practicing analytical chemists begin to employ these assays, the feed back should illustrate new applications of the technology as well as weak points in the technology than can be corrected by applying some of the approaches described in this text.

It is important that we examine the successes and some of the recent failures in clinical and veterinary diagnostics as we view the future of the environmental field. Immunochemistry is very widely used and well established in both of these fields. However we have a situation with many clinical immunoassays where epidemiologists have realized that the data are so variable that they have limited value in population studies. The laboratories claim that they are not at fault since they follow package inserts, the manufacturer claims Food and Drug Administration (FDA) approval, and FDA says that they are not responsible for operator error and quality control. In the veterinary field we see a still more complex situation where untrained users are expected to run immunoassays under field conditions for trace levels of drugs. In many cases the kits were not optimized for such work. The fact that the results of these legally mandated assays are often of poor quality may be taken as a failure of the technology rather than an inappropriate application. Both the successes and failures from other disciplines that have used immunoassays strongly encourage us to continue to have high standards of assay performance and rigorous procedures for validation.

The chemical industry will be a major user of immunochemical technology. By developing in house expertise in the technology rather than simply contracting the work to outside suppliers, the industry itself is assuring a high level of competence. The concept of supplying immunoassays as a component of product stewardship is very encouraging, and will likely expand the useful life of many pesticides by encouraging proper usage.

Computation and Validation of Results. The sensitivity, accuracy and precision that is expected in environmental chemistry will drive an increasing emphasis on the computation of immunoassay data. The data from most laboratory based assays are collected, stored and evaluated by microcomputers. Many of these microcomputers have great computational power only a fraction of which is being utilized. As outlined in this volume we will see increasingly sophisticated systems for the calculation of data as well as systems for the quality control and storage of analytical data hidden below user friendly interfaces. Even with powerful programs for the evaluation of data, most microcomputers associated with detector systems such as ELISA readers will lie idle most of the time. This computer power can be used to drive robotics systems to automate sample handling or artificial intelligence and expert systems to guide users through immunoassays. We will see increasingly powerful statistical packages and on line help imbedded transparently in the computer programs used for immunoassay.

Since the computer power needed to store and evaluate immunoassay data is under utilized the vast majority of the time, it is obvious that we can use this power as a teaching and quality control tool. Although immunoassays are generally very simple to perform a great deal of knowledge is required for trouble shooting. Manuals prepared in our laboratory for other users are often intimidating to them as we address numerous reasons for the rare failures of an assay. Suppliers of commercial assays are hesitant to even mention failures due to its negative impact on marketing. Whenever a user sees a massive manual on trouble shooting, immunochemical technology becomes intimidating. With an expert system on sample collecting and processing, assay performance, trouble shooting, quality control, data interpretation, resource allocation of ELISA wells among samples and standards, and archiving embedded in a transparent, user friendly interface one can more easily transfer immunoassay technology to other laboratories and insure a high degree of operator competence without relying on intimidating instruction manuals. Immunoassay technology may lead the way in more general application of expert systems to analytical chemistry thus allowing the analyst to bring increasingly sophisticated and diverse technologies to bare on problems with a high degree of competence.

Immunoassay is best as a single analyte system, but many immunoassay formats are being adapted to make use of different haptens and antibodies to identify, and in some cases estimate amounts of related compounds. Since the early days of RIA there has been the technology to use two assays to distinguish among two analytes by solving simultaneous equations. Statistical methods that have long been used to interpret instrumental data are now being applied to immunoassay. These methods give the analyst a means of quantifying how well the response is fitted by a mathematical or empirical model, and an estimate of the total error and its components. As multianalyte immunoassays gain acceptance, the analysis software will undoubtedly include more sophisticated statistical tools. In addition, the analyst will have to become better educated about the assumptions and potential errors inherent in these assays. Among the most sophisticated of these approaches will be the use of pattern recognition systems where an array of binding proteins are used to describe an analyte or a complex series of analytes both quantitatively and qualitatively.

Several chapters in this text illustrate that the development of an immunoassay is but one step in a process leading to validation and implementation of the technology. The development of an assay, however, is no longer an end in itself. We will see the assay compared with other technologies with real world samples and both technologies compared with standards in a blind fashion. Users of immunoassays are becoming more sophisticated in asking how much confidence they can place in their results. The low cost and speed of immunoassay compared to many classical technologies may allow a new standard to be set for evaluating environmental chemistry data resulting in our defining the source and magnitude of errors generated at each step from sample collection through the archiving of the data.

Use in Developing Countries

To date most immunoassay development and applications have been in developed countries. Immunoassays provide a powerful, sophisticated technology that can be brought to bear on complex environmental problems. However, much of the world lacks any adequate technology for the monitoring of chemicals in their food or their environment. Repeatedly international organizations have provided technical training and sophisticated analytical equipment to developing countries. Regardless of the dedication and sophistication of the analytical chemists involved, most of these efforts fail to provide the expected results because the technical infrastructure to keep complex machines operational is frequently lacking. As a result we often try to provide seemingly second class analytical technologies to developing countries in the name of appropriate technologies.

Immunochemistry offers solutions to these dilemmas. In contrast to many other analytical technologies, immunoassays are modern, sophisticated biotechnologies. In general the more sophisticated the immunoassay the easier the assay is to perform. In spite of this sophistication the assays depend upon the skill of the analyst and not the ability to maintain complex instrumentation, acquire a stable power supply, or obtain large amounts of high purity gases and solvents. Thus immunoassays can provide state of the art analytical chemistry to areas of the world where it is desperately needed to insure human and environmental health as well as to allow countries to export their food and fibers. Even the most sophisticated of the equipment needed for immunoassays is rugged and easily maintained while the more sophisticated assays require little equipment. Immediate use of immunoassay technology can be made internally in these countries. In many cases a common technology can be used to monitor pesticides for agriculture as well as for control of medical pests, disease organisms, and drugs. Thus the technology becomes attractive for joint development by agencies such as FAO, WHO, U.S. AID, and the World Bank. Of course if developing countries are to use such technology to verify the quality of products for export, immunochemical methods of analysis will have to be accepted in the developed countries as well. Similarly clear standards for the analysis must be established. A negative aspect is that immunochemical methods just like other analytical technologies can be used as trade barriers and will be used in this way if the technology is not available in exporting countries.

Many of technologies presented in this book certainly will have a positive impact on the use of immunochemistry in environmental analysis. Their synergistic application in the field is even more exciting, and the increasing lead that environmental chemists are taking in the development of these technologies is indicative of the health and vigor of the field.

Acknowledgments

This work was supported in part by NIEHS Superfund 2P42-ES04699, U.S. Environmental Protection Agency (EPA) Cooperative Research Grant CR819047, NIEHS Center for Environmental Health Sciences (at U.C. Davis) 1P30 ES05707; U.S. EPA. Center for Ecological Health Research at U.C. Davis CR819658 and U.S.D.A./P.S.W. 5-93-25 (Forest Service NAPIAP). Although the information in this document has been funded in part by the United States Environmental Protection Agency, it may not necessarily reflect the views of the Agency and no official endorsement should be inferred.

Literature Cited

(1) Ercegovich, C. D. In *Analysis of Pesticide Residues: Immunological Techniques*; American Chemical Society: Washington, D.C., 1971; pp. 162-177.

(2) Hammock, B.D., Mumma, R.O. In *Potential of Immunochemical Technology for Pesticide Analysis*; Harvey, J.; Zweig, G., Eds.; American Chemical Society; Washington, D.C., 1980; pp. 321-352.

(3) Williams, C.A.; Chase, M.W. *Methods in Immunology and Immunochemistry; Academic Press*: New York, NY, 1967.

(4) *Immunochemical Techniques, Part B*; Langone, J.J.; Van Vunakis, H., Eds.; Methods in Enzymology, Academic Press: New York, NY, 1981; Vol. 73.

(5) *Immunochemical Techniques, Part C*; Langone, J.J.; Van Vunakis, H., Eds.; Methods in Enzymology, Academic Press: New York, NY, 1981; Vol. 74.

(6) *Immunochemical Techniques, Part D;* Langone, J.J.; Van Vunakis, H., Eds.; Methods in Enzymology, Academic Press: New York, NY, 1982; Vol. 84.

(7) *Immunochemical Techniques, Part A*; Van Vunakis, H.; Langone, J.J.; Eds.; Methods in Enzymology, Academic Press: New York, NY, 1980; Vol. 70.

(8) Köhler, G.; Milstein, C. *Nature* **1975**, *256*, pp. 495-497.

RECEIVED October 13, 1994

RECOMBINANT ANTIBODIES AND ANTIBODY MIMICS

Chapter 2

Recombinant Antibodies Against Haptenic Mycotoxins

Heather A. Lee, Gary Wyatt, Stephen D. Garrett, Maria C. Yanguela, and Michael R. A. Morgan

Food Molecular Biochemistry Department, Institute of Food Research, Norwich Research Park, Colney, Norwich NR4 7UA, England

Recent developments in the molecular biology of antibody production have provoked much excitement among researchers interested in the generation of recombinant antibodies for immunodiagnostics. The speed of probe generation and the potential for manipulation of binding site properties would compare favourably with the difficulties (and cost) of monoclonal antibody production against haptens and other problematic targets. Unfortunately, translation of molecular biology methodology from other areas of research, in spite of the availability of kits, has not proved straightforward. This paper outlines some of the problems that have been encountered and some solutions employed in the production of short-chain Fvs to haptens. These targets present interesting problems to the immunochemist, particularly in situations where antibodies of broad specificity are sought.

Since early reports of the production of antibody fragments from cloned genes derived from lymphocytes and hybridomas (1), the technology of recombinant antibody production has developed rapidly and its potential has been recognized and welcomed by immunochemists in many areas. Several different cloning and expression systems have been developed, for example in yeast (2), insect cells (3), plants (4) and fungi (5), but most systems use *E.coli*. Many strategies exist for the production of different types of antibody fragments, heavy (VH) and light (VL) variable domains, single-chain variable domains (scFv) or Fab fragments, either directed towards the periplasm of *E.coli* (1) or displayed on the surface of filamentous phage particles. These phage-display systems are believed to be advantageous as they allow selection of clones expressing desirable antibody fragments from a large number of phage particles by affinity panning (6,7). Two systems have been developed: one expresses Fab fragments with the H and L

0097–6156/95/0586–0022$12.00/0

chain expressed separately (6), and the other expresses scFvs with the VH and VL chains linked by a synthetic polypeptide (7). The latter system is also available in a modified form as a kit from Pharmacia P-L Biochemicals Inc.

The use of recombinant antibodies (Rabs) in immunoassays has been eagerly awaited, but as yet has not materialised in a widespread manner. There have been a few reports of Rabs, derived mainly from hybridomas, which are capable of giving inhibition curves. These include Rabs to two proteins (8,9) and two haptens (10,11). Garrard et al. (8) measured the affinity of a phage-antibody (Fab) to the extracellular domain of the HER 2 receptor using iodine labelled protein and Hogrefe et al. (9) showed displacement of phage-antibodies (Fabs) to tetanus toxin in an ELISA with less than O.1nM protein. For the haptens, inhibition of binding in ELISAs was demonstrated using soluble recombinant antibodies (Fabs in both cases) to transition state analogues for the hydrolysis of amide bonds (10) and diuron (11).

Of the options available, we have chosen to use the phage-display system developed at the MRC Laboratory of Molecular Biology in Cambridge, U.K. (7). This paper will describe the progress made towards producing Rabs to two mycotoxins: diacetoxyscirpenol (DAS), a trichothecene, and aflatoxin M1 (AFM1). Mycotoxins, the secondary metabolite products of fungi, can contaminate food such as nuts and cereals via mould growth. Indirect contamination of animal tissue (through consumption of mycotoxins in animal feed) can also occur. Consequently, compounds such as the aflatoxins and the trichothecenes present a potential safety problem given their potent activities (12). Antibody-based diagnostics have been used widely to ensure the safety of both raw materials and finished products (13), particularly for aflatoxin B1. However, there remain analytical problems relating to specificity, sensitivity and speed of probe generation. If the full potential of the recombinant antibody approach is to be realised then the use of antibody libraries (rather than material from hybridomas) is essential. To date, the production of antibody domains of reasonable affinity has proved difficult when using combinatorial libraries; consequently we will highlight some of the difficulties we have experienced with the methods behind this new technology.

General Methods

Starting from the spleen cells of immunised Balb/c mice, (which gave good antibody titres in the tail bleeds) mRNA was extracted and cDNA prepared using commercially available kits. Antibody VH and VL genes were amplified by PCR from the mRNA:cDNA template using primers which match well to most of the mouse V gene families (14) and which do not contain restriction sites. The VH and VL genes (350 bp) were subsequently joined using a 100 bp DNA-linker fragment which codes for the polypeptide $(Gly_4Ser)_3$(15) by overlap PCR extension (16). Once assembled, restriction sites were added to the scFv gene fragment by further PCR with outer primers containing the *Not* I and *Sfi* I sites.

Three vectors were used in this work; pHEN 1 (7,14), pCANTAB and pCANTAB 5 E (the latter two are available commercially from Pharmacia Biotech). All three vectors are designed to produce the scFv as a gene 3 coat protein (g3p) fusion for display on the surface of the bacteriophage. In addition, pHEN 1 and pCANTAB 5 E also contain an amber mutation, allowing for the expression of soluble scFv antibodies which are directed to the periplasm in the *E. coli* non-suppressor strain HB2151 (17). The vector pCANTAB does not contain the amber mutation and can only be used to produce phage-antibodies. The immunoglobulin scFv libraries were cloned into one of these vectors; electroporated into *E.coli* and co-infected with helper phage. The antibodies were expressed fused to the g3p on the surface of the phage.

Phage expressing functional antibodies were selected by panning on hapten-conjugate-coated tissue culture flasks using the washing procedures discussed later. The bound phage were eluted by either triethylamine or hapten. After re-infection of *E.coli*, colonies containing the phagemid vector were selected and binding of expressed phage to the hapten-conjugate immobilised to ELISA plates was detected using an anti M13 horseradish peroxidase-labelled antibody. Enrichment by panning was repeated two or three times and then clones were selected for further characterisation.

Soluble antibodies were produced by transforming a non-suppressor strain (HB2151) with the selected plasmids. Colonies containing inserts were grown in culture and induced with IPTG (1mM) at 25^{0}C for 4-22h. The culture supernatant, periplasmic extract and whole cell extract were tested for antibody production by ELISA.

Specific problematic areas

Selection by panning. Following three rounds of panning, phage-antibodies (Ph-scFv) produced from spleen cells from a mouse immunised with DAS-bovine thyroglobulin (DAS-BTG) were selected for their binding in ELISA. The Ph-scFvs had been panned with DAS-BTG-coated tissue culture flasks and the same conjugate was used to coat the wells of the ELISA plates. Large batches of Ph-scFvs were prepared from overnight cultures (50ml) of transformed *E. coli*, co-infected with helper phage. The Ph-scFvs were precipitated with PEG/NaC1 and resuspended in 1 ml of Tris-EDTA buffer. Binding of these preparations to the 96 well plates was observed using an anti M13 horseradish peroxidase-labelled antibody down to a 1 : 500 dilution (18), but displacement of the Ph-scFvs from the plate with free DAS was very weak or non-existent. The specificity of the Ph-scFvs was investigated by measuring their binding to ELISA plates coated with other trichothecene-protein conjugates and to plates coated only with the carrier protein (Table I).

Table I.   **Binding of anti DAS Ph-scFvs. Ratio of binding to conjugate: binding to carrier protein.**

Conjugate	R36	R37	R44	R45
DAS-BTG	1.12	<u>1.57</u>	1.15	1.12
DON-BTG	1.29	1.47	1.28	1.39
T2-KLH	1.21	1.19	1.26	0.93
ADON-KLH	0.99	0.97	1.07	1.02

Ph-scFv at 1/50 dilution
Abbreviations :   DAS, diacetoxyscirpenol; DON, deoxynivalenol;
                       T2, T2 toxin; ADON, acetyldeoxynivalenol;
                       BTG, bovine thyroglobulin;   KLH,   keyhole limpet
                       haemocyanin.

There was considerable binding to the protein carriers alone, with a slight increase in binding to the DAS, DON and T2 conjugates for all four Ph-scFv preparations.

The polyspecific nature of these preparations lead us to investigate the panning procedure and in particular the washing and elution methods. Two wash procedures were compared (Table II); one employed multiple washes in PBS (10mM pH7.4) and PBST (PBS containing Tween 20,0.05%) (Method 1), and the other used fewer washes but gradually increased the pH with each wash (Method 2).

Table II. Comparison of two different wash procedures used during panning for Ph-scFv recombinant antibodies.

Method 1	Method 2
10 x 2 ml PBS	5 x 2 ml PBS
10 x 2 ml PBST	1 x 2 ml PBST
	1 x 2 ml Tris/NaCl pH 7.5
	1 x 2 ml Tris/NaCl pH 8.5
	1 x 2 ml Tris/NaCl pH 9.5
	1 x 2 ml NaHCO$_3$/NaCl pH 9.6
	(Elution with 3 ml PBS (15 min))

No. of colonies after re-infection with 10^{-1} dilution of eluted phage.

Method 1	Method 2
313	> > 1000

Far fewer phage were eluted non-specifically after the more extensive wash procedure of Method 1. Extending this procedure to a total of 60, 80 and 100 washes decreased the number of colonies obtained with non-specific elution to 23, 41 and 26 respectively. This background seems to be unavoidable, but at this low level quite acceptable.

Using the same library of Ph-scFvs to DAS-BTG, different elution procedures were investigated. The phage were eluted with PBS (3 ml), DAS (3 ml, 10 μg/ml) or triethylamine (3, ml 100mM) and mixed with Tris-HCl (IM, pH7.4, 0.5ml). All three elution procedures gave approximately the same number of colonies after re-infection of 200μl of TG1 cells with 100μl of phage at an equivalent dilution (1:10 or 1:100).

As the panning procedure using conjugate-coated flasks did not appear to be generating high affinity Ph-scFvs an alternative solid-phase was tested using DAS coupled directly to sepharose beads. The beads were washed with PBS (200 ml) and PBST (200 ml) then eluted with DAS solution (50 μg/ml). By comparison with a control gel (uncoupled sepharose) the number of colonies produced by re-infection with the eluted phage was increased about 15 fold for DAS-sepharose. However binding of selected eluted Ph-scFvs (after re-infection) showed the same low-affinity, polyreactivity as seen previously. Panning against DAS immobilised without any carrier protein had not improved the specificity of the Ph-scFvs.

Stability of phage-scFvs. From a library of scFv genes to aflatoxin M1, one phage-antibody (clone 7) was selected after only one round of panning with aflatoxin M1-bovine serum albumin (BSA) in tissue culture flasks. Following precipitation with PEG, binding of the Ph-scFv to an ELISA plate was demonstrated to a dilution of 1:50. On the subsequent day a certain amount of inhibition of binding of clone 7 with aflatoxin M1 was demonstrated. One ug/ml of AFMI gave 50% inhibition, although the standard deviation for each point of the standard curve was large (Figure 1a). When this was repeated the following day the binding had decreased to the background level obtained previously, and no inhibition was seen (Figure 1b). The instability of Ph-scFvs has been noted previously in our laboratory; storage in several different buffers has not improved the position, though PBST provides the best protection.

Soluble expression. Due to the unstable nature of Ph-scFvs and the difficulties encountered in displacement studies with free hapten, we decided to proceed to soluble expression of scFvs for characterisation of the antibody fragments after initial selection. Our initial attempts were with the pHEN 1 vector in *E.coli* HB2151 (18) and the DAS scFvs selected by three rounds of panning. No detectable scFvs were produced in the culture supernatant or the periplasm as detected by ELISA and immunoblotting using the anti C-terminal antibody 9E10 which recognises the c-myc tag (1,19). Subsequent work with AFM1 scFvs in pCANTAB 5 E has produced binding fragments in the periplasm which were detected by dot-blot and ELISA.

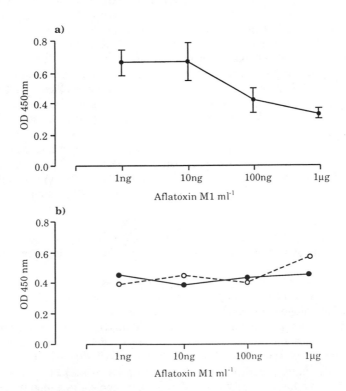

Figure 1. Inhibition of binding of a phage-scFv (clone 7) to aflatoxin M1-BSA by free aflatoxin M1 in an ELISA, a) Day 1 b) Day 2. Phage-scFv diluted $\frac{1}{20}$ (●) or $\frac{1}{40}$ (○).

Discussion

The panning procedures described here have not proved satisfactory for the production of high affinity, high specificity phage-scFvs from antibody gene libraries. Instead, generation of polyspecific antibody fragments which bind to a range of compounds was observed. The findings confirm problems experienced elsewhere, in that whereas specific scFvs can be accessed from hybridomas, much greater difficulty is experienced when tackling antibody gene libraries. A further frustration is that many results have been described in which affinity measurements for scFvs have been made for binding to immobilised hapten on solid phases. The immunochemist needs to see target recognition in solution. As with the production of antibodies to haptens by any method there is also the added problem of generating scFvs recognising the carrier protein or the linker group between the hapten and carrier. In order to overcome the problems we are now employing much more rigorous panning protocols. As well as antigen-coated tissue culture flasks, we shall use coated immuno-tubes (16) which are treated to improve their antigen-coating capacity; and streptavidin-coated beads with biotinylated antigen. If the hapten can be directly biotinylated this may prevent selection of antibody fragments recognising the carrier protein. In addition several different conjugates, made with different carriers and at different concentrations, will be used to increase the specificity of the panning procedure.

The underlying problem, however, would appear to be the difficulties of generating antibody fragments of reasonable affinity when selecting from libraries.

A second strategy we are employing is that of expression in filamentous fungi. Some species have been developed as hosts for secreted production of a wide variety of heterologous proteins (20). Fab fragments have, for example, been secreted from *Trichoderma reesei* at levels of 150 mg/l (3) as compared to levels of 0.2-10 mg/l generally found in *E.coli*. We have chosen to express the DNA encoding an scFv in *Aspergillus niger* using a gene fusion strategy which has been used previously in the production of heterologous proteins (21). In that work, the target gene was fused downstream of the gene encoding a highly secreted homologous protein and separated by a sequence encoding a Golgi - located endoproteolytic cleavage site. This approach ensures secretion of fully processed target protein.

Lastly, using a range of different hapten-conjugates for glyphosate and organophosphate pesticides, we propose to select scFvs from a large repertoire of phage antibodies made by transforming *E.coli* with a repertoire of heavy chains which are encoded on plasmids, and then infecting the same cells with a repertoire of light chains encoded on phage (22). The *lox* -Cre site specific recombination system of bacteriophage P1 has been employed to transfer the heavy chain genes from the plasmid into the phage so that the heavy and light genes can be expressed together. With a normal phage antibody repertoire produced (without immunisation) by random combinational linkage (6,16) the size of the repertoire, which is limited by the efficiency of transformation of *E.coli*, is 10^8. The range of antibody fragment affinities isolated from these libraries is 10^5 M^{-1}-10^7 M^{-1}, which are too low to be useful in diagnostic immunoassays. However with the new recombination library, the size of the library is dependent on infection of *E.coli* which is extremely efficient and theoretically could be as large as 10^{11} per litre. Used in combination with mutagenesis and chain-shuffling to facilitate affinity maturation, this much larger library will yield antibody fragments to a wider range of different antigens, with higher affinities.

The potential of recombinant antibody technology, including speed of probe generation and the ability to manipulate binding properties, will not be fully realised in our opinion unless selection from antibody libraries can be achieved. Much has been achieved with genetic material obtained from hybridomas, it seems important to us to access diversity in order to more closely mimic the immune system *in vivo*. Thus far it has proved difficult to isolate stable, high affinity antibody binding domains from libraries, in spite of much elegant molecular biology. Recent work has offered the possibility of generating larger antibody repertoires (22) than previously described. Immunochemists require rapid screening procedures, and to be able to select stable antibodies. Sufficient amounts of test sample must be available to perform preliminary characterisation before the need for scale-up and soluble expression. The challenge to molecular biologists will be to provide reproducible systems generally applicable to the needs of the user communities. Sufficient research has been reported to whet the appetite (application of recombinant antibodies in immunofluorescence studies (23) and in neutralisation of biological activity (24), for example) and expectations are high.

Literature cited.

1. Ward, E.S.; Gussow, D.; Griffiths, A.D.; Jones, P.T.; Winter G. *Nature*, 1989 *341*, 544-546.
2. Horwitz, A.H. *Proc. Natl. Acad. Sci.* USA, 1988, *85*, 8678-8682.
3. Hasemann, C.A.; Capra, J.D. *Proc. Natl. Acad. Sci. USA*, 1990, *87*, 3942-3946.
4. Hiatt, A,; Cafferty, R.; Bowdish, K. *Nature*, 1989, *342*, 76-78.
5. Nyyssönen, E.; Penttilä, M.; Harkki, A.; Saloheimo, A.; Knowles, J.K.C.; Keränen, S. *Bio/technology*, 1993, *11*, 591-595.
6. Barbas, C.F.; Kang, A.S.; Lerner, R.A.; Benkovic, S.J. *Proc. Natl. Acad. Sci. USA*, 1991, *88*, 7978-7982.
7. McCafferty, J.; Griffiths, A.D.; Winter, G.; Chiswell, D.J. *Nature*, 1990, *348*, 552-554.
8. Garrard, L.J.; Henner, D.J. *Gene*, 1993, *128*, 103-109.
9. Hogrefe, H.H.; Mullinax, R.L.; Lovejoy, A.E.; Hay, B.N.; Sorge, J.A. *Gene*, 1993, *128*, 119-126.
10. Partridge, L.J., Abstracts. The Biochemical Society Meeting No. 647, Sheffield, 1993, p25, F4.

11.   Karu, A.E.; Schothof. K-B.G.; Zhang., G.; Bell, C.W. Food Agric.
       Immunol. In Press.
12.   Krogh, P.; Mycotoxins in Food. 1987, Academic Press, London.
13.   Morgan, M.R.A.; Lee, H.A. In Development and application of
       immunoassay for food analysis: Rittenburg, J.H., Ed.: Elsevier Applied
       Science, London 1990. pp 143-170
14.   Clackson, T.; Hoogenboom, H.R.; Griffiths, A.D.; Winter, G. Nature,
       1991, 352, 624-628.
15.   Huston, J.S.; Levin, D.; Mudgett-Hunter, M.; Tai M-S.; Novotry, J.;
       Margolies, M.N.; Ridge, R J.; Bruccoleri, R.E.; Haber, E.H.; Crea,
       R.; Opermann, H. Proc. Natl. Acad. Sci, USA, 1988, 85, 5879-5883.
16.   Marks, J.D.; Hoogenboom, H.R.; Bonnert, T.P.; McCafferty, J;
       Griffiths, A.D.; Winter, G. J. Mol Biol, 1991, 222, 581-597.
17.   Carter, P.; Bedouelle, H.; Winter, G. Nucl. Acids Res., 1985, 13,
       4431-4443.
18.   Lee, H.A.; Wyatt, G.; Garrett, S.D.; Lacarra, T.G.; Alcocer, M.J.C.;
       Morgan, M.R.A. Food Agric. Immunol., In Press.
19.   Munro, S.; Pelham, H.R.B.; Cell, 1986, 46, 291-300.
20.   MacKenzie, D.A.; Jeenes D.A.; Belshaw, N.J.; Archer, D.B. J Gen.
       Microbiol, 1993, 139, 2295-2307.

21.   Jeenes, D.J.; Marczinke, D.; MacKenzie, B.; Archer, D.B. FEMS
       Microbiol, Letts, 1993, 107, 267-272.
22.   Waterhouse, P.; Griffiths, A.D. Nucl. Acids Res, 1993, 21, 2265-2266.
23.   Marks, J.D.; Ouwehand, W.H.; Bye, J.M.; Finnern, R.; Gorick,
       B.D.; Voak, D.; Thorpe, S.J.; Hughes-Jones, N.C.; Winter, G.
       Bio/technology, 1993, 11, 1145-1149.
24.   Froyen, G.; Ronsse, I.; Billiau, A. Mol. Immunol, 1993, 9, 805-812.

RECEIVED November 4, 1994

Chapter 3

Sequence Analysis of Individual Chains of Antibodies to Triazine Herbicides

Sabine B. Kreissig[1,2], Vernon K. Ward[1,3], Bruce D. Hammock[2], and Prabhakara V. Choudary[1,4]

[1]Antibody Engineering Laboratory and Department of Entomology, University of California, Davis, CA 95616–8584
[2]Departments of Entomology and Environmental Toxicology, University of California, Davis, CA 95616–8584

We rescued and sequenced the antibody genes from hybridomas secreting monoclonal antibodies with different cross-reactivity patterns against atrazine and terbutryn, members of the triazine family of herbicides, analyzed the sequence data by comparison with the sequences of antibodies against dioxin, and identified the contact amino acid residues. The antibody genes were isolated as cDNA copies coding for the light chains or for the Fd fragments of the heavy chains, using the baculovirus vector system. Using comparative sequence analysis and computer modeling, we concluded that the amino acid residues at positions H35B, H50 in the CDR1, H52 in CDR2 and H95 in CDR3 of the heavy chain, as well as L34 and L91 in the light chain CDRs may be contributory to the observed cross-reactivity patterns of the triazine herbicide antibodies studied. Further, the amino acid residues at these positions, because of their critical role in the constitution and functioning of the antigen-binding site, could be the initial targets for site-directed in vitro mutagenesis, aimed at generating recombinant triazine antibodies or fragments with improved binding properties.

Enzyme immunoassays (EIA's) have a long history of use in the clinical and pharmaceutical fields. In recent years, EIA's have been introduced as tools for environmental analysis (*1*). It has been shown that pesticide residues can be successfully quantified by enzyme-linked immunosorbent assays (ELISAs) (*2,3*).

[3]Current address: Department of Microbiology, University of Otago, P.O. Box 56, Dunedin, New Zealand
[4]Corresponding author

0097–6156/95/0586–0031$12.00/0

One of the bottlenecks of immunoassays, however, is the dependence on adequate supplies of suitable antibodies. Although polyclonal and monoclonal antibody (MAb) technologies yield antibodies in quantities sufficient for most analytical applications, it is expensive to produce them in large quantities, needed for widespread monitoring programs and detoxification efforts. The production of antibodies to a small molecule requires its prior conjugation to a larger carrier molecule, and involves time-consuming steps of large-scale immunization and screening. The smaller the size of the target molecule, the greater the difficulty in developing antibodies with desired specificities. The problems of handle-recognition (4) and cross-reactivity have to be overcome, and very often this makes it necessary to immunize a large number of animals or involves, as in the case of MAb's, large-scale screening procedures. To sustain the rapid pace of development of the field of immunoassay for small molecule recognition, these problems have to be solved.

The burgeoning recombinant Ab technology, in combination with novel screening strategies (5), provides powerful methods for circumventing some of these problems, especially with respect to large-scale production and development (6, and reviewed in 7). Further, it permits the addition of novel domains such as reporter enzymes (8), metal binding sites (9) and affinity tags to the antibody (Ab) molecule (Figure 1), aiding their detection and purification (10). Knowledge of sequence information of the Ab genes is critical to our understanding of the antigen-recognition (specificity) and the cross-reactivity patterns of antibodies to different pesticide residues and to our efforts to alter the binding properties of these Ab's by manipulation of the corresponding DNA sequences (11).

In this chapter, we provide a brief background of the recombinant Ab technology, describe the molecular tools and vectors used in cloning the genes encoding the light chain and Fd fragment of the heavy chain of the anti-terbutryn antibody, present the deduced amino acid sequences of the complementarity-determining regions (CDR's), and based on a comparative analysis of different Ab genes analyzed in our laboratory with published sequences and computer modeling of these data by Bell et al (12), predict the amino acid residues that appear to determine the specificity and affinity of the antigen-binding properties of the triazine Ab sequences.

Materials and Methods

Monoclonal Antibodies with Different Specificities and Cross-Reactivity Patterns. The hybridoma cell line secreting the MAb AM7B2.1 was a kind gift from Alexander Karu of the University of California Berkeley, the anti-terbutryn MAb K1F4 from Thomas Giersch and Bertold Hock of the Technical University of Munich, and the cDNA clones coding for anti-dioxin Ab DD1 and DD3 were from Larry Stanker and Adrian Recinos of Texas A&M University.

Triazine Antibodies. We isolated and sequenced cDNA fragments encoding the Ab genes from two different anti-triazine MAb's, with different specificities and cross-reactivity patterns (13; Kreissig, S.B.; Ward, V.K.; Hammock, B.D.; and Choudary, P.V., University of California at Davis, unpublished data). MAb

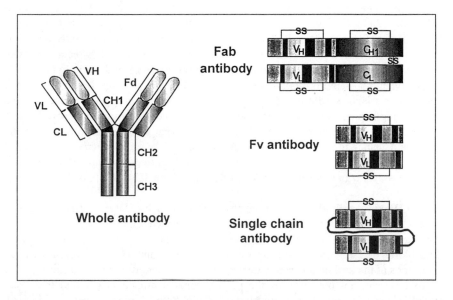

Figure 1. Schematic presentation of the structure of the antibody molecule and various fragments. V_H and V_L are the variable regions of the heavy- and light-chains, respectively. CH^1, CH^2, CH^3 represent the constant regions of the heavy chain, C_L, the constant region of the light chain, and Fd the truncated heavy chain. -ss- symbolizes disulfide bonds.

AM7B2.1 (*14*) binds propazine (2-chloro-4,6-bis (isopropylamino) s-triazine), cyanazine (2-(4-chloro-6-ethyl-amino-s-triazine-2-yl-amino)-2-methylpropio-nitrile) and atrazine (2-chloro-4-ethylamino-6-isopropylamino-s-triazine) with progressively decreasing affinity in that order, although the hapten used as immunogen for producing this Ab was atrazine mercaptopropionic acid, coupled to Keyhole Limpet Hemocyanin (KLH). Even though an ethyl group and an isopropyl group were present on the immunizing hapten distal to the site of the handle attachment, AM7B2.1 bound the analyte with two isopropyl groups (propazine) better than the analyte with one ethyl group and one isopropyl group (atrazine), suggesting the hydrophobic nature of the binding site of the Ab. In the case of the hybridoma K1F4, the immunizing hapten-protein conjugate was prepared by coupling ametryn sulfoxide (2-ethylamino-4-isopropylamino-6-methyl-sulfoxide-s-triazine) to BSA. MAb K1F4 (*15*) showed highest affinity for terbutryn (2-tert.-butylamino-4-ethylamino-6-methyl-thio-s-triazine), followed by significant cross-reactivity with prometryn (2,4-bis-isopropylamino-6-methyl-thio-s-triazine) and terbuthylazine (2-tert.-butylamino-4-chloro-6-ethylamino-s-triazine). The strong binding of the antibody to triazines with a tertiary butyl amino group, a group not present in the immunizing hapten, was a very unexpected result. The in vivo immune mechanism of the animal, independent of the influence of external stimuli, may be responsible for such a change in the specificity pattern. Details of the cross-reactivity patterns of both antibodies are summarized in Table 1. Both antibodies supported sensitive immunoassays, with a lower detection limit of approximately 0.05 ppb in each case.

Dioxin Antibodies. We used two different Mab's (DD1 and DD3) against dioxin (*16-18*) in this study. Both MAb's bind tetrachloro- and pentachloro-dibenzodioxins and -dibenzofurans, but don't bind to either non-chlorinated, octachloro- or 1,2,3,4,8,9-hexachloro-dibenzofurans. Thus, chlorine substitution on both rings of the analyte appears to be necessary for recognition by the antibody. The MAb DD3 did not bind any of the polychlorinated biphenyls (PCB's) that were tested, whereas MAb DD1 recognized the 3,3',4,4'-tetrachloro congener weakly (*16*). Recently, these two antibodies were studied further by computer modeling (*19*).

Molecular Cloning of the Antibody Genes. The procedures used for cloning the Ab genes from the hybridoma AM7B2.1 were reported (*13*). A step-wise procedure followed to clone the genes (cDNA fragments) encoding Fd fragment of the heavy chain and the light chain of the MAb K1F4 using the baculovirus system is schematically depicted in Figure 2.

Messenger RNA (mRNA) was extracted from hybridoma cells, using the Fast Track mRNA extraction kit (Invitrogen Corp., San Diego, CA). Multiple sets of oligonucleotide primers (Figure 3), designed to be complementary to antibody sequences and to allow cloning in the baculovirus transfer vector, were synthesized on an automatic DNA synthesizer Model 380A (Applied Biosystems Inc., Foster City, CA) by standard methods (*20*). The 3'-primers were designed to anneal at the 3'-end of the C_H1 or C_L domain of the mouse IgG heavy or light chain (*21*), and the 5'-primers at the 5'-end of the heavy or light chain. The primers were designed with

Table 1. Cross-reactvity patterns of the antibodies, AM7B2.1 and K1F4, toward different triazine herbicides (Common name and chemical stucture of each compound are shown). Cross-reactivities are shown as percent values relative to atrazine and terbutryn, respectively. n.t., not tested

		AM7B2.1	K1F4
Atrazine		100	14
Deethylatrazine		<1	1
Hydroxyatrazine		5.7	5
Simazine		31	3
Propazine		196	5
Terbuthylazine		n.t.	43
Terbutryn		21	100
Prometryn		30	89
Cyanazine		106	n.t.

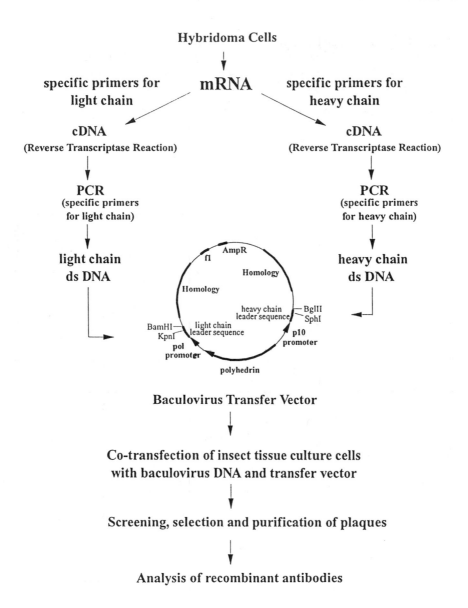

Figure 2. Schematic protocol for the production of recombinant antibodies in the baculovirus expression vector system. The baculovirus transfer vector carries the Origin of Replication (f1), ampicillin resistance gene (Amp^R) for selection of recombinants, and flanking segments the baculovirus DNA to facilitate homologous recombination in vivo. In the baculovirus system the antibody light chain is cloned under the control of the polyhedrin promoter, and the heavy chain (or fragment) under the control of the p10 promoter. In addition, the vector contains a functional polyhedrin gene. ds DNA: double stranded DNA.

Heavy chain N-terminal primers (5' to 3')

	SphI	Leader Sequence	H-Chain N-terminal Homology				
TTA CTC GCT GCC	GCA TGC	ATC CTT TCC	^{C}GAC GTG CAG	CTG CAG CAG ^{T}CCT GG			
TTA CTC GCT GCC	GCA TGC	ATC CTT TCC	CAG GTA CAG	CTC AAG GAG TCA GG			
TTA CTC GCT GCC	GCA TGC	ATC CTT TCC	GAG GTG ^{C}AAG	CTG CTG GAG TCT GG			

Heavy chain C-terminal primers (3' to 5')

H-Chain C$_{H1}$-domain C-terminal Homology			Stop Codons	BglII	EcoRI	
TG GT^{T}G CCA CCT GTT CTT T^{G}TA	AAC TGA CTC ACT	TCT AGA	CTT AAG	CGG		
A ACA CGG GTC CCT AAC	AAC TGA CTC ACT	TCT AGA	CTT AAG	CGG		
GG GAC AGG AGG TAC GTT	TAC TGA CTC ACT	TCT AGA	CTT AAG	CGG		

Light chain N-terminal primers (5' to 3')

	KpnI	Leader Sequence	L-Chain N-terminal Homology	
GGT TTC GCT ACC	GGT ACC	AGA TGT GAC	ATC ^{C}AAG ATG	ACC CAG
GGT TTC GCT ACC	GGT ACC	AGA TGT GAC	ATT ^{G}TTG ^{A}CTG	ACT CAG

Light chain C-terminal primer (3' to 5')

L-Chain C$_L$ domain C-terminal Homology	Stop	BamHI	
TCG AAG GTT TCC TTA CTC AGA	ACT ACT	CCT AGG	GCT AGC AAG AAG

Figure 3. Nucleotide sequences of the PCR primers used for cloning the antibody heavy and light chain genes. The restriction sites used for cloning are indicated in italics, and the three amino acid residues lost from the leader sequences during digestion with *SphI* and *BglII* (heavy chain) or *KpnI* and *BamHI* (light chain) but replaced by the primers are shown. Stop codons are indicated.

sufficient degeneracy to allow priming of the different subclasses of heavy and light chains (*13*). Using an appropriate 3' primer, aliquots of the mRNA were converted into cDNA by the reverse transcriptase reaction. In the next step, the cDNA was amplified in a polymerase chain reaction (PCR), using both 3' and 5' primers. Reactions for the heavy and light chains were performed separately.

A baculovirus transfer vector, suitable for the cloning (as well as expression) of the cDNA encoding both heavy and light chains of Ab, was constructed. The vector plasmid contained a leader sequence and cloning sites for the heavy chain after the p10 promoter, and after the polyhedrin promoter for the light chain, respectively (Ward, V.K.; Kreissig, S.B.; Hammock, B.D.; and Choudary, P.V., University of California at Davis, unpublished data). In addition, it contained a functional polyhedrin gene, which allows for easier screening of the recombinant baculoviruses, a f1 origin and an ampicillin-resistance gene for selection in *E. coli*. cDNA fragments encoding Ab heavy and light chain were ligated into the transfer vector, and the plasmid construct was introduced into *E. coli* by electroporation. Recombinant clones were identified by ampicillin-resistance. Recombinants carrying both heavy- and light-chain genes were identified by restriction digestion analysis of the plasmid DNA. *Spodoptera frugiperda* (Sf21) cells were cotransfected with recombinant transfer vector DNA, linearized baculovirus (AcNPV) genomic DNA, and lipofectin. Five days after transfection, the culture medium was collected and plaque assays were performed (*22*). Plaque picks were used to infect Sf21 cells growing in individual wells of 24-well plates. Ten days after infection, aliquots of the culture supernatant from the infected cell lines was collected and tested by ELISA for antigen-binding activity, essentially as described (*13*). The clones that appeared putative positive in the analyte-binding assay were analyzed by DNA sequencing.

Nucleotide Sequencing. The procedures used for sequencing the cDNA fragments from AM7B2.1 were reported by Ward et al (*13*). The cDNA clones from K1F4 were sequenced using standard dideoxynucleotide termination reactions containing 7-deaza dGTP, using [^{35}S]dATP as the label. The sequencing reactions were analyzed on 6% polyacrylamide wedge gels containing 8M urea. The primers synthesized for PCR amplification and additional internal sequencing primers were used in sequencing reactions.

Sequence Analysis and Alignments. The comparative analysis of the cDNA sequences and deduced amino acid sequences was performed using the HIBIO MENU (HIBIO DNASIS and PROSIS) of the Hitachi Software Engineering America Ltd (Brisbane, CA). In this study, we followed the amino acid numbering system of Kabat et al. (*21*). Wherever amino acid numbers from literature citations had to be converted, it is noted.

Computer Modeling of the Antigen-Binding Site in AM7B2.1. Computer modeling data was taken from Bell et al (*12*), who modeled the sequence data of AM7B2.1 reported by Ward et al (*13*), using the programs PROCHECK, ProExplore and INSIGHT II.

Results and Discussion

Molecular Cloning of the Anti-Terbutryn Antibody Genes. We cloned and sequenced the Ab genes from two different anti-triazine monoclonal antibodies (MAb's), AM7B2.1 and K1F4, with varying specificities and cross-reactivity patterns (*13;* Kreissig, S.B.; Ward, V.K.; Hammock, B.D.; and Choudary, P.V., University of California at Davis, unpublished data).

Although a variety of cloning/expression systems have been used for producing recombinant antibodies (reviewed in *7*), cloning in *Escherichia coli*, mediated by bacteriophage (*23, 24*) and plasmid vectors (*10, 13*), has been the method of choice, because of the simple and well-established technology. The rapid growth and the relatively simple fermentation of *E. coli* makes large-scale production of recombinant proteins in this host considerably easy and inexpensive. Consequently, *E. coli* has been also the host of choice used traditionally for the production of functionally-assembled antibody fragments.

In addition to *E. coli*, *Bacillus subtilis*, although not as well-characterized as the *E. coli* system, has been used as an alternative host for the production of functional single-chain antibodies (Figure 1) and individual variable domains of light (V_L) and heavy (V_H) chains, because of its greater efficiency in secreting recombinant proteins (*25*). The single-chain (scFv) antibody produced in *B. subtilis* was found to exhibit almost identical binding properties as the parent monoclonal antibody (*25*).

Since the carbohydrate moieties decorating the Ab are present mostly on the C_H2 domain of the Fc region (Figure 1), and are relatively rare in the Fv or CH^1 domains, functional Fv, Fab and scFv molecules have been synthesized efficiently in *E. coli and B. subtilis*. However, post-translational modifications, especially the complex glycosylations, of recombinant eukaryotic proteins produced in prokaryotic hosts are not the same as those occurring in vivo in the eukaryotic host. It is also thought to be very difficult to produce a functional whole antibody (Figure 1) in prokaryotic hosts, presumably because of the large size of the antibody, in addition to the absence of cellular machinery necessary for complex glycosylation of recombinant proteins produced in prokaryotes.

In light of the above limitations associated with prokaryotic expression systems, eukaryotic systems such as transgenic plants (*26, 27*), animals (*28*),Yeasts (*29*), and filamentous fungus, *Trichoderma reesei* (*30*) have been used successfully as hosts for the production of functional antibodies or fragments with binding properties comparable to those of the parent antibodies.

Another eukaryotic vector, based on the baculovirus - Nuclear Polyhedrosis Viruse (NPV), has proven to be a powerful expression system (*31*). In these viruses, intracellular virions (packages of virus DNA) are occluded in a crystalline matrix of the protein, polyhedrin. The polyhedrin promoter and another virus promoter, the p10 promoter, are usually used for the expression of foreign genes, because they both are strong and together can account for up to 50% of total cell or larval protein, although neither protein is essential for the function of the virus (*32*). The advantages of the baculovirus system lay in its easy handling, well developed technology, high expression levels, and correct folding and glycosylation of the recombinant proteins.

Insect cell cultures can be grown at room temperature and do not need CO_2 supply, unlike hybridoma cell cultures (*33*). The expression levels of foreign proteins are considerably high and large amounts of proteins are usually secreted into the culture supernatant (*34*) or can be recovered from the hemolymph of the infected larvae easily (*35*). Production of foreign proteins in larvae makes the method inexpensive. Using the baculovirus system, whole antibodies and fragments with appropriate carbohydrate residues have been produced (*36-38*).

Although the baculovirus system has been utilized conventionally for the production of recombinant proteins (*31*), it can be considered as a cloning system also, by taking advantage of its multiple attributes. These attributes include relatively easy cloning procedures, simple and relatively inexpensive insect cell culture systems that can be scaled up and the ability to store viruses indefinitely (*34, 35, 33, 31*). Although a panning procedure (*5*), analogous to the one available for the phage display systems may be a remote possibility for application with the baculovirus system, we have recently demonstrated the feasibility of constructing a cDNA expression library in the baculovirus and isolating cDNAs coding for functional antibody chain(s) by using conventional screening procedures (Ward, V.K.; Kreissig, S.B.; Hammock, B.D.; and Choudary, P.V., University of California at Davis, unpublished data). The baculovirus library yielded several cDNA clones producing functional antibodies. It is relatively straight-forward to develop procedures to screen thousands or even tens of thousands of clones. It is also easier to screen for baculovirus clones than hybridoma clones due to the properties of the cell-lines involved. When the goal is to clone an antibody from a hybridoma cell line or possibly from the spleen or lymphocytes of immunized animals, the limited screening capability associated with the baculovirus library may be off-set by the desirable expression attributes of the baculovirus library. Although the generation of a large library of random mutants of Ab genes in the baculovirus system is a daunting challenge at the current state of the technology, mutagenizing and screening an existing clone of an Ab gene for site-specific mutations is not a formidable task.

The Role of Different Amino Acids in the CDRs of Heavy and Light Chains in Antigen-Binding.   To achieve our long-term goal of producing genetically engineered Abs, especially triazine-binding molecules, with novel binding properties, knowledge of the molecular architecture of the binding site and identification of the contact (amino acid) residues that determine the binding properties is a pre-requisite. Accordingly, to identify the common structural features shared by antibodies to different pesticides, we compared the nucleotide sequence of the variable regions from the triazine antibodies and two different Mab's (DD1 and DD3) against dioxin (*16-18*) and analyzed our data in the context of other published studies.

Mian and coworkers (*39*) examined the structure-function correlates of the antibody binding sites in six different antibodies with known crystal structure of the antigen-Ab complex and found only 37 (plus 3 within the framework regions, L49, H30 and H47), out of a total of 85 residues present in the complementarity determining regions (CDRs) of the heavy and light chains, to be directly involved in the antigen-Ab interaction. Of all the amino acid positins examined, the light chain residue 91 (L91) was involved in all of the cases (100%), while L32, L96 and heavy chain residue 33 (H33) were involved in 83% of the cases.

An examination of the chemical and physical properties of the amino acids involved in antigen-Ab interaction reveal the following common requisite attributes:

- Amphipathic amino acids better tolerate the change from a hydrophilic to hydrophobic environment that occurs upon the formation of the antigen-antibody complex.
- Residues that are large and that can participate in a wide variety of van der Waals' and electrostatic interactions permit binding of the Ab to a greater range of antigens.
- Amino acids with flexible side-chains generate a structurally plastic region, that allows the Ab mould around antigens and thereby improve the complementarity to the interacting surfaces.

Because tyrosine and tryptophan meet the above criteria well, they would be expected to be most commonly present in the Ab-combining sites. Further, their ability to form hydrogen bonds, hydrophobic interactions and attractive electrostatic interactions between positively-charged groups and aromatic rings permits tyrosine and tryptophan to interact with structurally diverse antigens. A comparative analysis of the binding sites of antibodies with known crystal structures (21) confirms the frequent occurrence of tyrosine (25%) and tryptophan (10.2%) in Ab combining sites. Tyrosine and tryptophan frequently alternate with small amino acids such as glycine, alanine and serine, which allows maximum mobility of the tyrosine and tryptophan side-chains.

Ohnu and coworkers (40) reported that three short clusters of amino acids present in CDR1 and CDR2, placed in the immediate vicinity of the tryptophan loop, primarily determine the antigen preference of the v_H and therefore of the antibody. These amino acid clusters are 31 - 35 in CDR1, and 50 - 52 and 58 - 60 in CDR2. However, in these clusters, not all amino acids are hypervariable. At position 32, tyrosine occurs in 54.2% of all cases (23.3% are phenylalanine), while position 34 is usually occupied by methionine (74.1%), position 51 by isoleucine (69.0%) and position 59 by tyrosine (86.6%). This leaves a total of 7 variable amino acids in CDR1 and CDR2 that determine the antigen specificity of most antibodies. For an antibody against the hapten NIP (4-hydroxy-3-nitro-5-iodo-phenylacetyl), Reth et al (41) predicted that the amino acid residues in position 31, 35, 50, 52 and 99 of the heavy chain play a key role in hapten-binding. Brüggemann and coworkers (42) showed that an amino acid change in position 50 results in the loss of the hapten-binding and generates a new binding specificity. However other published data demonstrate that point mutations in CDR1 - H35 (43) and CDR3 of the heavy chain - H101 (44) and CDR3 of the light chain - L98 (45) can also change the Ab-specificity.

Computer Modeling of the Antigen-Binding site of the Antibody AM7B2.1.
Computer modeling of the Ab AM7B2.1 led Bell et al. (*12*) to the identification of
several amino acids in the heavy chain as well as in the light chain that seem to be
involved in the antigen-binding. The long CDR1 of the light chain and CDR3 of the
heavy chain point to the possibility of the AM7B2.1 having a deep antigen binding
pocket. Hydrophobic side chains seem to be involved in the antigen-binding. The
modeling study of Bell et al (*12*) further suggests that amino acids at positions 34
and 35B (CDR1), 50 and 52 (CDR2) and 95, 97 and 99 (CDR3) of the heavy chain
and the amino acids at positions 32 and 34 (CDR1) and 91 and 96 (CDR3) of the
light chain are involved in the binding of atrazine. The presence of the spacer arm of
the atrazine mercaptopropionic acid-protein conjugate in the R_1 position of the
triazine ring makes it likely that the isopropyl groups of atrazine or propazine would
be buried at the bottom of the binding pocket with the triazine ring sitting between
the many tryptophan and tyrosine residues. However, the AM7B2.1 light chain
synthesized by Ward et al (*13*) and used in the modeling study by Bell et al (*12*) is
unusual in two ways. The well-conserved cysteine at position 23 of the framework
region 1 (FR1) of the light chain that usually forms an intramolecular disulfide bond,
is replaced by a tyrosine. In addition, CDR3 is only 8 residues-long and ends in a
proline, instead of the usual threonine (*21*). These unusual features may at least in
part account for the decreased affinity of the F_{ab} fragment of AM7B2.1 produced in
E. coli (*13*) for the analyte.

Amino Acid Sequences of the Heavy Chain CDR's. Amino acid sequences of the
CDR's of the heavy chains of four different Ab's against pesticides are shown in
Figure 4. AM7B2.1 (*14*) and K1F4 (*15*) are hybridomas secreting Ab against triazine
herbicides, and DD1 and DD3 (*16-18*) against dioxin. K1F4 and DD3 belong to the
IgG1 subclass, and AM7B2.1 and DD1 to IgG2b (*21*).

Table 2 shows that AM7B2.1 and K1F4 vary in all but one positions
involved in the binding of atrazine to AM7B2.1. This is position 34, usually
occupied by tyrosine (54.5%; see also *Ref 40*, position 32). The positions that vary in
heavy chain CDR's of all the four antibodies (AM7B2.1, K1F4, DD1, DD3) are H50,
H56 and H62 in CDR2, and H95 and H99. Three out of these five residues are
involved in the antigen binding of AM7B2.1. Some of these amino acids, H50 (*41;
40*) and H99 (*41*), are also reported in the literature as being important for binding.
Changes of amino acids at these positions could possibly result in antibodies with
different binding specificities or cross-reactivity patterns. It is intriguing that the
heavy chain of K1F4 contains additional residues (100A-100E) in CDR3. The role of
these extra amino acids in the binding of the antigen is not determined yet; however,
the presence of two tyrosines within the neighborhood of small amino acids (glycine
and serine) points to their possible involvement in the antigen-binding (*39*).

Amino Acid Sequences of the Light Chain CDR's. A comparison of the amino
acid sequences of the CDR's of light chains from 3 different Ab's is shown in Figure
5. All the light chains considered here belong to the kappa (κ) subclass. The CDR3
sequence is quite similar in the two dioxin antibodies (DD1 and DD3) but differs

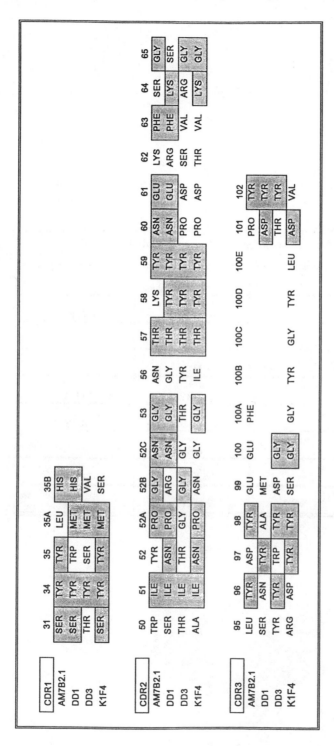

Figure 4. Amino acid sequences of the CDR's of the heavy chains of four different antibodies. AM7B2.1 (IgG2b) and K1F4 (IgG1) are MAb's directed against the triazine herbicides, and DD1 (IgG2b) and DD3 (IgG1) against dioxins. Amino acid residues are numbered according to Kabat et al. (1991). The shading indicates amino acids that are common to two or more antibodies.

Table 2. The amino acid residues and their positions in the sequences of
AM7B2.1 and K1F4 that seem critical for antigen-binding

Position	Amino Acid	
	AM7B2.1	K1F4
H34	TYR	TYR
H35B	HIS	SER
H50	TRP	ALA
H52	TYR	ASN
H95	LEU	ARG
H97	ASP	TYR
H99	GLU	SER

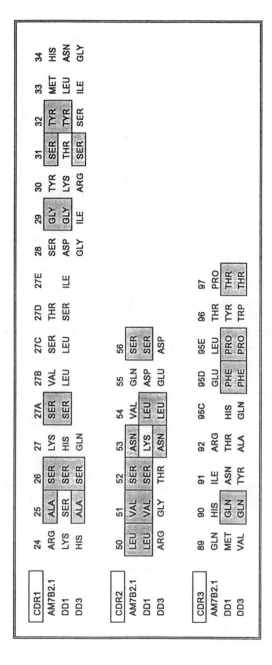

Figure 5. Amino acid sequences of the CDR's of the light chains of three different MAb's. AM7B2.1 is directed against the triazine herbicides, and DD1 and DD3 were raised against dioxins. All three light chains belong to the κ subclass. Amino acid residues are numbered according to Kabat et al (1991). Shading indicates the amino acids that are common to two or more antibodies.

completely from the CDR3 of AM7B2.1. However, the CDR2 sequence is similar in all the three light chains. Further conclusions concerning the light chain sequences are difficult to reach at this stage because of the non-availability of sequence information on additional anti-triazine Ab light chains. In addition, the AM7B2.1 light chain seems to be quite different from the other light chains. Work on the K1F4 light chain sequence, as well as the sequence of another triazine Ab with cross-reactivity pattern similar to that of AM7B2.1 is currently in progress.

Putative Role of Framework Regions in Antigen-Binding. It is well known that changes in the framework regions, in addition to those in the CDR's, may affect the conformation of the binding site and consequently the nature of the binding, a possibility that is particularly likely to be relevant in the case of anti-hapten antibodies. Foote and Winter (*46*) showed that mutations in a triad of heavy chain residues (H27, H29 and H71) as well as a phenylalanine to tyrosine substitution of the light chain residue L71 in a reshaped (humanized) antibody against lysozyme increased the free binding energy of the antibody-antigen complex. These substitutions were made in the ß-sheet framework regions, closely underlying the complementarity determining regions, but did not participate in the direct interaction with the antigen. Mian et al. (*39*) found three amino acids outside the CDRs (L49, H30, H47) that were influencing the antigen binding of the antibodies. However, the positions that were identified would be unlikely to be involved in the cases presented here. In position L49 there is a tyrosine in all three cases. The residue at H30 in AM7B2.1 (threonine) differs from K1F4 (serine), but the dioxin antibodies, which belong to the same IgG subclass, contain in both cases the same amino acid. We found a tryptophan at position H47 in AM7B2.1, whereas K1F4 contains a serine, and a tryptophan in DD1 and DD3. In IgG2b, this position is invariably occupied by tryptophan (*21*). However, these findings do not exclude the possibility that residues at this or other positions in the framework regions may have an influence on the conformation or stability of the binding pocket and thus on the binding specificity.

Conclusions

Recombinant DNA technology facilitates the design of antibodies with novel properties, unforeseen by the use of conventional technologies of polyclonal and monoclonal antibodies. A number of gene cloning and expression systems facilitate large-scale production of engineered antibodies. The ability of the baculovirus system to produce whole antibodies that are correctly glycosylated purports it to be a superior expression system for some applications, notwithstanding its utility as a direct cloning system.

The analysis of the structure-function relationship between DNA sequence and antigen-binding can be used for identification of the contact residues that determine specificity and cross-reactivity patterns of antibodies to pesticides. Once identified, these residues can be changed by the use of genetic manipulation methods to produce Ab's with desired, previously unavailable specificities, including Ab's that are completely monospecific or that bind analytes from a family of structurally similar antigens with equal affinity. Based on a combination of sequence comparisons and modeling data, we predict that the amino acids at positions H35B, H50 in the CDR1, H52 in CDR2 and H95 in CDR3 of the heavy chain, as well as L34 and L91 in the CDRs of the light chain would likely provide suitable targets for mutagenesis in our efforts to alter their binding properties in the case of MAb, AM7B2.1. In addition, the possibility of synthesizing antibody chains fused to reporter enzymes, metal binding sites and affinity tags can prove very useful for a wide range of applications such as EIA's and biosensors.

A judicious combination of all these converging technologies, including the manipulation of antibody genes to achieve new properties, expression of functional antibodies in various heterologous hosts, affinity purification of engineered antibodies and the development of increasingly sensitive immunoassay procedures or biosensor applications will rapidly advance the field of small molecule analysis and have a significant impact on our efforts to apply this burgeoning technology in efficient detection and possible remediation of environmental contaminants.

Acknowledgments

We thank Alexander Karu, Bertold Hock and Thomas Giersch for the anti-triazine monoclonal antibodies, Larry Stanker and Adrian Recinos for the anti-dioxin antibody sequences, and Christopher Bell and Alexander Karu for providing the computer modeling data of AM7B2.1. Funding for this work was provided by grants from the Office of Research of the University of California Davis, Environmental Protection Agency of the USA (CR-819047-01-0), Superfund Basic Research Program (2P42 ES 04699), Center for Affects of Agrochemicals (IP30 ES 05707), Center for Ecological Health Research (CR 819658) and Center for water Resources (W-840).

Literature Cited

1. Vanderlaan, M.; Watkins, B.E.; and Stanker, L. *Environ. Sci. Technol.* 1988, **22**, 247-254.
2. Jung, F.; Gee, S.J.; Harrison, R.O.; Goodrow, M.H.; Karu, A.E.; Braun, A.L.; Li, Q.X.; and Hammock, B.D. *Pestic. Sci.* 1989, **26**, 303-317.
3. Hall, J.C.; Deschamps, R.J.A.; and McDermott, M.R. *Weed Technology* 1990, **4**, 226-234.
4. Harrison, R.O.; Goodrow, M.H.; Gee, S.J.; and Hammock, B.D. In *Immunoassays for Trace Chemical Analysis*; Vanderlaan, M.; Stanker, L.H.; Watkins, B.E.; and Roberts, D.W., Eds.; *ACS Symposium Series*, **451**, 14-27; American Chemical Society: Washington DC, 1991.
5. Parmley, S.F.; and Smith, G.P. *Gene* 1988, **73**, 305-318.
6. Carter, P.; Presta, L.; Gorman, C.M.; Ridgway, J.B.B.; Henner, D.; Wong, W.L.T.; Rowland, A.M.; Kotts, C.; Carver, M.E.; and Shepard, H.M. *Proc. Natl. Acad. Sci. USA* 1992, **89**, 4285-4289.

7. Winter, G.; Griffiths, A.D.; Hawkins, R.E.; and Hoogenboom, H.R. *Annu. Rev. Immunol.* 1994, **12**, 433-455.
8. Ducancel, F.; Gillet, D.; Carrier, A.; Lajeunesse, E.; Ménez, A.; and Boulain, J.-C. *Bio/Technology* 1993, **11**, 601-605.
9. Pessi, A.; Bianchi, E.; Crameri, A.; Venturini, S.; Tramontano, A.; and Sollazzo, M. *Nature* 1993, **362**, 367-369.
10. Schmidt, T.G.M.; and Skerra, A. *Protein Engineering* 1993, **6**, 109-122.
11. Brüggemann, M.; and Neuberger, M.S. In: *Monoclonal Antibody Therapy. Prog. Allergy*; **45**, 91-105; Waldmann, H., Ed.; 1988.
12. Bell, C. Maulik, S.; Roberts, V.A.; Ward, V.K.; and Karu, A.E. In *Antibody Engineering; Research and Application of Genes Encoding Immunoglobulins*; Keystone Symposia: Lake Tahoe, CA, March 7-13, 1994,
13. Ward, V. K.; Schneider, P.; Kreissig, S. B.; Karu, A.; Hammock, B. D.; and Choudary, P. V. *Protein Engineering* 1993, **6**, 981-988.
14. Karu, A.E.; Harrison, R.O.; Schmidt, D.J.; Clarkson, C.E.; Grassman, J.; Goodrow, M.H.; Lucas, A.; Hammock,, B.D.; Van Emon, J.M.; and White, R.J. In *Immunoassays for trace chemical analysis*; Vanderlaan, M.; Stanker, L.H.; Watkins, B.E.; and Roberts, D.W., Eds.; *ACS Symp. Ser.* **451**, 59-77; American Chemical Society: Washington DC, 1991.
15. Giersch, T.; and Hock, B. *Food Agric. Immunol* 1990, **2**, 85-97.
16. Stanker, L.H.; Watkins, B.; Rogers, N.; and Vanderlaan, M. *Toxicology* 1987, **45**, 229-243.
17. Recinos, A.; Silvey, K.J.; Ow, D.J.; Jensen, R.H.; and Stanker, L.H. *EMBL database* accession No. **Z21788 and Z19575**, 1989a.
18. Recinos, A.; Silvey, K.J.; Jensen, R.H.; and Stanker, L.H. *EMBL database* accession No. **X58884 and X59052**, 1989b.
19. Stanker, L.H.; Recinos III, A.; and Linthicum, S., this volume.
20. Sambrook, J.; Fritsch, E.F.; and Maniatis, T. *Molecular Cloning*; Vol 1-3; Cold Spring Harbor Laboratory, NY, 1989.
21. Kabat, E.A.; Wu, T.T.; Perry, H.M.; Gottesman, K.S.; and Foeller, C. In *Sequences of proteins of immunological interest*; NIH Publication No. 91-3242; US Department of Health and Human Services, Public Health Service, National Institutes of Health: Bethesda, MD,1991.
22. O'Reilly, D.R.; Miller, L.K.; and Luckow, V.A. In *Baculovirus Expression Vectors: A Laboratory Manual*; W.H. Freeman and Company, New York, NY,1992.
23. Nissim, A.; Hoogenboom, H.R.; Tomlinson, I.R.; Flynn, G.; Midgley, C.; Lane, D.; and Winter, G. *EMBO J.* 1994, **13**, 692-698.
24. Barbas III Jr, C.F. This Volume, pp.
25. Wu, X.-C.; Ng, S.C.; Near, R.I.; and Wong, S.-L. *Bio/Technology* 1993, **11**, 71-76.
26. Hiatt, A, Caffertey, R.; and Bowdish, K. *Nature* 1989, **342**, 76-78.
27. Tavladoraki, P.; Benvenuto, E.; Trinca, S.; De Martinis, D.; Cattaneo, A.; and Galeffi, P *Nature* 1993, **366**, 469-472.

28. Duchosal, M.A.; Eming, S.A.; Fisher, P.; Leturcq, D.; Barbas III, C.F.; McConahey, P.J.; Caothein, R.H.; Thornton, G.B.; Dixon, F.J.; and Burton, D.R. *Nature* 1992, **355**, 258-262.
29. Horwitz A.H.; Chang, C.P.; Better, M.; Hellstrom, K.E.; and Robinson, R.R. *Proc. Natl. Acad. Sci. USA* 1988, **85**, 8678-8682.
30. Nyyssönen, E.; Pentitilä, M.; Harkki, A.; Saloheimo, A.; Knowles, J.K.C.; and Keränen, S. *Bio/Technology* 1993, **11**, 591-595.
31. Luckow, V. A. In *Recombinant DNA Technology and Applications*; Prokop, A.; Bajpai, R.K.; and Ho, C.S.; Ed.; McGraw Hill: New York, NY, 1991, pp 97-152.
32. Davies, A.H. *Bio/Technology* 1994, **12**, 47-50.
33. Summers, M.D.; and Smith, G.E. *Tex. Agric. Exp. Stn. Bull.* 1987, **1555**.
34. Maeda,S. In: Mitsuhashi, J. (ed) *Invertebrate Cell System Applications* 1989a, **1**, 167-181.
35. Maeda, S. *Ann. Rev. Entomol.* 1989b, **34**, 351-372.
36. Reis, U.; Blum, B.; von Specht, B.-U.; Domdey, H.; and Collins, J. *Bio/Technology* 1992, **10**, 910-912.
37. Hasemann, C.A.; and Capra J.D. *Proc. Natl. Acad. Sci. USA* 1990, **87**, 3942-3946.
38. Putlitz, J.z.; Kubasek, W.L.; Duchene, M.; Marget, M.; Specht, B-U.v; and Dumdey, H. *Bio/Technol*, 1990, **8**, 651-654.
39. Mian, S.; Bradwell, A.R.; and Olson, A.J. *J. Mol. Biol.* 1991, **217**, 133-151.
40. Ohnu, S.; Mori, N.; and Matsunaga, T. *Proc. Natl. Acad. Sci. USA* 1985, **82**, 2945-2949.
41. Reth, M.; Kelsoe, G.; and Rajewsky, K. *Nature* 1981, **290**, 257-259.
42. Brüggemann, M.; Müller, H.-J.; Burger, C.; and Rajewsky, K. *EMBO J.* 1986, **5**, 1561-1566.
43. Diamond, B.; and Scharff, M.D. *Proc. Natl. Acad. Sci. USA* 1984, **81**, 5841-5844.
44. Cook, W.D.; Rudikoff, S.; Giusti, A.M.; and Scharff, M.D. *Proc. Natl. Acad. Sci. USA* 1982, **79**, 1240-1244.
45. Azuma, T.; Igras, V.; Reilly, E.B.; and Eisen, H.N. *Proc. Natl. Acad. Sci. USA* 1984, **81**, 6139-6143.
46. Foote, J.; and Winter, G. *J. Mol. Biol.* 1992, **224**, 487-499.

RECEIVED November 10, 1994

Chapter 4

Recombinant Antibodies to Diuron

A Model for the Phenylurea Combining Site

Christopher W. Bell[1], Victoria A. Roberts[2], Karen-Beth G. Scholthof[3],
Guisheng Zhang[1,4], and Alexander E. Karu[1]

[1]Hybridoma Facility, University of California, Berkeley College of Natural Resources, 1050 San Pablo Avenue, Albany, CA 94706
[2]Department of Molecular Biology, The Scripps Research Institute, 10666 North Torrey Pines Road, La Jolla, CA 92037
[3]Department of Plant Biology, University of California, Berkeley, CA 94720

Recombinant antibody technology and computational modeling make it possible to deduce intermolecular interactions between antibody and antigen, and genetically alter antibody affinity and selectivity. A recombinant antibody (Fab 481.1) that competitively bound diuron was cloned from a diuron-specific hybridoma. Fabs that bound diuron hapten conjugate but not free diuron were also selected from semi-synthetic combinatorial antibody libraries. A computer model of Fab 481.1 was constructed based on structural homology with known antibodies. The steps in the modeling process and subsequent refinements are described. From the model we identified a putative diuron binding site and the atomic interactions of diuron with amino acids in the site. Inferences from the model will be tested by site-directed mutagenesis to produce Fabs with altered binding properties.

The ability to manipulate antibody genes inaugurated a third generation of antibody technology. Polyclonal antisera and, more recently, monoclonal antibodies (MAbs) have been valuable enzyme immunoassay (EIA) reagents. MAb technology made it possible to select and immortalize antibody producing cells and obtain antibodies of defined affinity and specificity (*1*). Standard molecular biology techniques now allow the amplification, selection, and alteration of antibodies expressed in *E. coli* and other cell types. Using techniques similar to immunoassay, bacteriophage that express antibody fragments (Fab) as coat protein fusions can be rapidly selected (*2*). In addition, molecular modeling and site-directed mutagenesis of cloned antibodies may provide an alternative to the synthesis of haptens and the derivation of new antibodies.

[4]Current address: Department of Transplantation Surgery, University of Pittsburgh School of Medicine, Pittsburgh, PA 15261

The phenylurea herbicide diuron [3-(3,4-dichlorophenyl)-1-1-dimethylurea] has been important for studying photosynthetic electron flow in the thylakoid membrane. Diuron inhibits photosystem II electron transport in plants and other photosynthetic organisms by displacing plastoquinone from the Q_B binding site of the active-center protein D1 (*3*). In agriculture, the phenylureas are widely used leachable herbicides and regulatory agencies must monitor the residues. They also have numerous analogs, metabolites, and breakdown products, some of which are difficult to analyze. These characteristics are parameters for a model system to test the versatility of recombinant antibodies. The asymmetric structure and charge distribution of phenylureas make them interesting compounds to orient in the combining site of the antibody. The interactions in binding diuron may be useful for understanding how other macromolecules bind organochlorines. To our knowledge, this is the first model of the interaction between an organochlorine and a competitively binding antibody.

Immunoglobulins (antibodies) are proteins composed of light (L) and heavy (H) chains folded into globular domains and stabilized by disulfide bridges. Their unique antigen-binding specificity is mediated by three hypervariable loops, the complementarity determining regions (CDR) at the variable (V) amino terminus of each chain. The CDRs are separated by more highly conserved sequences known as framework regions (FR) in the order FR1—CDR1—FR2—CDR2—FR3—CDR3—FR4 (*4*). The CDRs project from one end of the V_L/V_H complex and are supported on the frameworks, which form a conserved "β-barrel" fold (*5*). The lengths, conformation, and sequence of the CDR loops vary considerably between antibodies, but the general architecture is the same.

Attempts to engineer antibodies have focused on manipulating the CDR gene sequences. The main strategies used are: alteration of amino acids (*6*), direct substitution of the CDRs (*7*), and introduction of a new ligand recognition property (e.g., metal ion binding) into the antibody (*8*). Several groups have also prepared semi-synthetic combinatorial antibody libraries in which at least one of the CDRs is randomly varied (*9, 10*). Such libraries have a large potential diversity, allowing selection of antibodies to human self-antigens or small organic molecules that are difficult to prepare by conventional immunization.

This chapter describes the first step toward preparing antibodies with altered specificities for the phenylurea herbicides. Recombinant Fabs were isolated from cloned L and H chains from a high-affinity MAb (481.1) to diuron and by selection from semi-synthetic combinatorial Fab gene libraries (*9*). Amino acid sequences were derived from the Fab L and H genes. An initial computational model for the V domain of MAb 481.1 was constructed with the antibody modeling package AbM. This model was analyzed and refined by direct comparison with crystallographically determined antibody structures. The revised model was used to identify amino acids that may be critical for binding diuron. Experiments are planned to test the model by preparing and characterizing mutant antibodies with specifically altered amino acids. These engineered antibodies could have useful properties for immunodetection and immunoaffinity recovery of phenylureas and their metabolites.

METHODS.

Derivation of Recombinant Antibodies. Antibody genes were cloned from a diuron-specific hybridoma (MAb 481.1) (*11*) into the M13 phagemid vector pComb3 (*2*). In addition, nine semi-synthetic combinatorial libraries (*12, 13*) prepared from a human tetanus toxoid antibody, p3-13TT, modified to have a randomized CDRH3 and/or CDRL3 were screened for diuron hapten binding. Details of the cloning and selection of these recombinant antibodies and of their response to phenylureas in EIAs are given elsewhere (Karu, A.E., *et al.*, *Food & Agric. Immunol.*, 1994, in press; Scholthof, K-B.G., Zhang, G., and Karu, A.E., in preparation).

DNA Sequencing. Phagemid DNA was isolated from cultures of *E. coli* XL-1 Blue by alkaline lysis and purified by polyethylene glycol precipitation (*14*). The DNA was sequenced on both strands by the dideoxy nucleotide chain termination method using α-[^{35}S]dATP and a kit (Sequenase 2.0 - USB, Cleveland, OH). The variable region sequences from both strands of clones 112 and 224 (Scholthof, K-B.G., Zhang, G., and Karu, A.E., in preparation) derived from MAb 481.1 were obtained using synthetic oligonucleotide primers (Operon Technologies, Alameda, CA). The nucleotide sequences and open reading frames were collated and analyzed using MacVector and AssemblyLIGN software (IBI). The sequences of clone 112 were submitted to GenBank (accession nos. U04352 and U04353, H and L chain respectively). We subsequently refer to this clone as Fab 481.1, because its behavior in EIAs was identical to that of Fab obtained by proteolysis of MAb 481.1 (Scholthof, K-B.G., Zhang, G., and Karu, A.E., in preparation). In the synthetic combinatorial clones, the CDRL3 sequence was obtained with the primer SQCLHuKX (5'-GAAGTTATTCAGCAGGCACAC-3') and the CDRH3 sequence with the primer SQCH1HuGX (5'-GGGAAGTAGTCCTTGACCAGG-3'). These primers bind in the C$_L$ and the C$_{H1}$ regions, respectively, of the combinatorial clones. The remainder of the sequence of p3-13TT was provided by C. Barbas (The Scripps Research Institute).

Molecular Modeling. Initial models of the antibodies were constructed from their deduced amino acid sequence with AbM 1.2 software (Oxford Molecular, Palo Alto, CA) on a Silicon Graphics R-4000 work station. AbM identifies CDR loops that have a canonical structure (see below) and selects coordinates of the CDR with the highest sequence identity from an antibody database. For CDR loops that do not have a canonical shape, it searches a protein loop database for candidate structures. In the search, the loop constraint distances between alpha carbons (Cα), in Å, are those from the first Cα to every other Cα in the loop (DP) and from the last Cα to every other Cα in the loop (DM). For the CDRH3 loop of 481.1, it was necessary to relax the default constraints in order to find any candidates. The following values were used: DP1=3.65 3.98, DP2=3.32 8.18, DP3=3.56 12.15, DP11=4.18 5.09, DP12=5.23 6.84, DP13=6.94 8.31, DM1=3.59 3.99, DM2=5.17 6.64, DM3=8.52 9.94, DM11=9.63 13.89, DM12=6.21 10.35, DM13=6.94 8.31, with all other constraints set to zero. The model for Fab 481.1 was completed in 4 hours of cpu time. The final model is a file of atomic coordinates in Brookhaven Protein Data Bank (PDB) format (*15*). Amino acids in models generated by AbM are numbered sequentially beginning with the light chain. To facilitate subsequent analysis, we renumbered the CDR and framework amino acids to match the system of Kabat *et al.* (*16*).

The stereochemical quality of the AbM model was analyzed with the program PROCHECK (*17*). The model was visualized and subsequent changes (loop grafting and adjustment of side chain dihedral angles) were made with the graphics program INSIGHT 2.2.0 (Biosym Technologies, San Diego, CA) on an IBM RS/6000 Powerstation 320H.

The Fab 481.1 model was analyzed by superimposing it upon variable region domains of known antibody crystal structures from the antibody structural database (ASD) (*8, 18*). The coordinates for the antibodies McPC603 (1MCP) and 50.1 (1GGI) were obtained from the PDB. Coordinates for antibody glb2 (Gloop2 (*19*)) are not available in the PDB, but are provided with AbM. For replacement of loops in the model, a template loop was grafted from crystallographically determined antibody structures. The template side chains were replaced by those of the 481.1 sequence and their conformation was retained unless there were steric clashes with other amino acids. This procedure is described in more detail by Roberts *et al.* (*18*).

Models of Diuron and Haptens. For reasons described in the Results, the crystallographic coordinates for diuron were obtained from the Cambridge database of small molecules.

RESULTS

Diuron-binding Fab 481.1. Soluble Fabs from clones 112 and 224 competitively bound free diuron in competition EIAs with the same sensitivity and specificity as Fab fragments made by papain digestion of MAb 481.1 (Scholthof, K-B.S., Zhang, G., and Karu, A.E., in preparation). The DNA sequences of clones 112 and 224 were identical except for a single nucleotide difference in the codon for Kabat amino acid 203 near the C terminus of the L chain constant region (A in clone 112 and G in clone 224). This difference was silent (i.e., it codes for the same amino acid) and was probably a substitution made by the *Taq* polymerase used in polymerase chain reaction amplification of the genes. Thus, 112 and 224 were identical functional clones of MAb 481.1, even though they were selected by binding to different diuron haptens. When clones 112 and 224 were derived, an *Spe I* restriction enzyme site was found within the L chain sequence. This site was confirmed from the DNA sequence as being within the codon for amino acid L28 in CDRL1.

The light chain V—C_{L1} region was 664 bp in length. From the amino acid sequence (Table I) it belongs to the mouse kappa L chain subgroup III (*16*). The CDRL1 loop is relatively long at 15 amino acids. Long CDRL1 loops have been observed in other hapten-binding antibodies: 16 amino acids in the fluorescein binding Fab 4-4-20 (*20*) and 17 amino acids in the anti-phosphorylcholine Fab McPC603 (*21*).

The H chain V—C_{H1} region was 669 bp and the amino acid sequence of the CDRs is shown in Table I. It belongs to the mouse H chain subgroup III(D) (*16*). The 14 amino acid long CDRH3, which is the most variable CDR, is in the upper size range for mouse CDRH3s (*22*). As with CDRL1, long CDRH3s have also been reported in antibodies that bind small haptens: 12 amino acids in the *p*-azophenylarsonate-binding antibody 36-71 (*23*) and 15 amino acids in the *p*-azobenzenearsonate Fab R19.9 (*24*).

Diuron-binding Synthetic Combinatorial Fabs. Forty-two Fabs that bound diuron-alkaline phosphatase conjugates were isolated from semi-synthetic human combinatorial antibody libraries. None of these, however, competitively bound free diuron (Karu, A.E., *et al., Food & Agric. Immunol.*, 1994, in press). Soluble Fab prepared from the tetanus toxoid antibody clone (p3-13TT) used to derive the libraries did not bind diuron conjugates, nor did soluble protein from *E. coli* transformed with the pComb3 vector (Table II). Therefore the diuron hapten conjugate binding must be due to the CDRH3 and/or CDRL3 regions, which are the only regions modified from the p3-13TT Fab.

Sequences of the CDRH3 and CDRL3 regions of the 16 strongest diuron conjugate binders were determined (Table II). Due to constraints in the library construction (*12, 13*), clones had either 5, 10 or 16 amino acid CDRH3s with Asp H101 fixed (except for library H5) and 8 or 9 amino acid (libraries K9/E and K9/F) or 10 amino acid (libraries K10/E and K10/G) CDRL3s. Thus, the library from which each Fab originated could be identified from the sequence and length of L3 and H3 (Table II). From analysis of the atomic structure of antibodies, Chothia & Lesk (*25*) demonstrated that a relationship existed between the amino acid sequence and the 3-dimensional conformation of five of the six CDRs (*25-27*). The presence of key amino acids identified the loop as having a particular conformation or "canonical" class. Table III includes the lengths and canonical classes of the CDRs in Fab 481.1 and the combinatorial Fabs. The combinatorial Fabs varied in length and sequence of

     IMMUNOANALYSIS OF AGROCHEMICALS

Table I. CDR and key framework (FR) sequences of the murine antibody to diuron (Fab 481.1).

L1

Kabat No	24	25*	26	27	27a	27b*	27c	27d	27e	27f	28	29	32	33*	34	FR 2*	FR 71*
Fab 481.1	R	A	S	E	S	V	E	Y	Y	G	T	S	L	M	Q	L	F
1ggi	R	A	S	E	S	V	D	D	D	G	N	S	F	L	H	I	F

L2

Kabat No	50	51	52	53	54	55	56	FR 48*	FR 64*
Fab 481.1	G	A	S	N	V	E	S	I	G
glb2	A	A	S	T	L	D	S	I	G

L3

Kabat No	89	90*	91	92	93	94	95*	96	97†
Fab 481.1	Q	Q	S	R	K	V	P	A	T
glb2	L	Q	Y	L	S	Y	P	L	T

H1

Kabat No	31	32	33	34*	35	FR 26*	FR 27*	FR 29*	FR 94*
Fab 481.1	D	Y	G	M	H	G	F	F	R
3D6	D	Y	A	M	H	G	F	F	K

H2

Kabat No	50	51†	52	52a	53	54*	55	56	57	58	59	60	61	62	63	64	65	FR 71*
Fab 481.1	Y	I	S	S	G	S	S	T	I	Y	Y	A	D	T	V	K	G	R
3D6	G	I	S	W	D	S	S	I	G	Y	Y	A	D	S	V	K	G	R

H3

Kabat No	95	96	97	98	99	100	100a	100b	100c	100d	100e	100k	101	102	FR 94
Fab 481.1	W	D	T	T	V	S	G	H	Y	Y	V	M	D	Y	R
McPC603	N	Y	Y	G	S	T	W	-	-	-	Y	F	D	V	R

* Key amino acids that determine the canonical shape of CDR loops (25).

† CDR side chains that may also determine CDR structure due to extensive contacts with FRs (18).

Table II. Amino acid sequences of CDR3 regions of Fab clones selected from synthetic combinatorial antibody libraries, which bound diuron hapten conjugate in direct EIA.

Clone	CDRH3 (H95-H102)	Size	CDRL3 (L89-L97)	Size	EIA rate[a] vs		Library[b]
					DIII	DI	
p3-13TT	GDTIFGVTMGYYAMDV	16	QQYGGSPW	8	2	<1	Tet tox Fab
pComb3	–	-	–	-	2	0	vector
10.4	SIFDP	5	QQYGAQMGAT	10	42	6	K10/G
12.4	STSWG	5	QQYGGSPW	8	15	3	H5
g1-23-8	GGWWRMLLDY	10	"	"	21	2	F
g1-23-9.2	VGRLWYPRDF	10	"	"	16	23	"
g1-24-9.2	SRPNRWWSDK	10	"	"	10	0	"
g1-28-1.1	GWGLLAKRDP	10	"	"	32	28	"
g1-16-6	GSWL(S/R)FLLDS	10	"	"	32	13	"
g1-23-6	GPRSFRGFFWGWLLDL	16	"	"	59	26	E
g1-26-5	SGHYRFHHAWPRVLDW	16	"	"	37	7	"
g1-28-1.3	GQGRRLYDLWRGYFDF	16	"	"	21	47	"
g1-15-4	GGRPFRGLRVVHALDY	16	"	"	53	23	"
g1-24-9.1	SWPLWGRFWSLWGQDN	16	"	"	25	50	"
g1-26-8.2	GSRAMFLGWMARPFDD	16	nd	"	8	56	"
g2-25-1	GNNVGSWMGWRRRFDV	16	QQYWGAGGLT	10	18	3	K10/E
g2-19-10.3	GGRVWHRLGGATWADI	16	QQYZVWLGVT	10	62	46	"
g3-25-7.1	SLMWILRLRWGWFPDL	16	QQYRAGGRT	9	41	66	K9/E
112/224	WDTTVSGHYYVMDY	14	QQSRKVPAT	9	54	37	Fab 481

[a] Rate = ΔA_{405nm} min^{-1} x 10^{-3} of a 1:10 (1:100 for 112/224) dil of soluble Fab in a direct EIA on wells coated with 500ng of anti-human Fab (anti-mouse for 112/224) and detected with diuron-III or diuron-I alkaline phosphatase conjugate (*11*).

[b] Library designations from C. Barbas, The Scripps Research Institute and (*12, 13*).

CDRH3 and/or CDRL3 with a predominance of hydrophobic amino acids, but none had the same CDRH3 or CDRL3 as mouse Fab 481.1 (Table II).

The mouse and combinatorial Fabs were similar in that both had 7-amino acid CDRL2s and 5-amino acid CDRH1s of the same canonical class. However, CDRs L1, L3, and H2 differ in length and/or canonical structure or lack thereof, in each. The majority of combinatorial clones were from libraries E and F, which have the same CDRL3 sequences as the parent tetanus toxoid Fab. This CDRL3 is unusual because it is only 8 amino acids long and is therefore unlikely to have the canonical shape of the 481.1 CDRL3 (Table III).

Table III. Length and canonical class[a] of the CDRs in the mouse and synthetic combinatorial antibodies for diuron.

Antibody	CDRL1 size	class	CDRL2 size	class	CDRL3 size	class	CDRH1 size	class	CDRH2 size	class	CDRH3 size
Fab 481.1	15	—	7	1	9	1	5	1	17	3	14
Comb. Fabs[b]	12	—	7	1	8,9,10	—	5	1	17	—	5,10,16

[a] As defined by Chothia and Lesk (25-27).
[b] The combinatorial Fabs all have the same framework and CDR sequences (except CDRs H3 and/or L3).

In addition, the side chain of the last amino acid in the tetanus toxoid CDRL3, Trp L97 (Table II), may disrupt the loop structure because it is much larger than the Thr or Val side chains that usually occur at this position (16). The lack of competitive binding of free diuron by the combinatorial Fabs suggests that interactions with the attached protein and/or the linking group may contribute to strong binding. The differences between CDRs L1, L3, and H2 of the combinatorial Fabs and Fab 481.1 raise the question of whether antibodies with high-affinity diuron binding could be obtained from the libraries we used. Although hapten contacts with CDRs L3 and H3 are usually more extensive than with other CDRs, the predetermined shape of L1, L3 and H2 may preclude strong hapten binding. Models of three of the combinatorial clones were created with AbM (data not shown), but how they bind diuron conjugate was not determined.

Molecular Modeling of Fab 481.1. Computational modeling of the 3-dimensional structure of Fab 481.1 was used to evaluate how diuron may be bound. This may also provide insight into how diuron conjugate but not free diuron was bound by the combinatorial Fabs. Although X-ray crystallography would provide the most accurate determination of the Fab structure and binding interactions, a close approximation can be obtained by computer modeling as a result of the extensive conformational similarities in antibody framework regions and some antigen binding loops (25, 26). Initial molecular models were constructed with the antibody modeling package AbM, which brings together features of knowledge-based and conformational search modeling in a combined algorithm (28, 29).

Construction of the Framework Domains. The first step in modeling makes use of known V_H and V_L amino acid sequences to generate suitable framework region structures. The template for the light chain FR is assigned from an antibody crystal structure database by comparison of sequence homology. The L chain FR for Fab 481.1 was taken from McPC603 (2MCP), which has 64% sequence identity. The H chain FR template was selected on a slightly different basis. Two conformational classes of CDRH3 have been defined; one class has Tyr or Val and the other has Ser or Gly at the C-terminal position (amino acid 102) in CDRH3 (30). The H chain FR was assigned by sequence identity with known structures of the same class. Fab 481.1

V_H had 72% identity to Fab 17/9 (1HIL) of class 1 (*30*), which was selected for the H chain framework.

After AbM found the appropriate templates, the angle between the H and L chains was taken from that of the light chain template (*30*). The FR amino acids in the templates were replaced with those from the diuron Fab, with sidechain torsion angles $(X_1, X_2,$ etc.) chosen so that the conformations matched those of the template.

Building the Canonical CDRs. The loops of the CDRs, which are more varied in structure than the FRs, were modeled using two strategies. For the first of these, AbM used the canonical structure rules (*25-27*) to search the 481.1 sequence, assigning loops with key amino acids to specific canonical structures. Four of the six CDRs (L2, L3, H1, H2 in Table III) in Fab 481.1 were found to have canonical structures. The backbone coordinates for the template loop with the highest overall sequence identity was used for the loop in the model. The sequence from the diuron Fab was then substituted into the canonical loop and the side chain conformation was built by conformational search (*31*). It should be pointed out that the CDRs defined by Chothia and coworkers (*25-27*) and in AbM overlap. However they do not entirely correspond (CDRH2 in particular) to the hypervariable regions defined from sequence variation by Kabat *et al.* (*16*).

CDRL2 is composed of amino acids L50 to L56 and connects adjacent strands in the framework ß-sheet. The presence of Ile (L48) and Gly (L64) in the FR of Fab 481.1 identified CDRL2 as a member of canonical class 1, with highest sequence identity to CDRL2 of antibody glb2 (Gloop2) (*19*) (Table I). The 9-amino acid CDRL3 loop of Fab 481.1 (L89-L97 in Table I) was placed in canonical class 1, due to the Gln at L90 and Pro at L95, and had highest sequence identity to glb2 CDRL3 (Table I). The shape of CDRL3 is maintained by hydrogen bonds formed between the side chain of L90 and the main chain atoms of L92, L93 and L95 (*25*). The conservation of the amino acids at L90 and L97 may also be important, because their side chains make several contacts with the framework (*18*). The predicted orientation of L89, L91 and L96 in this canonical class directs the side chains towards the H chain, forming part of the antigen-binding pocket.

The shape of CDRH1 (amino acids H31-H35) is constrained by FR H29 (Phe), which acts as a hydrophobic anchor. This side chain is predicted to pack against the side chain of H34 (Met) and the main chains of amino acids of H72 and H77. The Phe amino acid at H27 is partially buried in a cavity next to H94 (Arg) (*25*). Along with Gly at H26, these amino acids place the CDRH1 of Fab 481.1 in canonical class 1 (*27*). The Tyr at H32, which is conserved in many antibodies, contacts the VH framework of 481.1 and may also be important for the conformation of CDRH1 (*18*). The CDRH1 from Fab 3D6 (1DFB) (*32*) had the closest sequence identity with that of 481.1 and was used as the template (Table I).

CDRH2 (H50-H65 in Table I) of Fab 481.1 has conserved amino acids that define it as a class 3 canonical loop. These are Ile H51, Ser H54, and FR Arg H71, which contact the framework (*25, 26*). The size of the amino acid at FR H71 helps define the shape of CDRH2 (*33*). The presence of Arg at H71 indicates that the 481.1 CDRH2 has a well-defined structure with amino acid H52a (Ser) solvent exposed. AbM used the CDRH2 of 3D6 (1DFB) for the model (Table I).

Modeling Non-Canonical Loops. AbM uses a second strategy, a combined algorithm, for modeling loops that do not fit into a specific canonical structural class. A database of protein loop Cα distance constraints derived from the PDB is searched to match specific constraints for the loop to be built. For loops of seven amino acids or less, there are sufficient candidates to cover all the conformational space. For those eight or more amino acids long, there are usually too few candidate loops, so part of the loops is rebuilt using the conformational search routine CONGEN (*34*). This latter method was used for the long loops CDRs L1 and H3 of 481.1. Although database exploration combined with conformational searching works well for loops up to seven amino acids, the assumption that enough examples of longer loops will be well represented in protein crystal structure databases has been questioned (35). We found this to be the case with the 15 amino acid CDRL1 (L24-L34), as only a single loop was retrieved with the default constraints from 167,487 entries and this happened to be the CDRL1 loop from the antibody 50.1 (1GGI). AbM then selected the CDRL1 of the model from 1,316 resultant conformations. Modeling the CDRH3 (14 amino acids - H95-H102) required relaxing the default database search constraints within the program to return any candidates. When this was done, 9 loops were identified and AbM selected a final shape from 539 conformations.

Evaluation and Refinement of the Fab 481.1 Model. The initial model was evaluated with the programs in PROCHECK (17), which compare a series of parameters in the model to those of crystallographically determined structures for over-all stereochemical quality. The programs provide several indices of the local, amino-acid-by-amino-acid reliability of the model. Figure 1 is a Ramachandran plot that compares the polypeptide backbone torsion angles ϕ and ψ in the H and L chains of the Fab 481.1 model with the range of observed values in crystallized proteins. Although 97% of the amino acid backbones were in the most favored and allowed ranges, three amino acids were in the generously allowed and three in the disallowed regions. Five of these occurred within CDRs and the sixth (L15) was in the framework of the L chain. The main-chain properties of the percentage of amino acids in core favored regions and the standard deviations (S.D.) of peptide bind planarity (ω angle) were better in the model when compared to other well-refined structures. The geometry of main-chain hydrogen bonds (H-bond), measured as S.D.s from the expected H-bond energy, was just outside the mean, at one S. D. However, the number of non-bonded or bad contacts (where amino acids are ≤ 2.6 Å from each other) and the α-carbon tetrahedral distortions (zeta S.D.) were significantly worse (respectively, 25.5/100 amino acids compared with a typical value of 4.2/100 and 13.5 compared with a typical value of 3.1). Stereochemical properties of the side chains of the initial model were all better or within one S. D. of values from well-refined structures. It was noted that significant errors in stereochemical quality, namely a-carbon chirality and H-bond energy deviation, were all clustered in the CDRL1 and H3 regions, indicating that these areas required further attention.

This analysis did not reveal more global dissimilarities, such as CDR positioning. Therefore the model was visually compared with antibodies from a database of superimposed antibody structures, the ASD (8, 18). This showed that the loops modeled by conformational search in AbM (CDRL1 and CDRH3), differed significantly in structure from the way these loops appear in known antibodies.

Correction of CDRL1. The conformation of the AbM-built CDRL1 loop was found to deviate significantly from other antibody structures when superimposed onto the ASD. Met L33, which was one of the amino acids identified in the disallowed region of the Ramachandran plot (Figure 1), was incorrectly oriented so that highly conserved main-chain hydrogen bonds were not formed and the side chain did not properly extend into the V_L framework. The orientation of this amino acid is critical to formation of the binding pocket. In addition, the highly conserved main-chain conformation of the first six residues of CDRL1 (26) was not retained in the conformationally built loop.

The 481.1 CDRL1 (15 amino acids, Table I) has key amino acids in most of the positions that would define it as canonical class 2: Leu L2, Ala L25, Val L27b, Met L33, and Phe L71, but it is longer than other CDRL1s in this class. The extra amino acids would most likely be accommodated as a loop or turn on the antibody's solvent-

Upper Half

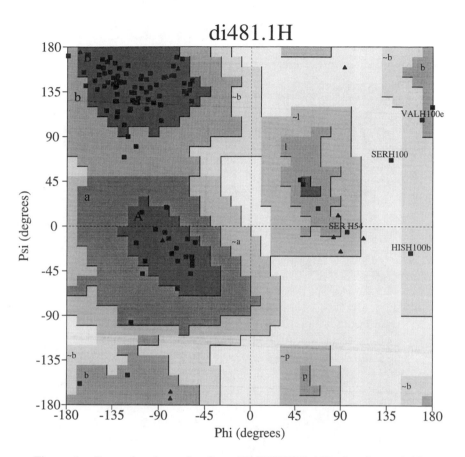

Figure 1. Ramachandran plot from PROCHECK (*17*) for the variable regions of the light chain (upper) and the heavy chain (lower) of the initial AbM-built model of Fab 481.1. Side chains of all amino acids are marked with a square except for glycines, which are shown as triangles. Those side chains in the generously allowed and disallowed regions have been labeled.

Lower Half

di481.1L

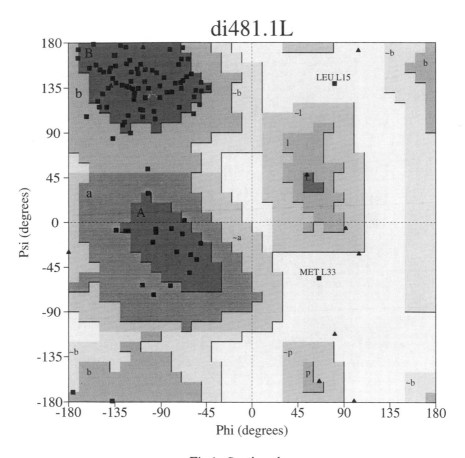

Fig.1. *Continued*

accessible surface. A recent antibody structure (Fab 50.1 - 1GGI (*36*)) with a 15 amino acid CDRL1 (53% sequence identity with Fab 481.1) was added to the PDB. The CDRL1s of 50.1 and Fab 481.1 have the same key structural amino acids, so the lower halves of these loops should share the conserved features. Accordingly, we replaced the AbM-built CDRL1 loop of 481.1 (L24-L34) with that of Fab 50.1 by least squares superposition and corrected the sequence. The difference in shape was evident when the Cα backbones of the two loops were superimposed (Figure 2).

Resolving a Steric Clash in CDRs H1 and H2. At the bottom of the binding pocket, the van der Waals radii of the side chains of His H35 in CDRH1 and the first amino acid in the CDRH2 (Try H50) overlapped. In several antibodies, the H35 side chain is oriented so that it creates a hydrogen bond with the conserved framework amino acid Trp H47 (*18*). We therefore moved the side chain of His H35 to a position that allows hydrogen bonding between Nδ of the His imidazole ring and the indole nitrogen atom of Trp H47, which allowed reorientation of Tyr H50.

The amino acid at H35 has both a structural and an antigen-binding role in numerous antibodies. In HyHEL-5 Glu H35 forms a salt bridge with the antigen lysozyme (*37*). His H35 packs against the phenyl ring of phenyloxazolone in antibody NQ10/12.5 and is preserved in all other antibodies that bind this antigen (*38*). Mutation of His H35 in the phosphotyrosine antibody Py20 (*39*) or the catalytic antibody 43C9 (*40*) destroyed binding and function. In the phenylarsonate-binding antibody 36-71, Asn H35 contacts the antigen and is conserved in 81 other phenylarsonate antibodies (*41*).

Adjustment of CDRH3. The CDRH3 loop is formed by recombination and is subject to additional variation by other mechanisms. This makes CDRH3 the most difficult CDR to model, especially in Fab 481.1, where it is 14 amino acids long. However, some structural conservation does exist at the termini of this loop. A salt bridge between framework amino acids Arg H94 and Asp H101 of CDRH3 is conserved in numerous other antibodies (*42*) and probably stabilizes the CDRH3 structure in 481.1. Many antibodies share a conserved shape for the first two and the last four amino acids, with the main-chain nitrogen of H96 (Asp in 481.1) forming a hydrogen bond with the main-chain carbonyl oxygen of the fourth amino acid from the end of CDRH3 (Val H100e in Fab 481.1; see for example McPC603 (*25*)). This structural motif then defines the positions of the side chains of the first (Trp H95) and last three amino acids of CDRH3 (Met-Asp-Tyr — H100k-H102 in 481.1). In addition, a hydrogen bond exists between the main-chain carbonyl oxygen of the last amino acid in CDRH3 (Tyr H102) and the main-chain nitrogen of Arg H94. None of these motifs were preserved in the CDRH3 conformation modeled by AbM. Figure 3 compares the CDRH3 loop of Fab 481.1 from the initial AbM model with that of McPC603, which has the conserved features. On the basis of these similarities, we substituted the backbone conformations of amino acids H93-H96 and H100e-H103 from McPC603 into the 481.1 model, and then replaced the necessary side chains. The upper part of the loop (H97-H100d) was retained since it is exposed to the solvent and likely to be flexible and may adjust its conformation upon antigen binding.

We concentrated on the base of CDRH3 as a recent study of known antibodies indicated that the shape of CDRH3 is strongly influenced by the conformation near the base of the loop (*35*). There are two classes of base conformations -- kinked and unkinked. An extensive conformational modeling search with the CDRH3 loop of a carcinoembryonic antigen (CEA) antibody, also 14 amino acids long, resulted in three equally plausible structures. These structures all had the kinked pattern, with an Arg H94-Asp H101 salt bridge and a Met at H100k (same amino acid as in 481.1) filling a solvent-inaccessible cavity (*35*). The CDRH3 of McPC603 also had this kinked pattern at the base. The three CEA loops were very similar to each other at the base

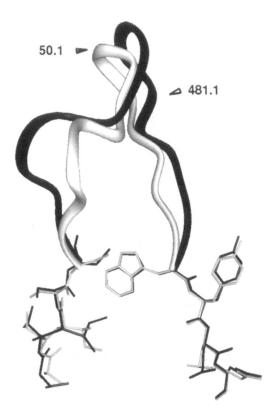

Figure 2. The CDRL1 loops of 50.1 (1GGI) and Fab 481.1, as built by AbM, are shown in ribbon outline for the Cα backbone. The loops were superimposed by least squares fitting of the backbone atoms of the two framework residues on either side of the loop.

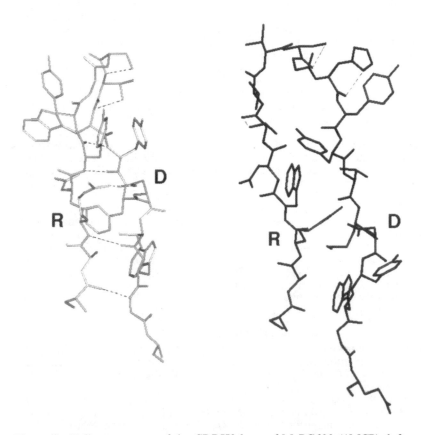

Figure 3. Full atom traces of the CDRH3 loop of McPC603 (1MCP), left, and Fab 481.1 as modeled by AbM, right. They were superimposed by least squares fitting of the backbone atoms of two framework residues on either side of the loop, then moved apart. The conserved Arg H94 and Asp H101 are labeled, and hydrogen bonds, salt-bridge, and other close interactions are identified with dotted lines.

but diverged significantly at the top. Such conformational flexibility is expected for large loops and is often seen as a region of poorly defined electron density in crystallographic data.

Accuracy of the Revised Model. Re-evaluation of the corrected model with PROCHECK (*17*) showed an improvement in its stereochemical quality. Exchange of the CDRL1 and H3 loops significantly reduced the deviations observed in α-carbon chirality and H-bond energy for amino acids L24, L25, L33, H95 and H102. The disallowed Ramachandran plot status of L33 was eliminated. Although the status of L27e was changed to disallowed, this amino acid does not contribute to the binding pocket. The non-bonded contact between H35 and H47 was also greatly reduced. Since the model was not subjected to subsequent minimization, some worsening of values for bond lengths and non-bonded contacts around the positions of loop-grafting was observed. Overall there was a good improvement in the 481.1 model's quality, with a lower H-bond energy S. D. and one more amino acid in the most favored regions of the Ramachandran plot.

Analysis of the Diuron Binding Site. After refining the model, diuron was placed within the combining site. We took into account the structure of diuron and the diuron haptens, the cross-reactivity of Fab 481.1 with various phenylureas in competition EIAs, and the predominant types of side-chain contacts and interactions of small molecules with crystallographically solved antibodies.

The crystal structure from the Cambridge database of small molecules indicated that diuron and related molecules can have four possible conformations, with a dihedral angle of about 30° between the planes through the urea group and the dichlorophenyl ring (Figure 4). The crystallographic structure of diuron was used for docking studies because models built in Chem3D (Cambridge Scientific Computing) or Nemesis (Oxford Molecular) differed significantly from the crystallographic coordinates. Energy minimization with the modified MM2 or the COSMIC force fields in these programs moved the urea and dichlorophenyl groups into the same plane, causing a steric clash between the urea group and hydrogens on the ortho position of the dichlorophenyl ring.

It is likely that diuron would fit into the binding site in the same orientation as that of the diuron conjugates used to select the Fab and original MAb. The spacer arm was attached to either the internal nitrogen (diuron-I) or the urea nitrogen (diuron-III in Figure 4) (*11*), positioning the hapten with the carbonyl oxygen pointing downwards.

MAb 481.1 and Fab 481.1 bind monuron (a single chlorine on the phenyl ring) only 3% to 10% as well as diuron in competition EIAs. Fenuron, which lacks any chlorines, is not competitively bound (*11*). These data indicate that the chlorinated phenyl moiety is the primary recognition feature, suggesting it has extensive contacts with the antibody.

The unusually long CDR L1 and H3 loop of the revised Fab 481.1 model gave the antigen binding site a deep pocket (Figure 5a). The binding site is a cleft with CDRL3 and CDRH3 on one side and CDRH1 and CDRH2 on the other, as seen in other hapten-binding antibodies (*43*). Within this cleft is a hydrophobic area, into which extend the side chains of several nonpolar amino acids, including Ala L96 (CDRL3), Tyr H32 (CDRH1), Tyr H50 (CDRH2), and Trp H95 and Tyr H100d from CDRH3. The dichlorophenyl ring of diuron was docked by computer graphics into this hydrophobic region, resulting in extensive hydrophobic interactions of the dichlorophenyl ring and the methyl groups on diuron with the antibody. Similar interactions are seen in the binding of phenylarsonate by Fab 36-71 (*41*), 2-phenyloxazolone by Fab NQ11.7.22 (*44*), and fluorescein by Fab 4-4-20 (*20*). In addition, this placement centers the dichlorophenyl ring in the antigen-binding pocket, consistent with its important role in recognition.

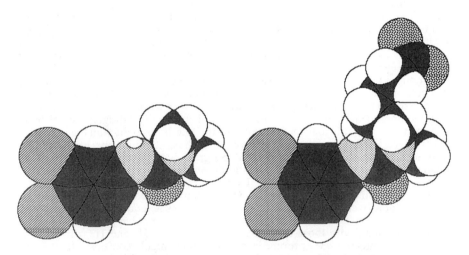

Figure 4. Diuron (left) and diuron-III hapten (right) (*11*) are shown in space filling display.

The last amino acid in CDRH1 (His H35) is at the bottom of the pocket in a position to hydrogen bond with the carbonyl oxygen of diuron. The chlorines on diuron have a diffuse electronegative charge that could form an electrostatic dipole with the somewhat electropositive nitrogen atoms on vicinal side chains in the binding pocket. The side chains of Gln L34 on CDRL1, Gln L89 and Ser L91 of CDRL3, and possibly Met 100k on CDRH3 are in positions to form such a dipole with diuron. These amino acids and a proposed orientation for diuron are shown in Figure 5b. In our model there are no amino acids of CDRL2 that contact diuron.

DISCUSSION

Antibodies are a specialized class of binding proteins in which sequence diversity in about 10% of the molecule gives rise to the ability to bind a vast number of different antigens. Antibody modeling strategies usually draw upon knowledge of the conserved nature of antibody structure gained from X-ray crystallography. A computational model is a valuable tool for antibody engineering because it can provide information about the amino acids involved in ligand binding and help define a rational strategy for antibody design and mutational analysis (6, 18). We analyzed the binding of the phenylurea herbicide diuron to a high affinity antibody because it may serve as a prototype for interactions between organochlorines and macromolecules.

An initial model of Fab 481.1 was built from the deduced amino acid sequence with the computer program AbM. The conformations of CDRs L2, L3, H1 and H2 were based upon the presence of structurally conserved amino acids (25-27). When we superimposed the AbM-built Fab 481.1 model onto crystallographic antibody structures using the ASD, the conformations of the four loops built with the canonical rules were very similar to those of known antibodies, although one loop required adjustment of the side chains. The AbM program was much less successful in modeling the two longer CDRs for which it did not find canonical structures, CDRs L1 and H3. AbM modeled these CDRs by conformational searching, loop reconstruction, and energy screening (28, 31), but important conserved features at the ends of the loops were not retained in the model. The structurally conserved regions at the end of CDRs L1 and the beginning of H3 are particularly critical, as these amino acids, along with those at the base of CDRs L3, H1, and H2, often either contact antigen directly (45) or form a layer between the highly conserved framework and antigen-contacting residues (46). Positioning these side chains as accurately as possible is essential for building a properly shaped antigen-binding pocket. Fortunately, comparison of the determined crystallographic structures suggests that both the main-chain (25, 26) and side-chain (18) positions for most of these residues are highly conserved. Therefore, we chose to replace all or part of the initially built L1 and H3 loops with corresponding regions from known antibody structures.

The entire CDRL1 loop was replaced by the CDRL1 coordinates from a recently determined antibody, 50.1. This antibody has the same length CDRL1 as 481.1 and shares the same key residues for determining loop conformation. Even though AbM selected this 50.1 CDRL1 loop during its Cα loop database search, we would have expected that the loop should still retain important structurally conserved features during the subsequent conformational modeling process. In this case, the AbM program did not do so.

The CDRH3 loop presents a much more difficult modeling task because it is sensitive to the surrounding environment. CDRH3 loops of the same length in different antibodies often have significantly different conformations, and this loop can even show substantial conformational changes upon antigen binding in the same antibody (47). The base of CDRH3, however, does show strong structural conservation, which was not preserved in the AbM-built 481.1 loop. Therefore, the base of the 481.1 CDRH3 was rebuilt from antibody crystallographic coordinates of

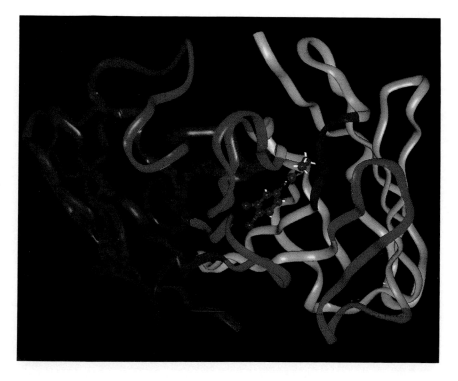

Figure 5. A. Cα backbone ribbon trace of the revised model of Fab 481.1, viewed looking down into the antibody combining site. The light chain (purple) is on the left and the heavy chain (light blue) is on the right. The CDRs are colored red (CDRs L1 and H1), yellow (CDRs L2 and H2) and green (CDRs L3 and H3). Diuron is positioned within the site and is shown in ball and stick representation (colored by atom type)

antibodies with increased off-rates for hapten and antigen binding may be desirable in applications such as self-regenerating biosensors for analyzing small toxic analytes.

Acknowledgments

This work was supported in part by the UC Toxic Substances Research and Teaching Program and a DOE Office of Technology Development subcontract LLL-B244824 to AEK and by NIH grant GM 48877 to VAR. AEK is an Investigator in the Environmental Health Sciences Center at UCB (NIEHS Grant ES01896). K-BGS is an NIH Postdoctoral Fellow (Grant AI-08710). We thank Carlos Barbas for providing pComb3 and the synthetic combinatorial libraries, Andy Jackson for advice and laboratory facilities, Tim Robinson (Graphics Facility, UCB College of Chemistry) for help with computer graphics, Dave Rockhold and Bill Belknap (USDA-ARS, Albany, CA) for use of the sequence analysis programs, and Sunil Maulik (Oxford Molecular) for help with AbM.

Literature Cited

1. Köhler, G.; Milstein, C. *Nature* **1975**, *256*, 495-497.
2. Barbas, C. F.; Kang, A. S.; Lerner, R. A.; Benkovic, S. J. *Proc. Natl. Acad. Sci. USA* **1991**, *88*, 7978-7982.
3. Jansen, M. A. K.; Mattoo, A. K.; Malkin, S.; Edelman, M. *Pest. Biochem. Physiol.* **1993**, *46*, 78-83.
4. Wu, T. T.; Kabat, E. A. *J. Exp. Med.* **1970**, *132*, 211-250.
5. Alzari, P.; Lascombe, M.-B.; Poljak, R. *Ann. Rev. Immunol.* **1988**, *6*, 555-580.
6. Roberts, S.; Cheetham, J. C.; Rees, A. R. *Nature* **1987**, *328*, 731-734.
7. Reichmann, L.; Clark, M.; Waldmann, H.; Winter, G. *Nature* **1988**, *332*, 323-327.
8. Roberts, V. A.; Iverson, B. L.; Iverson, S. A.; Benkovic, S. J.; Lerner, R. A.; Getzoff, E. D.; Tainer, J. A. *Proc. Natl. Acad. Sci. USA* **1990**, *87*, 6654-6658.
9. Barbas, C. F.; Bain, J. D.; Hoekstra, D. M.; Lerner, R. A. *Proc. Natl. Acad. Sci. USA* **1992**, *89*, 4457-4461.
10. Hoogenboom, H. R.; Winter, G. *J. Mol. Biol.* **1992**, *227*, 381-388.
11. Karu, A. E.; Goodrow, M. H.; Schmidt, D. J.; Hammock, B. D.; Bigelow, M. W. *J. Agr. Food Chem.* **1994**, *42*, 301-309.
12. Barbas, C. F., III; Rosenblum, J. S.; Lerner, R. A. *Proc. Natl. Acad. Sci. USA* **1993**, *90*, 6385-6389.
13. Barbas, C. F., III; Amberg, W.; Simoncsits, A.; Jones, T. M.; Lerner, R. A. *Gene* **1993**, *137*, 57-62.
14. Sambrook, J; Fritsch, E. F.; Maniatis, T. *Molecular Cloning, A Laboratory Manual*; 2nd Edn; Cold Spring Harbor Laboratory: Cold Spring Harbor, NY, 1989; Vol. 1-3.
15. Bernstein, F. C.; Koetzle, T. F.; Willliams, G. J. B.; Meyer, E. F., Jr; Brice, M. D.; Rodgers, J. R.; Kennard, O.; Shimanouchi, T.; Tasumi, M. *J. Mol. Biol.* **1977**, *112*, 55-542.
16. Kabat, E. A.; Wu, T. T.; Perry, H. M.; Gottesman, K. S.; Foeller, C. *Sequences of Proteins of Immunological Interest*; Public Health Service, N. I. H.: Washington, D. C., 1991.
17. Laskowski, R. A.; MacArthur, M.; Moss, D. S.; Thornton, J. M. *J. Appl. Crystallogr.* **1993**, *26*, 283-291.
18. Roberts, V. A.; Stewart, J.; Benkovic, S. J.; Getzoff, E. D. *J. Mol. Biol.* **1994**, *235*, 1098-1116.
19. Darsley, M. J.; Rees, A. R. *EMBO J.* **1985**, *2*, 383-392.
20. Herron, J. N.; He, X. M.; Mason, M. L.; Voss, E. W., Jr; Edmundson, A. B. *Proteins: Struct. Funct. Genet.* **1989**, *5*, 271-280.

21. Segal, D. M.; Padlan, E. A.; Cohen, G. H.; Rudikoff, S.; Potter, M.; Davies, D. R. *Proc. Natl. Acad. Sci. USA* **1974**, *71*, 4298-4302.
22. Wu, T. T.; Johnson, G.; Kabat, E. *Proteins: Struct. Funct. Genet.* **1993**, *16*, 1-7.
23. Rose, D. R.; Strong, R. K.; Margolies, M. N.; Gefter, M. L.; Petsko, G. A. *Proc. Natl. Acad. Sci. USA* **1990**, *87*, 338-342.
24. Lascombe, M.; Alzari, P.; Boulot, G.; Saludjian, P.; Tougard, P.; Berek, C.; Haba, S.; Rosen, E.; Nisonhoff, A.; Poljak, R. *Proc. Natl. Acad. Sci. USA* **1989**, *86*, 607-611.
25. Chothia, C.; Lesk, A. M. *J. Mol. Biol.* **1987**, *196*, 901-917.
26. Chothia, C.; Lesk, A. M.; Tramontano, A.; Levitt, M.; Smith-Gill, S. J.; Air, G.; Sheriff, S.; Padlan, E. A.; Davies, D.; Tulip, W. R.; Colman, P.; Spinelli, S.; Alzari, P. M.; Poljak, R. J. *Nature* **1989**, *342*, 877-883.
27. Chothia, C.; Lesk, A. M.; Gherardi, E.; Tomlinson, I. M.; Walter, G.; Marks, J. D.; Llewelyn, M. B.; Winter, G. *J. Mol. Biol.* **1992**, *227*, 799-817.
28. Martin, A. C. R.; Cheetham, J. C.; Rees, A. R. *Proc. Natl. Acad. Sci. USA* **1989**, *86*, 9268-9272.
29. Martin, A. C. R.; Cheetham, J. C.; Rees, A. R. *Meth. Enzymol.* **1991**, *203*, 121-153.
30. *AbM Users' Guide*; Oxford Molecular Ltd., Oxford, 1992.
31. Bruccoleri, R. E.; Haber, E.; Novotny, J. *Nature* **1988**, *335*, 564-568.
32. He, X.-M.; Rüker, F.; Casale, E.; Carter, D. C. *Proc. Natl. Acad. Sci. USA* **1992**, *89*, 7154-7158.
33. Tramontano, A.; Chothia, C.; Lesk, A. M. *J. Mol. Biol.* **1990**, *215*, 175-182.
34. Bruccoleri, R. E.; Karplus, M. *Biopolymers* **1987**, *26*, 137-168.
35. Mas, M. T.; Smith, K. C.; Yarmush, D. L.; Aisaka, K.; Fine, R. M. *Proteins: Struct. Funct. Genet.* **1992**, *14*, 483-498.
36. Stura, E. A.; Stanfield, R. L.; Fieser, G. G.; Silver, S.; Roguska, M.; Hincapie, L. M.; Simmerman, H. K. B.; Profy, A. T.; Wilson, I. A. *Proteins: Struct. Funct. Genet.* **1992**, *14*, 499-508.
37. Sheriff, S.; Silverton, E. W.; Padlan, E. A.; Cohen, G. H.; Smith-Gill, S. J.; Finzel, B. C.; Davies, D. R. *Proc. Natl. Acad. Sci. USA* **1987**, *84*, 8075-8079.
38. Alzari, P. M.; Spinelli, S.; Mariuzza, R. A.; Boulet, G.; Poljak, R. J.; Jarvis, J. M.; Milstein, C. *EMBO J.* **1990**, *9*, 3807-3814.
39. Ruff-Jamison, S.; Glenney, J. R., Jr *Protein Eng.* **1993**, *6*, 661-668.
40. Stewart, J. D.; Roberts, V. A.; Thomas, N. R.; Getzoff, E. D.; Benkovic, S. J. *Biochemistry* **1994**, *33*, 1994-2003.
41. Strong, R. K.; Campbell, R.; Rose, D. R.; Petsko, G. A.; Sharon, J.; Margolies, M. N. *Biochemistry* **1991**, *30*, 3739-3748.
42. Chien, N. C.; Roberts, V. A.; Giusti, A. M.; Scharff, M. D.; Getzoff, E. D. *Proc. Natl. Acad. Sci. USA* **1989**, *86*, 5532-5536.
43. Wilson, I. A.; Stanfield, R. L. *Curr. Opin. Struct. Biol.* **1993**, *3*, 113-118.
44. McManus, S.; Reichmann, L. *Biochemistry* **1991**, *30*, 5851-5857.
45. Glockshuber, R.; Stadlmüller, J.; Plückthun, A. *Biochemistry* **1991**, *30*, 3049-3054.
46. Chen, C.; Roberts, V. A.; Rittenberg, M. B. *J. Exp. Med.* **1992**, *176*, 855-866.
47. Rini, J. M.; Schulze-Gahmen, U.; Wilson, I. A. *Science* **1992**, *255*, 959-965.
48. Lerner, R. A.; Kang, A. S.; Bain, J. D.; Burton, D. R.; Barbas, C. F. *Science* **1992**, *258*, 1313-1314.
49. Goodsell, D. S.; Lauble, H.; Stout, C. D.; Olson, A. J. *Proteins: Struct. Funct. Genet.* **1993**, *17*, 1-10.
50. Friedman, A. R.; Roberts, V. A.; Tainer, J. A. *Proteins: Struct. Funct. Genet.* **1994**, *20*, 15-24.
51. Reichmann, L.; Weill, M.; Cavanagh, J. *J. Mol. Biol.* **1992**, *224*, 913-918.
52. Cheung, P. Y. K.; Kauvar, L. M.; Engqvist-Goldstein, A. E.; Ambler, S. M.; Karu, A. E.; Ramos, L. S. *Analyt. Chim. Acta* **1993**, *282*, 181-192.

RECEIVED August 15, 1994

Chapter 5

Antidioxin Monoclonal Antibodies

Molecular Modeling of Cross-Reactive Congeners and the Antibody Combining Site

Larry H. Stanker[1], Adrian Recinos III[2], and D. Scott Linthicum[3]

[1]Food Animal Protection Research Laboratory, Agricultural Research Service, U.S. Department of Agriculture, 2881 F&B Road, College Station, TX 77845
[2]Lawrence Berkeley Laboratory, Donner Laboratory, University of California, Berkeley, CA 94720
[3]Department of Veterinary Pathobiology, Texas A&M University, College Station, TX 77845

A series of modeling experiments were undertaken to help clarify the factors controlling binding of a set of monoclonal antibodies that bind polychlorinated dibenzo-p-dioxins. Minimum energy conformations were generated for a number of polychlorinated dibenzo-p-dioxins, furans and PCBs congeners. These models suggest that antibody binding is a complex process but that the size, position of chlorines and planarity of the molecules are critical for antibody binding. Similar experiments with the hapten used to generate these monoclonal antibodies suggest that both structural and electronic alterations introduced in order to facilitate conjugation to carrier protein are recognized by the antibodies. The amino acid sequence for these antibodies also is presented as well as models of the antibody combining site.

The polychlorinated dibenzo-p-dioxins (PCDDs) are a group of highly toxic, environmentally significant compounds. Since the toxicity of the 75 dioxin congeners varies widely, analysis requires the use of sophisticated analytical methods such as gas chromatography and mass spectroscopy (1). We have developed a series of monoclonal antibodies (2) that bind specific dioxin and dibenzofuran congeners. These antibodies form the basis of an enzyme-linked immunosorbent assay (ELISA) that is capable of detecting dioxin, at part-per-billion levels, in a variety of environmental and industrial matrices (3,4).

0097–6156/95/0586–0072$12.00/0

Antibodies are complex molecules consisting of two identical heavy-chain (H-chain) and two identical light-chain (L-chain) polypeptides. Amino acid sequence analysis of a large number of antibodies revealed that both the H-chain and the L-chain have only a single domain, referred to as the variable domain or V-region, in which significant amino acid variation is observed. Antibody V-regions are located at the amino end of the H-chain and L-chain peptides (5). Association of the H-chain and L-chain variable regions forms the antibody combining site, (the antigen binding pocket). Figure 1 is a schematic representation of a typical IgG immunoglobulin showing these domains. Analysis of the amino acids in the V-regions of antibody molecules revealed areas of hypervariability, referred to as complementarity determining regions (CDR) that are separated by the more constant framework regions (5).

The ability of antibodies to bind small molecules has been well documented. Antibody binding can exhibit extraordinary specificity, in some cases binding specific molecules (6), or an antibody may recognize an entire class of compounds (7). The specificity of the antibody-antigen binding reaction is governed by precise interactions between the antigen and individual amino acid residues located within the variable region of the antibody molecule. These include ionic interactions (salt bridges), van der Waals and hydrogen bonding, hydrophobic and short range electronic interactions, but not covalent bonds.

In an effort to better understand the nature of the chemical interactions controlling binding of different polychlorinated dibenzodioxin and dibenzofuran molecules with our anti-dioxin monoclonal antibodies, we undertook a series of modeling experiments. First, energy-minimized molecular models for the polychlorinated dioxins, furans and PCBs were generated. These models are idealized gas-phase models that do not take into account interactions with solvent. This is one of the major limitations of these models. Next, the antibody combining sites were modeled for two different anti-dioxin monoclonal antibodies (Mabs), DD-1 and DD-3. The results from these studies are discussed here.

Methods

Antibody Production and Gene Sequencing.

Monoclonal Antibody Production. Production of the hybridoma cell lines DD-1 and DD-3 producing monoclonal antibodies DD-1 and DD-3 were previously described (2). Cross-reactivity studies using a competition ELISA were previously reported (2).

Construction and screening of cDNA libraries: Construction of cDNA libraries and sequencing of the H-chain and L-chain genes for DD-1 and DD-3 were reported earlier (Recinos, A.; Silvey, K. J.; Ow, D. J.; Jensen, R. H.; Stanker, L. H. *Gene* **1994,** in press). Briefly, total RNA was purified from approximately 5×10^8 hybridoma cells by the guanidine thiocyanate

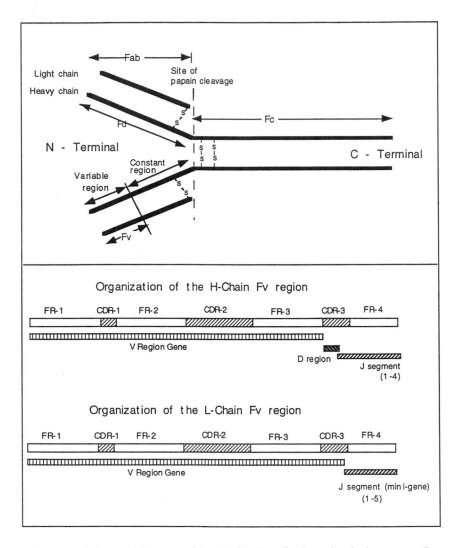

Figure 1. Schematic diagram of a typical IgG antibody molecule (upper panel) showing the variable and constant regions of the heavy and light chains. Organization of the variable portion of the L-chain and H-chain regions (lower panel) into complementarity determining regions (CDR) and framework regions (FR).

methods of Okayama *et al.* *(10)* as modified by Pharmacia LKB, (Piscataway, NJ). Presence of desired mRNAs for cDNA synthesis was verified by Northern analyses using [^{32}P]-end-labeled synthetic oligonucleotides (24-mers) specific for H and L chain constant region sequences as probes (data not shown). Approximately 50 µg of total RNA was subjected to cDNA synthesis according to Pharmacia LKB reagent kit protocols adapted from Gubler and Hoffman *(11)*. Reaction products were ligated into plasmid pUC 18, and the resultant constructs were used to transform competent *E. coli* DH5aF'IQ (GIBCO BRL, Gaithersburg, MD). Transformants were plated on selective medium, transferred to Whatman 541 paper disks and screened by *in situ* colony hybridization with the oligonucleotide probes noted above *(12)*. Colonies positive by hybridization for pUC 18 cDNA constructs containing either H or L chain inserts were cultured and "Geneclean" minipreped (BIO 101, Inc., La Jolla, CA) for plasmid DNA. Full-length cDNA clones were verified by restriction analyses. Plasmid DNAs were then further purified ("Midi prep", QIAGEN Inc., Studio City, CA), and the DNA sequences of the variable regions of the H and L chains in pUC plasmids were determined by double-stranded plasmid methods as specified with Promega's T7 polymerase/7 deaza dGTP kit on a Bio Rad (Richmond, CA) Sequi-Gen wedge gel apparatus.

Molecular Modeling

Determination of Minimum Energy Conformation. Molecular modeling studies were performed using a CAChe WorkSystem running on a Macintosh Quadra 700 equipped with a RP88 coprocessor board and a CAChe stereoscopic display (CAChe Scientific, Inc.; Beaverton, OR). Minimum energy conformations for selected dioxins, dibenzofurans, PCB's and related molecules were calculated using Allinger's standard MM2 force field *(13)* augmented to contain force field parameters for cases not addressed by MM2 (CAChe Scientific, Inc.). Following the initial optimization, a sequential search for low energy conformations was performed by rotating all dihedral angles through 360 degrees in 15 degree increments. The structures resulting from all computations were viewed and superimposed using the CAChe Visualizer+ application.

Determination of Electronic Properties. The electronic wave function for all compounds was calculated by solving the Schrödinger equation using the Extended Hückel approximation *(14)*. The wave function data were converted into three-dimensional coordinates for visualizing electron densities and electrostatic potentials using the CAChe Tabulator application. The electron probability density value was set at 0.01 electrons/Å^3 for all calculations. Electrostatic potentials were calculated in reference to an incoming positive charge and represent repulsive energies.

Modeling of Antibody Combining Site. The combining sites of Mab DD-1 and DD-3 were modeled using the knowledge based modeling algorithm AntiBody structure GENeration (ABGEN) developed by one of us

(D.S.L.) and previously described (*15-17* & Chhabinath, M.; Anchin, J. M.; Subramaniam, S.; Linthicum, D. S. ABGEN: A knowledge-based automated approach for antibody structure modeling, **1994**, submitted). Briefly, the V-region amino acid sequence of DD-1 and DD-3 were compared to the sequences of immunoglobulins whose three-dimensional structure has been solved. Features of ABGEN algorithm include analysis of invariant and strictly conserved residues, structural motifs of known Fabs, canonical features of hypervariable loops, torsional constraints for residue replacement and key inter-residue interactions. The H-chains for DD-1 and DD-3 were modeled on to the anti-hen egg white lysozyme antibody Hy-HEL-5 *(18)* and the anti-influenza virus hemagglutinin antibody FAB 17/9 (19) respectively. Likewise, the L-chain from antibody 4-4-20 *(20)* (an anti-fluorescein antibody) and Gloop 2 *(21)* an anti-hen egg white lysozyme antibody were used for modeling of DD-1 and DD-3 L-chains respectively. In all cases no segment grafting was needed nor were any insertions or deletions needed. The best "hypervariable loop-scaffoldings" were then obtained using sequence comparisons with loop segments of known structures as well as using several different computational algorithms and energy minimization methods (Chhabinath, M.; Anchin, J. M.; Subramaniam, S.; Linthicum, D. S. ABGEN: A knowledge-based automated approach for antibody structure modeling, **1994**, submitted).

Results and Discussion

Binding Studies. Competition ELISA experiments using DD-1 and DD-3 revealed that only a subset of dioxin and furan congeners were recognized by DD-1 and DD-3 (2). These studies (summarized in Table I) clearly demonstrated that DD-1 and DD-3 bound a restricted, but similar set of dioxin and furan congeners. Nonchlorinated dioxin and nonchlorinated dibenzo-p-furan were not recognized, nor were the octachloro congeners of these chemicals. Antibody binding was observed to tetrachloro, and pentachloro isomers as well as to some of the hexachloro isomers. However, most of the hexachloro isomers tested were not bound by these antibodies. Thus DD-1 and DD-3 preferentially bind dioxin and furan congeners of intermediate chlorination, especially those congeners having the chlorines located in the lateral positions. This is not unexpected since the hapten (1-amino-3,7,8-triCDD) contained chlorines at carbon numbers 3, 7, and 8, and was conjugated to carrier protein via a nitrogen on the number 1 carbon (2) (Figure 2). Inspection of the competition ELISA data, however, reveals that some differences in binding do occur between DD-1 and DD-3. The most dramatic differences were observed with 1,2,3,4,7,8-Hexa-CDD and with the PCB's. Competition ELISA experiments using PCB's revealed that DD-1 and DD-3 did not bind any of the PCB's tested, except for weak binding with the coplanar 3,3',4,4'-TCBP. However, DD-1 has almost a 10-fold greater relative affinity for 3,3',4,4'-TCBP than does DD-3 (Table I) (2). Likewise, competition ELISA analyses using other chlorinated chemicals (Table I) revealed that DD-1 had a stronger binding than did DD-3 to a number of

Table I: Binding Characteristics of DD-1 and DD-3

Competitors	IC$_{50}$ (ppb)	
	DD-1	DD-3
Dibenxodioxin (DD)	>2000	>2000
1-CDD	>2000	>2000
2,7-diCDD	7.5	10
1,2,4-TriCDD	2000	>2000
1,2,3,4-TCDD	2000	>2000
2,3,7,8-TCDD	20	25
1,3,7,8-TCDD	1.2	4.5
1,2,3,4,7-Penta-CDD	27	32
1,2,3,7,8-Penta-CDD	3.2	8
1,2,3,4,7,8-Hexa-CDD	15	>2000
Octachloro-DD	>2000	>2000
1-amino-3,7,8-tri-CDD	2.5	0.4
1-nitro-3,7,8-tri-CDD	0.4	1
1-N-(adipamino)-3,7,8-tri-CDD	1.1	1.2
Dibenzofuran (DF)	>2000	>2000
2,8-DiCDF	60	2000
2,3,7,8-TCDF	8	7
2,3,4,7,8-Penta-CDF	50	3
1,2,3,4,8,9-Hexa-CDF	>2000	>2000
Octachloro-DF	>2000	>2000
PCBs		
2,2',4,6-TCBP	>2000	>2000
3,3',4,4'-TCBP	250	2000
2,2',3,4,4-Penta-CBP	>2000	>2000
2,2',3,4,4',5-Hexa-CBP	>2000	>2000
2,2',3,4,5,5"-Hepta-CBP	>2000	>2000
2,2',3,3',4,4',6-Hepta-CBP	>2000	>2000
2,2',3,3',4,5,6,6'-Octa-CBP	>2000	>2000
2,2',3,3',4,4',5,5'-Octa-CBP	>2000	>2000

Continued on next page

Table I Continued

Competitors	IC$_{50}$(ppb)	
	DD-1	DD-3
Other Chlorinated Chemicals		
2,4-Dichloro-6-nitrophenol	>20,000	>20,000
2,4-dichlorophenol	20,000	>20,000
2,4-dichlorophenoxyacetic acid	>20,000	>20,000
2,5-dichloronitrobenzene	900	20,000
4,5-Dichlorocatechol	16,500	>20,000
2,2,2-Trichloroethanol	>20,000	>20,000
2,4,5-Trichlorophenol	6,500	>20,000
2,4,5-Trichlorophenoxyacetic acid	20,000	>20,000
Aldrin	>20,000	>20,000
Chlordane	>20,000	>20,000
Chlorobenzene	>20,000	>20,000
DDT	>20,000	>20,000
Endosulfan	>20,000	>20,000
Endrin	>20,000	>20,000
Heptachlor	>20,000	>20,000
Hexachlorocyclohexane (mixed Isomers)	3,000	>20,000
Kepon	>20,000	>20,000
Pentachlorophenol	>20,000	>20,000
Toxaphene	>20,000	>20,000

Data from Stanker et al.(2).

chemicals (e.g., 2,5-dichloronitrobenzene, and 2,4,5-trichlorophenol). However, binding to these chlorinated chemicals was greatly reduced (200- to 900-fold) as compared to binding with of the polychlorinated dioxins and furans. These observations strongly suggest that the antibodies require a molecule with a planar geometry, but that DD-1 has a less stringent requirement for planarity (i.e., will tolerate a looser "fit") than does DD-3.

Molecular Modeling. In an effort to more clearly understand the binding patterns observed above we generated a series of molecular models for specific dioxins, furans, PCBs and for the hapten (1-(adipamino)-3,7,8-triCDD) used to produce DD-1 and DD-3. Energy minimized models and electron density calculations were performed for all of the dioxin molecules listed in Table I. The energy minimized conformations of 2,3,7,8-TCDD, the hapten, 2,3,7,8-TCDF, and 2,2',4,6-TCBP are shown in Figure 2. Panels A, D, and G are ball-and-stick, space-filled (atoms represent 100% of their van der Waals radius) models of 2,3,7,8-TCDD, the hapten, and 2,3,7,8-TCDF (Figure 2). Electron density calculations for 2,3,7,8-TCDD, and the hapten are shown in Figure 2, panels B-C and E-F respectively. These electron density calculations are colored by the electrostatic potential (white is the most positive and cyan the most negative). These models demonstrate the symmetry and planar nature of 2,3,7,8-TCDD. All of the dioxin congeners studied resulted in similar models with respect to planarity. Likewise, the electron density surfaces were similar for all of the dioxin congeners studied and varied only as a result of their chlorination number. An identical set of calculations was performed for the hapten, and the results are shown in panels D-F. Clearly, the portion of the hapten distal to the linkage site (Figure 2D, arrow indicates linkage site), "the dioxin portion", maintained the structural and the electron density distribution features of non conjugated 2,3,7,8-TCDD. Thus, our earlier conclusions (based solely on cross reactivity) suggesting that these antibodies preferentially bind tetrachloro and pentachloro dioxins and furans, especially when the lateral positions are chlorinated, are not surprising in light of these models. Energy minimization and electron density calculations also were performed for all of the furans tested and for selected PCB's. Representative examples of these calculations are shown in Figure 2 (panels G-I) for 2,3,7,8-TCDF and (panels J-L) for 2,2',4,6-TCBP. These models demonstrate the planar nature of the polychlorinated furans and dioxins versus the PCB's in which the rings are free to rotate. The models for 2,2',4,6-TCBP shown in Figure 2, panels J-L are ball and stick models where the atom size represents 50% of their van der Waals radii. Panels J and K represent two different views of the lowest energy conformation. In this conformation the rings are almost completely out of phase. In panel L, a higher energy conformation is shown. These models serve as examples of the ability of the PCBs to assume different degrees of planarity, even for a specific isomer. Clearly the higher energy forms are not favored under standard conditions.

Partial charge calculations were made for the dioxins and furans listed in Table I. Typical results, for the hapten, 2,3,7,8-TCDD, and octaCDD, are shown in Figure 3. Areas of positive charge are represented by red spheres and negative charge by yellow spheres. The diameter of the sphere is proportional to the magnitude of the charge. The models suggest that the chlorines contribute little to the partial charge on the molecule and that the greatest contribution is from the oxygens. The oxygens could possibly be involved in hydrogen bonding with amino acid side chains in the antibody combining site. However, the large size of the chlorines may sterically hinder such interactions especially in the higher chlorinated congeners. In any event, both oxygens clearly are not necessary for antibody binding since these antibodies bind polychlorinated dibenzofurans as well as they bind dioxins.

Inspection of the results from our cELISA experiments and the above models suggest that antibody binding is controlled by a number of factors. Binding appears to require a molecule with the correct geometry, i.e., it must be planar and of the correct size. Chlorines must be present, and are probably necessary for proper positioning of the molecule within the antibody combining site. It appears that chlorines must be present on both rings, preferably on the lateral positions. The failure of the antibody to bind congeners that have chlorines only on one ring may be explained by inspection of a series of models representing electrostatic potential isosurfaces. The electrostatic potential isosurface is represented by red and blue surfaces indicating positive and negative potentials, respectively (Figure 4). In these calculations the potential is not constrained by structure or electron density. Figure 4 shows the electrostatic potential isosurface calculated for 1,2,3,4-TCDD (panel B), a molecule not bound by either DD-1 or DD-3, and for 2,3,7,8-TCDD (panel A). Notice that the symmetrical distribution of the negative potential regions in 2,3,7,8-TCDD is completely disrupted when all the chlorines are placed on one ring. The negative region is allowed to slide over the center of the benzene rings.

The exact features of the analytes that govern antibody binding are difficult to elucidate, in part because these molecules are highly similar in their properties. The lack of binding to the PCB's is readily explained by their lack of planarity. This is clearly suggested by the models studied and the ELISA binding data. However it is more difficult to explain the lack of binding to any specific dioxin or furan congener. Binding to these molecules is most probably a result of multiple factors including correct positioning of the molecule within the binding pocket which may be sterically hindered by additional chlorines. Likewise, the lack of binding to a specific congener could be the result of inappropriate electrostatic potential distributions, (a feature that appears to be heavily influenced by the chlorination pattern) or the inability to form hydrogen bonds or other non covalent interactions.

The Antibody Combining Site. Modeling the various properties of the analytes bound by DD-1 and DD-3 represents one approach to understanding the binding characteristics of these antibodies. However, a full explanation of the factors responsible for the differential binding of various dioxin and furan congeners can only be gained by understanding the three-dimensional features of the antibody combining site. The amino acid sequences for DD-1 and DD-

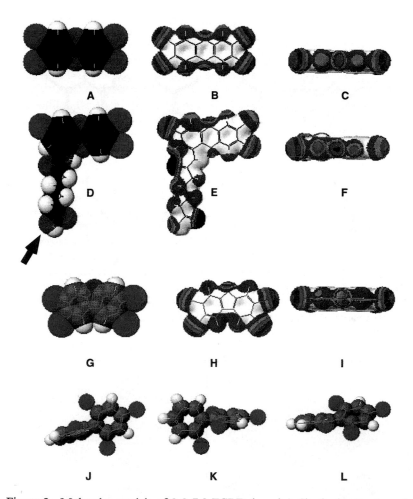

Figure 2. Molecular models of 2,3,7,8-TCDD (panel A-C), the hapten (panel D-F), 2,3,7,8-TCDF(panel G-I), 2,2',4,6-TCPB (panel J-L). Ball and stick, space-filled models are shown in panels A, D, G, and J-L (The colors indicate the following elements: gray-carbon, white-hydrogen, red-oxygen, and green-chlorine). Electrostatic potentials are displayed on the electron density surfaces in B, C, E, F, H, and I for the lowest energy conformations. The electron probability density value was set at 0.01 electrons/Å^3 for all calculations. The energy values in atomic units (1 a.u. = 627.503 kcal/mole) of the color boundaries are: white/red +0.09, red/yellow +0.03, yellow/green +0.01, green/cyan 0.00, cyan/blue -0.01, blue/violet -0.03, and violet/charcoal -0.06. The arrows indicates the position of linkage of the hapten to the carrier protein. Panels J and K are different views of the lowest energy conformation (-ll.47 kcal/mole) for 2,2',4,6-TCBP while panel L is a higher energy (30.67 kcal/mole) conformation.

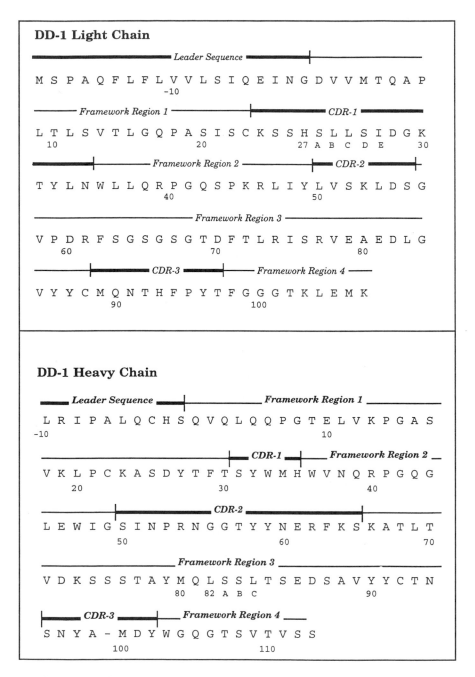

Figure 5. Amino acid sequence of the L-chain (top) and H-chain (bottom) of DD-1. The protein sequence was determined from the corresponding gene sequences (9).

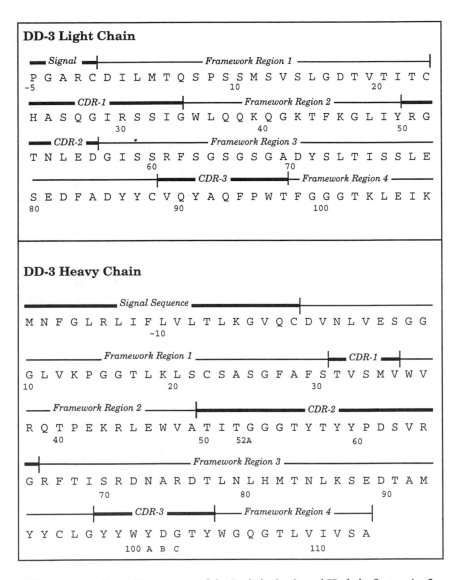

Figure 6. Amino acid sequence of the L-chain (top) and H-chain (bottom) of DD-3. The protein sequence was determined from the corresponding gene sequences (9).

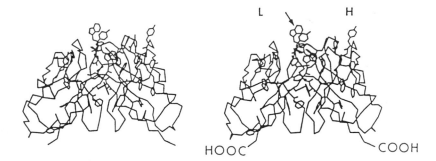

Figure 7. Stereoscopic view of the combining site of DD-3. The H-chain is on the right and the L-chain on the left. Arrow points to L-chain tryptophane-97. Note that the L96 and L98 amino acids are Tyr and Phe, respectively forming a hydrophobic cluster.

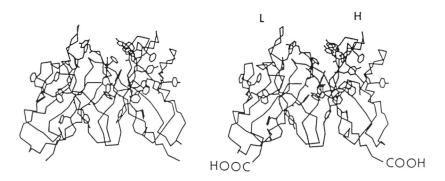

Figure 8. Stereoscopic view of the combining site of DD-1. The H-chain is on the right and the L-chain on the left.

binding and may be required in order for the analyte to become correctly positioned in the antibody combining site. After correct positioning, specific interactions with amino acid side chains (e.g., π-π interactions, ionic interactions, hydrogen bonding, and hydrophobic interactions) can occur. It is these specific interactions that are most likely controlling the differences in the relative binding affinities observed for different dioxin and furan congeners.   Molecular modeling of the hapten demonstrated that both structural and electronic alterations as compared to 2,3,7,8-TCDD were introduced by incorporating the amino group and linking via a six-carbon chain to the carrier protein.   These modifications appear to be most pronounced about the region of the number two Chlorine and are probably responsible for the greater relative binding affinity observed in cELISA experiments using the hapten or the hapten containing the six-carbon linker versus 2,3,7,8-TCDD as the competitor. These data clearly demonstrate that molecular modeling of haptens is a useful design tool for determining optimum linkage strategies prior to an extensive synthetic chemistry effort.

The amino acid sequences for the H-chains and L-chains of DD-1 and DD-3 were previously determined from their gene sequences (Recinos, A.; Silvey, K. J.; Ow, D. J.; Jensen, R. H.; Stanker, L. H. *Gene* **1994,** in press). Inspection of these sequences revealed typical antibody molecules with identifiable CDR and framework regions.   All of the invariant residues associated with immunoglobulin molecules (*5*) are present in DD-1 and DD-3. A large number of aromatic amino acids (Phe, Tyr, and Trp) capable of a variety of molecular interactions are present in the CDR regions of both antibodies.   Three-dimensional models of the combining sites for both DD-1 and DD-3 were generated using these sequences.   The combining site models are significantly different for the two antibodies.   DD-3 appears to have a more cleft-like pocket where as DD-1 has a more open, bowl-like combining region.   The latter structure may explain the broader specificity of DD-1.

Future studies are aimed at modifying specific amino acids in the antibody combining sites.   Such studies should lead to a clear understanding of the nature of analyte binding.   Similarly, experiments to measure charge transfer should aid in determining whether Trp residues are involved with antigen binding.

Acknowledgements

Mention of a trade name, proprietary product, or specific equipment does not constitute a guarantee or warranty by the U S Department of Agriculture and does not imply its approval to the exclusion of other products that may be suitable.   D.S.L. supported in part by research grant GM46535 from the National Institute of Health.   This work was performed in part under the auspices of the U.S. Department of Energy by Lawrence Livermore National Laboratory under contract No. W-7405-ENG-48.

Literature Cited

1.   Crummett, W. B. *Chemosphere* **1983,** *12*, 429.

2. Stanker, L.; Watkins, B.; Vanderlaan, M.; Budde, W.L.   1987.
    Chemosphere **1987**, *16*, 1635-1639.
3. Vanderlaan, M.; Stanker, L. H.; Watkins, B. E.; Petrovic, P.; Gorbach. S.
    Environ. Toxicol. Chem. **1988**, *7*, 859-870.
4. Watkins, B. E.; Stanker, L. H.; Vanderlaan, M. 1989. An Immunoassay
    for Chlorinated Dioxins in Soils. *Chemosphere* **1989**, *19*, 267-270.
5. abat, E. A.; Wu, T. T.; Perry, H. M.; Gottesman, K. S.; Foeller, C.
    Sequences of Proteins of Immunological Interest, 5th edition Vol. I-III,
    1991, U.S. Department of Health and Human Services, NIH Publication
    No. 91-3242.
6. Rowe, L. D., Beier, R. C., Elissalde, M. H., and Stanker, L. H. *J. Agric.
    Food Chem.* **1994**, 1132-1137.
7. Stanker, L. H., Watkins, B., Vanderlaan, M., Ellis, R., and Rajan, J., In:
    *Immunoassays for Monitoring Human Exposure to Toxic Chemicals in
    Foods and the Environment.* Vanderlaan, M., Stanker, L. H., Watkins,
    B., and Roberts D. W., Eds.   ACS Symposium Series 451, **1991**,
    American Chemical Society, Washington DC, pp 108-123.
8. Davies, D. R.; Padlan, E. A.; Sheriff, S. *Ann. Rev. Biochem.* **1991**, *59*,
    439-473.
9. Stanker, L.H.; Branscomb, E.; Vanderlaan, M.; Jensen R. H.   *J.
    Immunol.* **1986**, *136*, 615-622.
10. Okayama, H.; Kawaichi, M.; Brownstein, M.; Lee, F.; Yokota, T.; Arai,
    K. *Methods Enzymol.* **1987**, *1554*, 3-28.
11. Gubler, U.; Hoffman, B. J. *Gene* **1983**, *25*, 263-269.
12. Recinos, A.; Loyd, R. S. *Biochem. Biophys. Res. Commun.* **1986**, *138*,
    945-952.
13. Allinger, N. L. *J. Am. Chem. Soc.* **1977**, *99*, 8127-8134.
14. Hoffmann, R. *Chem. Phys.* **1963**, *39*, 1397-1412.
15. Kussie, P. H.; Anchin, J. M.; Shankar, S.; Glasel, J. A.; Linthicum, D. S.
    J. Immunol. **1991**, *146*, 4248-4257.
16. Anchin, J. M.; Subramaniam, S.; Linthicum, D. S. *J. Mol. Rec.* **1991**, *4*,
    7-15.
17. Anchin, J. M.; Linthicum, D. S. *J. Clin. Immunoassay* **1992**, *15*, 35-41.
18. Smith-Gill, S. J.; Mainhart, C. R.; Lavoie, T. B.; Rudikoff, S.; Potter, M.
    J. Immunol. **1984**, *132*, 963-967.
19. delaPaz, P.; Sutton, B. J.; Darsley, M. J.; Rees, A. R. *EMBOL. J.* **1986**,
    5, 415-425.
20. Bedzyk, W. D.; Johnson, L. S.; Rioran, G. S.; Voss, E. W. Jr. J. *Biol.
    Chem.* **1989**, *264*, 1565-1569.
21. Griest, R. E.; Jeffrey, P. D.; Taylor, G. L.; *J. Mol. Biol.* **1992**, *223*, 381-
    382.
22. Kabat, E. A.; Wu, T. T.; Reid-Miller, M.; Perry, H. M.; Gottesman, K. S.
    In *Sequences of Proteins of Immunological Interest,* **1977**, US
    Department of Health and Human Services, Washington, DC .
23. Mian, I. S.; Bradwell A. R.; Olson A.J. *J. Mol. Biol* **1991**, *217*, 133-151.

RECEIVED December 2, 1994

Chapter 6

Antibody Mimics Obtained by Noncovalent Molecular Imprinting

Lars I. Andersson, Ian A. Nicholls, and Klaus Mosbach[1]

Department of Pure and Applied Biochemistry, Chemical Center,
University of Lund, P.O. Box 124, S-221 00 Lund, Sweden

Molecular imprinting is becoming increasingly recognized as a technique for the ready preparation of polymeric materials containing recognition sites of predetermined specificity. In many instances molecularly imprinted polymers show binding affinities approaching those demonstrated by antigen-antibody systems. The imprints are highly specific, with selectivity profiles comparable to those of antibodies often being observed. Imprinted polymers, with their extremely easy preparation from simple chemical components and ligand-selective recognition, may be regarded as effective and efficient mimics of biological antibodies. The use of polymer based antibody mimics in immunoassay-like techniques is discussed.

Animals can make antibodies against virtually any foreign chemical group. Pauling's 1940 (*1*) explanation for the formation of such a diversity of antibodies involved imprinting of the antigen in the immunoglobulin molecule during its folding around the foreign antigen molecule. Although later proven incorrect his students began to explore this theory as a strategy for making tailor-made compounds with predetermined adsorptive properties (*2*). Whilst early attempts were of limited success, in recent years highly selective ligand binding systems, for not only mimicking those seen in biology, but also for the construction of novel ligand-receptor recognition systems, have been developed. Antibody-like affinities and selectivities are now achievable with molecularly imprinted systems. Molecular imprinting has already been utilized in a range of other applications requiring selective ligand binding, such as in the areas of selective detection, separation and purification, directed synthesis and catalysis, and in the study of ligand-receptor interactions (*3,4*).

Molecularly imprinted polymer (MIP) preparation involves the polymerization of functional monomers around an imprint species (Figure 1). The monomers used are capable of engaging in reversible non-covalent (*3,4*) or reversible covalent (*5,6*) interactions with specific functionality present in the imprint molecule. During this reaction 'cavities' or 'clefts' are formed in the polymer matrix which reflect the size and shape of the imprint molecule. After completion of the polymerization, these interactions between complementary functionalities present between the imprint molecule and the monomer(s) are conserved through the rigidity of the cross-linked polymer network. In this way functional groups of the monomer residues become

---

[1]Corresponding author

0097–6156/95/0586–0089$12.00/0

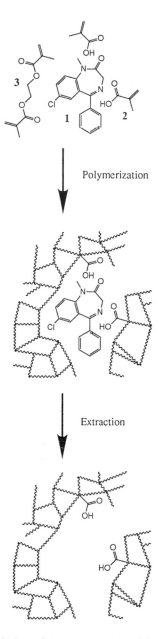

Figure 1. Schematic depiction of the preparation of imprints against diazepam (**1**) using methacrylic acid (**2**) as the functional monomer and ethylene glycol dimethacrylate (**3**) as the cross-linking monomer. It is understood that the polymer chains are interconnected to form a continous network.

spatially positioned around the cavity in a pattern which is complementary to the chemical structure of the imprint molecule. Subsequent removal of the imprint species exposes the imprints within the polymer matrix. This process constitutes a formation of permanent 'memory' for the original imprint molecule in terms of both shape complementarity and chemical functionality. These recognition sites or 'stamped memories' enable the polymer to later selectively rebind the imprint molecule from a mixture of closely related compounds.

Artificial Antibodies

It was recently demonstrated that imprinted polymers can indeed be substituted for antibodies in immunoassay protocols (7). Molecular imprints against clinically significant drugs were successfully used in a competitive radio-ligand binding assay, Molecularly Imprinted sorbent Assay (MIA) for the accurate determination of drug levels in human serum. In the first reported study of this type, two chemically unrelated drug compounds, theophylline and diazepam, were studied (7). Theophylline, a bronchodilating drug commonly used in the prevention and treatment of asthma, has a narrow therapeutic index requiring careful monitoring of serum concentrations. Diazepam (*i.e.* Valium) is a member of the benzodiazepine group of drugs widely used as antidepressants, tranquilizers and muscle relaxants. Benzodiazepines are one of the substances most commonly implicated in drug overdose situations and their detection in body fluids is very useful in clinical and forensic toxicology. Both drugs could be determined in clinically significant concentrations with an accuracy comparable with that obtained using a traditional immunoassay technique. Specifically, the MIA assays for theophylline and diazepam were linear over the ranges 14-224 μM and 0.44-28 μM with detection limits of 3.5 μM and 0.2 μM, respectively, which in both cases are satisfactory for therapeutic monitoring of the drugs. Prior to the actual assay, performed under optimized incubation conditions, the analyte is extracted from the serum using standard protocols. A comparison of the results obtained using a commercial immunoassay technique and the MIA competitive binding assays for the determination of theophylline in patient samples showed excellent correlation between the two methods (Figure 2).

We anticipate that in the near future molecularly imprinted polymers will begin to play a role in pharmaceutical and environmental analysis. An important issue and one already of great concern is the detection and analysis of contaminants in ground and fresh water supplies. The wealth of potential hazards available demands a fast and cheap strategy for assay development. Anti-atrazine (Figure 3) MIPs are already at hand (Siemann, M., Andersson, L. I. and Mosbach, K., unpublished results) and may potentially be useful in the determination and/or specific removal of this pesticide.

Both the theophylline and diazepam MIA methods showed cross-reactivity profiles for their major metabolites and structurally related drugs similar to those as is reported using commercial biological antibody based immunoassays. Anti-theophylline MIPs, for example, showed excellent selectivity for theophylline (1,3-dimethylxanthine) in the presence of the structurally related compound caffeine (1,3,7-trimethylxanthine) (Table I). Despite their close resemblance (they differ by only one methyl group) (Figure 3), caffeine showed less than 1% cross-reactivity. Like polyclonal antibodies, the polymers contain a heterogeneous population of binding sites with a range of affinities for the imprint molecule. Thus, multiple equilibrium dissociation constants (K_{Diss}), varying from high to low affinity, are obtained on analysis of binding data. Apparent K_{Diss}-values down to 10^{-9} M have been obtained, which compare favourably with the 10^{-6}-10^{-14} M range (although K_{Diss}-values better than 10^{-10} M are exceptions) typical for antibodies.

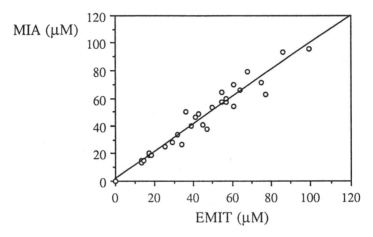

Figure 2. Comparison of the MIA and Enzyme-Multiplied Immunoassay Technique (EMIT) competitive binding assays for determination of theophylline serum concentration in patient samples.

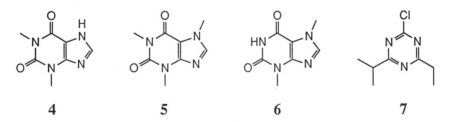

Figure 3. Structures of theophylline (4), caffeine (5), theobromine (6) and atrazine (7).

Table I.   Cross-reactivity observed for the theophylline MIP

Competitive ligand	Cross-reaction (%)	
	MIP	Antibody[a]
Theophylline (1,3-dimethylxanthine)	100	100
3-Methylxanthine	7	2
Caffeine (1,3,7-trimethylxanthine)	< 1	< 1
Theobromine (3,7-dimethylxanthine)	< 1	< 1
Xanthine	< 1	< 1
Hypoxanthine	< 1	< 1
Uric acid	< 1	< 1

[a]Data from Poncelet, S. M. et al. *J. Immunoassay*, **1990**, 11, 77-88.

The preparation of antibodies against small organic compounds (so called haptens) necessitates hapten conjugation to a carrier protein before injection into the animal. Such derivatization often significantly alters the properties of the antigen presented to the immune system and the resultant antibodies may be directed to a different structure than desired. The MIP preparation avoids the need for derivatisation of haptenic antigens, which may result in superior specificity of the artificial system, as is found using anti-morphine MIPs (8). The very closely related structure codeine interferes to a lesser extent with morphine binding to the imprinted polymers than to most of the anti-morphine antibodies (including monoclonal antibodies) reported to date. This finding is significant in the context of codeine being a notoriously difficult cross-reactant for biological anti-morphine antibodies.

Imprinted polymers may also show high binding affinity and selectivity in aqueous buffers, which was demonstrated only very recently (8). This is an important breakthrough which greatly extends the working area of imprinted antibody mimics since the ligand binding assays can be performed under conditions compatible with biological systems. In aqueous buffers the detection limit is, in general, only around one order of magnitude higher than under optimized incubation conditions using organic solvents. The affinity of morphine MIPs would in principle be sufficient for their use in less demanding assay situations such as in screening programmes for drugs of abuse. For analyte extraction protocols prior to the subsequent assay, as described above, determination of clinically significant levels of morphine in body fluids may be possible. In the same study, methacrylic acid based imprinted polymer recognition systems were also successfully applied to the preparation of imprints against the endogenous neuropeptide Leu[5]-enkephalin (8). The MIPs expressed excellent selectivity for free Leu[5]-enkephalin over other tetra- and pentapeptides, with unrelated amino acid sequences. The imprints were sufficiently well defined to allow discrimination between the imprint structure and the D-amino acid containing analogues D-Ala[2]-Leu[5]-enkephalin and D-Ala[2]-D-Leu[5]-enkephalin.

The findings reviewed here demonstrate the ability to use highly specific chemically prepared macromolecules, instead of the traditional biomolecules such as antibodies, as receptors in competitive ligand binding assays. MIPs provide a combination of polymer mechanical and chemical robustness with highly selective molecular recognition comparable to biological systems. The specificity of a MIP is predetermined by the choice of imprint species used during its preparation. Before use they can be stored in the dry state at ambient temperatures for several years without loss of recognition capabilities. Further advantages of the MIPs are their cheap, simple and rapid preparation and, furthermore, the MIA approach does not involve the use of laboratory animals, nor any material of biological origin. These antibody combining site mimics are routinely prepared against small haptenic (relatively low-molecular weight) compounds, against which biological antibodies, in many instances, are difficult to elicit with current methodology. In our opinion molecular imprinting may, before long, be considered as a useful complement to immunological and combinatorial library technologies.

Other Application Areas

In parallel to the successful generation of catalytic antibodies, endeavors to develop catalysts employing molecular imprinting are being explored. Although only a handful of reports have to date been made, several groups around the World are already highly active in this particularly challenging area. The MIP derived substrate selective catalytic polymers (or enzyme mimics) reported possess high substrate specificity (4), although catalytic rates worthy of direct comparison with naturally occurring enzymes have yet to be achieved. The potential of enzyme-like MIPs lies not only in their use as mimics of enzymes present in nature, but also for carrying out reactions either not observed in natural systems, nor possible due to the presence of water.

The increasing awareness of the sometimes very different physiological effects of asymmetric compounds (enantiomers) in biological systems have led to intense research efforts into the development of enantio-selective synthesis, analysis and preparative scale purification of enantiomers. Recent years have seen the commercialization of a plethora of chiral stationary phases (CSPs) for analytical and preparative chromatographic separations. Most of these, although being highly efficient, are restricted in their use to limited classes of compounds. Alternative to surveying the traditional type CSPs available, molecular imprinting of enantiomers offers a strategy by which a specific sorbent can be made for each particular separation problem. The relative ease with which the specific recognition sites may be produced and the long term stability of the polymer systems make them most suitable for use within this field. A distinct advantage offered by imprinted CSPs, in contrast to most commercial CSPs, is the predictable enantiomer elution order which is predetermined by the choice of imprint molecule enantiomer. For a series of closely related chemical structures, i.e. enantiomers and diastereoisomers, the imprint species will be the last to elute. Developmental work into this area has primarily used amino acid (3) and sugar based (5) structures, these being of fundamental interest in biological systems. Imprint systems showing very impressive separative capabilities have now become available (9), with separation factors of up to 18 for the enantiomers of the dipeptide Ac-Phe-Trp-OMe having been reported (10). In parallel, the development of MIP-derived CSPs for the resolution of some important pharmaceutical structures like the β-adrenergic blocking agent timolol (11) and the nonsteroidal anti-inflammatory drug naproxen (12) have been pursued.

Ligand-selective sensory devices, ideally, interact specifically with a predetermined compound or compounds from amongst a complex mixture and provide a signal which may be monitored externally. This concept has been widely realized in the area of biosensors where a biomolecule, such as an antibody or enzyme, has been used in conjunction with an electronic transducer. It was conceived that MIPs may be employed in place of such biomolecules. The greater inherent physical and chemical stability of MIPs make them ideally suited to this rôle. In principle, the capacity exists to produce recognition sites for any ligand, reinforcing the potential for MIPs in this area. The implementation of MIPs with field-effect type sensors for substrate-selective determinations is presently being explored (13).

Imprint Preparation

Of the two molecular imprinting approaches the non-covalent strategy (*3,4*) is more easily employed than its covalent (*5,6*) counterpart as the imprint molecule is simply allowed to pre-arrange with the monomers in solution prior to initiation of the polymerization (Figure 1), rather than requiring pre-derivatization with functional monomers. Furthermore, a higher number of compounds are amenable to non-covalent imprinting and the final imprinted material is more versatile, at least for the application range covered by this review.

To date, bulk polymerization, followed by grinding and particle sizing, has generally been the most often used technique for imprinted polymer preparation (Figure 1). These imprinted polymer systems utilize a very high mole percentage of cross-linking monomer and one or more functional monomers. Ethylene glycol dimethacrylate is thus far the most extensively utilized cross-linking monomer, due to its mechanical and thermal stability, ease of removal of imprint molecule and the high selectivities that have been observed. Recently, several interesting more highly methacrylate substituted cross-linkers have come into use in non-covalent imprinting applications and show promising properties (*14*). Several functional monomers, carrying chemical functionality suitable for interacting non-covalently with the imprint molecule, in particular methacrylate, acrylate or vinylic type polymerizable functionality, have thus far been employed in MIP preparation and are compatible with the cross-linking strategies discussed above. Methacrylic acid and vinyl-pyridine, in particular, have proven to be extremely versatile. Combinations of these monomers can sometimes yield superior results (*15*). Typically, these polymer systems utilize conventional aza-compound derived thermal or UV free radical initiation. More recently, emulsion, suspension and grafting techniques have been explored and in one report the bulk polymer was used *in toto* for subsequent chromatographic ligand binding experiments (*16*). The application of new polymerization technology to MIP preparation should pave the way to polymers with as yet uninvestigated physical characteristics, *e.g.* thin-films (*13*), microbead, highly macroporous coatings and inorganic-organic polymer composites.

Conclusions

Molecular imprinting provides a powerful new tool by which recognition sites of predetermined specificity can easily be made for compounds of a diverse array of chemical classes. Although to date the technique has been restricted to relatively small molecules, preliminary studies on the imprinting of large biomolecules and their aggregates, *e.g.* large peptides, proteins and enzymes, have already furnished interesting results. Imprinting of nucleotide base-like structures yielded excellent materials for chromatographic separations of the purine and pyrimidine bases (*17,18*). MIPs are interesting not only for basic studies on molecular recognition, but are versatile materials useful in numerous analytical, preparative, catalytic applications. Molecular imprinting should find further areas of application in Science and Technology.

Literature Cited.
1. Pauling, L. *J. Am. Chem. Soc.* **1940**, *62*, 2643-2657.
2. Tailor-made compounds predicted by Pauling *Chem. Eng. News*, **1949**, *27*, 913.
3. Mosbach, K. *Trends Biochem. Sci.* **1994**, *19*, 9-14.
4. Andersson, L. I.; Nicholls, I. A.; Mosbach, K. In *Highly Selective Separations in Biotechnology*; Street, G., Ed.; Blackie: Glasgow, UK, 1994; pp. 206-224.
5. Wulff, G. In *Polymeric Reagents and Catalysts*, Ford, W. T., Ed., ACS Symp. Ser. vol. 108, 1986, pp. 186-230.
6. Shea, K. J. *Trends Polym. Sci.* **1994**, 2, 166-173.

7. Vlatakis, G., Andersson, L. I., Müller, R. and Mosbach, K. *Nature* **1993**, *361*, 645-647.
8. Andersson, L. I., Müller, R., Vlatakis, G. and Mosbach, K. (1994) *Proc. Nat. Acad. Sci. (USA)* submitted.
9. Andersson, L. I. and Mosbach, K. *J. Chromatogr.* **1990**, *516*, 313-322.
10. Ramström, O., Nicholls, I. A. and Mosbach, K. *Tetrahedron: Asymmetry* **1994**, *5*, 649-656.
11. Fischer, L., Müller, R., Ekberg, B. and Mosbach, K. *J. Am. Chem. Soc.* **1991**, *113*, 9358-9360
12. Kempe, M. and Mosbach, K. *J. Chromatogr.* **1994**, *664*, 276-279.
13. Hedborg, E., Winquist, F., Lundström, I., Andersson, L. I. and Mosbach, K. *Sensors and Actuators A* **1993**, *37-38*, 796-799.
14. Kempe, M. and Mosbach, K. Manuscript.
15. Ramström, O., Andersson, L. I. and Mosbach, K. *J. Org. Chem.* **1993**, *58*, 7562-7564.
16. Matsui, J., Kato, T., Takeuchi, T., Suzuki, M., Yokoyama, K., Tamiya, E. and Karube, I. *Anal. Chem.* **1993**, *65*, 2223-2224.
17. Akashi, M., Waki, K., Miyauchi, N. and Mosbach, K. *HEFEI International Microsymposium on Funktional Polymers* **1987**, September 8.
18. Shea, K. J., Spivak, D. A. and Sellergren, B. *J. Am. Chem. Soc.* **1993**, *115*, 3368-3369.

RECEIVED October 5, 1994

NEW HAPTEN DESIGN
AND ANALYTICAL METHODS

Chapter 7

Type Reactivity for Analyte Profiling

Lawrence M. Kauvar and Peter Y. K. Cheung[1]

Terrapin Technologies, Inc., 750-H Gateway Boulevard,
South San Francisco, CA 94080

Olfactory recognition of small organic molecules has been shown to
involve comparisons of each molecule's unique combination of
binding strengths against a fixed panel of receptor proteins. We have
previously demonstrated adaptation of this molecular fingerprinting
principle for use in chemical detection *in vitro*. Families of proteins
found to be suitable for fingerprinting a large variety of molecules
include primary repertoire antibodies, described here, as well as
enzymes from the cellular toxic chemical defensive network.
Classification of chemicals on the basis of their binding fingerprints
to a compact panel of reference proteins can provide a generalized
detector for chemicals as well as a means of indexing chemicals for
use in drug design.

Immunoglobulins are chemical binding proteins which can be used as analytical
reagents independent of their function in the body [1]. One fundamental property of
any protein-based detector system is that the combinatorial explosion of possible
primary sequences can generate immense panels [2]. Considering the ~50 residues
considered to be part of the hypervariable regions in immunoglobulins [3], with 20
possible amino acids at each, there are 20^{50} mathematically conceivable structures,
an astronomically large number compared to the ~10^{12} lymphocytes present in the
body at one time [4] or even to the number of bacteria or phage that can be
reasonably handled in the laboratory.

Examination of each member of the set is clearly prohibited in practice by the
sheer size of the whole set. For characterizing the large variety of small organic
molecules, the full set is presumably sufficient, but given the practical limitations, it
is important to ask what fraction of the full set is necessary. Several lines of
evidence, most notably the operation of the olfactory system [5,6], indicate that a
much smaller set of reagents can be used to characterize a large variety of analytes.
Similar issues also arise in using chromatography as an analytical tool [7]. From
these precedents, it appears that use of pattern recognition techniques can provide
adequate discrimination with a primary detection panel that is far smaller than the
number of analytes to be detected.

[1]Current address: Genelabs, 505 Penobscot Drive, Redwood City, CA 94080

If a small number of reagents is to suffice, however, it is critical to select appropriate elements for the primary screening panel. The prevailing model for the immune system includes just such a primary repertoire, composed largely of immunoglobulins of the IgM subclass, generated autonomously in the absence of foreign antigens [8]. These reagents are often viewed as difficult to work with in the laboratory, due to their large size (approximately equal to five copies of the more familiar IgG class antibody, attached to a common core region). A study of the specificity characteristics of the primary repertoire is likely to be useful nonetheless as an indicator of what is possible to achieve in a more convenient embodiment, such as recombinant antibodies.

Early estimates of the germ line encoded primary immunoglobulin synthetic capability, as revealed by recombinant DNA analysis, yielded numbers in the millions [9] after taking into account the combinatorial mixing of gene fragments to assemble a final immunoglobulin. Analysis of preferred combinations for splicing of Joining segments with Constant and Variable domains has suggested that one to two orders of magnitude fewer constructs are actually present *in vivo* [10]. In addition, some heavy and light chain combinations will not form a stable immunoglobulin, further limiting the actual size of the repertoire [11]. Furthermore, many of the immunoglobulins appear to have redundant binding characteristics when tested *in vitro* against a wide range of antigens, including haptens. The results of such studies have typically yielded positive binding clones at a frequency between 1 in 100 and 1 in 1,000 of the resting B-cells [12-14]. The minimum size of the functional primary repertoire in mammals thus appears to be about a thousand unique specificities, comparable to the size of the olfactory receptor system in mice [5]; lower animals get by with even fewer primary defensive proteins.

The small number of unique chemical specificities in the primary repertoire contrasts sharply with the much lower frequency of clones that proliferate in response to antigenic stimulation *in vivo*, and implies that there are additional constraints on clonal response beyond chemical specificity, presumably relating to avoiding anti-self responses [15]. Along with the class switch from IgM to IgG, a process of somatic mutation is initiated [16] as a means to generate higher affinity antibodies; evolution has apparently chosen to restrict the number of clones able to proceed down this path, with its inevitable attendant risk of creating an auto-immune disease.

Previous work on documenting the frequency of clones recognizing particular epitopes as chemical entities has been preliminary in nature, and primarily directed to understanding the mechanisms which restrict clonal expansion. The present study confirms the earlier qualitative results by providing a careful quantitative analysis of a panel of 450 hybridomas randomly selected from unimmunized mice. Following presentation of the data, the implications are discussed with regard to generalized chemical detection and classification.

Materials and Methods. General buffer reagents and media were purchased from Sigma as were kassinin, KLH (keyhole limpet hemocyanin), and BSA (bovine serum albumin); phage f1 was a gift from Walter Soeller and Tom Kornberg (UC San Francisco). Fluorescein and simazine conjugates to BSA were prepared following published procedures [17,18]; Isotyping kits and other immunological reagents: from Zymed; iodinated antibody: from Amersham.

Immortalizing the primary repertoire. Following stimulation *in vitro* with bacterial lipopolysaccharide for 4.5 days, 30-50% of resting B-lymphocytes, from dissected spleens of 2-3 week old Balb/c mice, began to proliferate and thereby became suitable partners for hybridoma formation using 50% PEG as fusion agent and SP2/0 cells as the myeloma partner. Standard HAT selection procedures and limiting dilution subcloning were employed to establish the fused cells as viable clones [19]. Stable producers were expanded into 24-well and then 6-well plates before freezing cells in liquid nitrogen and supernatants at -20°C for storage. Clones from 5 independent fusions are included in the panels described here. All clones studied were verified by isotyping assays as being of the IgM class.

Assaying primary repertoire binding properties. Antigen was deposited from 96-well microplates onto Immobilon-P membrane (Millipore) using a custom made 96-well transfer device which consists of stainless steel pins with a shallow groove on the bottom end (a variant of the CloneMaster™ transfer device available from Fisher). Approximately 1μL is retained in the groove by capillary action, thereby allowing the antigen to be deposited reliably onto replicate membranes in the 96-well format with each deposit forming a dot 2-3mm in diameter. Pin to pin reproducibility and cycle to cycle reproducibility with this instrument are both high, with coefficients of variation below 5% using mouse IgM as a test antigen and a radiolabelled secondary antibody for detection. Linear dose response curves were obtained across a range of test antigen concentrations from 0.05 to 5.0 mg/mL. Actual antigens were used at 1 mg/mL. Excess protein binding sites on the membrane were then blocked with 1mM ethanolamine and 1% casein solution for 30 minutes at room temperature. For the actual antigens of interest, the 96-pin device was again used to deposit culture supernatants from 96-well microplates over the antigen dots. Duplicate droplets of each concentrated culture supernatant were overlaid and incubated at 37°C for 1 hour. In this manner, a large number of replicate experiments could be performed using far less culture supernatant than needed in the otherwise quite similar standard dot blot formats. Unbound primary antibody was washed off, and the bound antibody quantified following probing with a secondary antibody labelled either with an enzyme or ^{125}I, following standard procedures.

Washing of the membranes was achieved in a novel manner described in detail elsewhere [20]. Briefly, the membrane was positioned on a dry piece of 3MM filter paper (Whatman) and overlaid with a second filter paper saturated with PBS/Tween (phosphate buffered saline with 0.1% Tween-20). Liquid from the overlay filter paper was squeezed through the membrane and trapped in the underlying filter paper by application of pressure using a simple levered clamp device. With this washing technique, three washes of the membrane were sufficient to fully remove the unbound proteins, and the washes could be accomplished in under a minute. The method is particularly convenient for working with large numbers of membranes.

Two different labels attached to the secondary antibody were used in these experiments. For the experiments described in Figures 1 and 2, a radioactive iodine goat anti-mouse IgM (mu chain specific) label was used with quantification in a gamma radiation counter (Abbott). For the experiments summarized in Figure 3, the label attached to the goat anti-mouse IgM secondary antibody was alkaline phosphatase working on the substrate BCIP/NBT (bromo-chloro-indolyl phosphate/nitro blue tetrazolium). Quantification of colored precipitate on the membrane was achieved via a laser scanning densitometer (Zeineh Instruments SLR/2D-1D) which directly transferred data to computer storage (IBM compatible personal computer).

Prior to large scale assay, the immunoglobulin content of each supernatant was normalized to within a factor of two of 1 μg/mL. For some of the supernatants, dilution in serum free media was used to reach this level, but for most of the samples, concentration in a Centriprep (Sartorius) centrifugal concentrator was needed. Actual immunoglobulin content of each sample was then determined by directly spotting the sample onto a membrane with the 96-well transfer device and visualizing it with labelled secondary antibody.

Assay results are reported in units denoted "binding coefficient," which is proportional to the antibody's observed avidity for the immobilized antigen which in turn is related to the intrinsic solution phase affinity for the antigen of each individual paratope. Due to the 10-fold repetition of the paratope on multimeric IgM, the binding coefficient is an overestimate of intrinsic affinity, particularly for low affinities [21]. Because it is difficult to make a precise adjustment for this factor, which depends on antigen density on the solid phase as well, we have not attempted to correct for it. The coefficient was calculated as the average response of duplicate assays, in units of counts per minute or optical density, minus antigen-independent background signal, with the results then normalized to constant IgM concentration (requiring less than a factor of 2 in all cases due to the prior physical normalization of Ig content). In the experiments using radioactive iodine label, the binding coefficient was also corrected to constant specific activity of the labelled secondary antibody. For each series of experiments, all reagent conditions were held constant across the set of antigens, allowing direct comparison of binding coefficients, on which all the reported analyses are based. Overall reproducibility of the entire assay procedure yielded a coefficient of variation of 20% in the binding coefficient.

Results. The binding of three quite different antigens to a panel of murine primary repertoire monoclonal antibodies was measured and analyzed statistically. An explanation of a statistical feature found in the data was then confirmed by a further empirical analysis of two additional antigens.

Binding profiles of simple and complex antigens. Figure 1 provides a representative sampling of our binding data for a simple peptide antigen against a panel of murine primary repertoire monoclonal IgM antibodies. These data were collected using a carefully controlled variation of standard dot-blot methods, as described in Methods. The results illustrated demonstrate high reproducibility of the data on each antibody compared to the variation across the panel. The assay technique was designed for rapid processing with minimal wash solution in order to preserve weak binding signals, including those attributable to multi-dentate avidity effects. Such effects contribute to the functional binding characteristics of the IgMs and are intentionally preserved in the analysis, with results reported in units of an accurately measureable binding coefficient rather than a poorly extrapolated conversion of this coefficient to an estimate of intrinsic affinity. The binding coefficients plotted have already had the blank well assay background subtracted. Even the weakest signals are reproducible and represent some degree of specificity, not random background binding via totally non-specific mechanisms such as mechanical factors or variable denaturation of the proteins. As also described in Methods, the technique minimizes consumption of the supernatants allowing accurate replicates to be performed from even small scale cultures.

The binding of three antigens of different complexity was similarly measured against a panel of 335 antibodies. All clones were of the IgM class, drawn non-selectively from unimmunized mouse splenic cell cultures stimulated briefly with a mitogen to make them suitable partners for laboratory growth by fusion to an established myeloma cell line. The population thus represents a random subset of the primary repertoire. The antigens chosen for this experiment span a wide range in the variety of epitopes exposed to the antibody probes. Kassinin is a 12 amino peptide; f1 is a filamentous phage which displays the same coat protein, pVIII composed of 50 amino acids, hundreds of times in close proximity; KLH is a large, >500,000 dalton, protein with at least twelve distinct subunits [22]. Each antigen was able to capture a substantial fraction of the antibody types, with varying strengths. Comparing kassinin to KLH, a major difference in the profiles was the higher average binding coefficient in the case of the more complex antigen. This

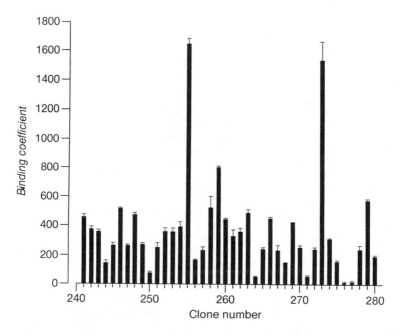

Figure 1. The variation in binding coefficients of a panel of monoclonal IgM antibodies to the immobilized antigen kassinin is much larger than the assay error. Mean and standard deviations for 40 representative clones are illustrated; units are cpm of radiolabelled second antibody.

difference is not accounted for simply by epitope density, leading to an exaggerated multi-dentate avidity effect for the larger antigen, since phage f1 did not have as high an average binding as KLH.

Figure 2 plots the frequency of occurence of clones with low to high binding coefficients for the three antigens. The distribution for the simplest antigen, kassinin, is heavily biased towards low binding strength. This result is not due to loss of antibody integrity, by denaturation of the proteins for example, since the distribution for the complex antigen, KLH, is fairly flat indicating that most antibodies are functional. The phage coat is intermediate in both complexity and distribution profile.

Other qualitative investigations of the frequency of antigen-reactive antibodies in very small samples of the primary repertoire have also noted a proportionality to molecular weight [11], implying that our results are not due to peculiarities of the particular antigens studied here. The simplest rationalization for the different frequency profiles for our test antigens is that the more complex antigen provides a larger number of possible epitopes to which a given antibody can bind at high strength. With a comparable distribution of binding coefficients for each particular epitope, the larger number of chances for a strong interaction would yield the observed higher average binding strength.

Non-correlation of hapten profiles. If our interpretation of the elevated frequency of high binding clones for complex antigens is correct, then the binding profiles of unrelated simple antigens should be largely uncorrelated. To test this prediction, we measured binding of two chemically unrelated haptens to the panel of monoclonal antibodies used in Figure 1 plus 115 additional clones prepared similarly in independent fusions, with the results displayed in Figure 3.

The haptens used were fluorescein, conjugated to BSA as an isothiocyanate derivative [17], and simazine, conjugated to BSA through a hexanoic acid linker [18]. Both are hydrophobic ring structures, but are otherwise unrelated. Since some of the binding to these antigens is presumably attributable to the BSA carrier, we measured that binding as well, and then renumbered the clones based on binding coefficient for BSA. After disregarding the first 25-50 clones whose binding is probably due primarily to interaction with the carrier BSA, the resulting patterns of binding strengths to the two haptens (panels A and B) are quite uncorrelated. The results are also consistent with the previously obtained frequency distribution of clones with regard to binding coefficient against a simple antigen. For BSA, the frequency of stronger binding clones is lower than expected, possibly due to an intrinsic bias built into the repertoire against the common serum albumin protein.

These results extend to a quantitative level prior observations on large hybridoma panels from unimmunized mice, first tried independently by two groups shortly after the invention of hybridoma methods [12,13]. Unfortunately, these early studies did not include careful control over parameters of the experiment shown to be important by subsequent experience working with hybridomas. For example, in the earlier work the majority of the hybridoma lines were not monoclonal due to initial plating at a high density which yielded essentially no blank wells. Furthermore, no attempt was made to normalize the immunoglobulin content of the wells before screening, although ensuing experience has shown that the immunoglobulin content in primary culture supernatants can vary by several orders of magnitude. Finally, the earlier studies did not perform replicate assays. Nevertheless, the qualitative conclusion from the earlier studies, namely that the primary repertoire contains numerous clones recognizing a randomly chosen hapten, was correct. Several other studies have also confirmed this general observation [14], although on much smaller samples of the primary repertoire than in the present study and again without quantification of degrees of binding.

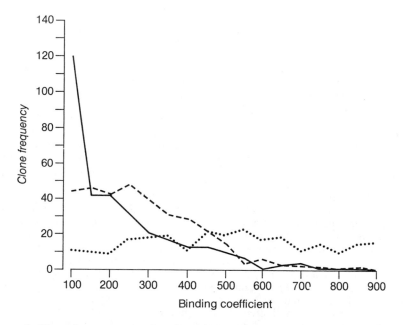

Figure 2. Clone frequency as a function of binding coefficients reflects complexity of the antigen. Data are for 335 monoclonal IgM antibodies from the murine primary repertoire, assayed as in Figure 1, binding to: kassinin (—), phage f1 major coat protein (– –), and KLH (· ·). Beginning with the interval of 0-100 in binding coefficient and continuing at intervals of 50 units, the number of clones in the interval is plotted above the end value of the interval.

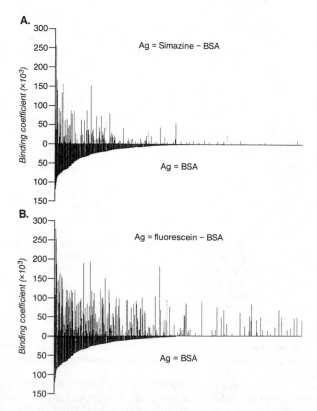

Figure 3. Unrelated haptens generate distinctive patterns of binding to a sample of 450 IgM monoclonal antibodies from the primary repertoire. Panels A and B show binding against the haptens simazine and fluorescein respectively, both coupled to BSA as a carrier for membrane immobilization. To better display the lack of correlation in the binding arising solely from the haptens, the clone numbering (abscissa) has been arranged to put the high BSA-binding clones first, with binding to BSA alone plotted below the axis

Discussion. The reduced size of the functional primary repertoire compared to the full set of combinatorial possibilities, along with the high frequency of antigen binding cells within that repertoire for a variety of antigens, suggests that monoclonal antibodies are less specific chemical reagents than initially expected from experiences with polyclonal sera drawn from hyper-immune animals. Early observations on monoclonal antibodies, prior to the hybridoma era, suggested that moderate specificity at moderate affinity was a general phenomenon [23]. Mapping of antibody specificities against panels of peptides has now dramatically established the conclusion that even high affinity IgG monoclonal antibodies can show significant cross-reactivities to a large number of simple epitopes [24].

In recent years, a variety of antibody-like libraries of binding agents have been described including combinatorial heavy and light chain co-expression libraries in bacteria [25] or phage [26], single chain fusions of heavy and light chain variable domain DNA fragments [27,28], and purely synthetic varieties of binding pockets that can be constructed using modern peptide synthesis technology [29]. Depending on the cloning vector, recombinant proteins can be prepared which provide either high or low avidity effects, for example by using the major (pVIII) or minor (pIII) filamentous phage coat protein as fusion site [30]. In the case of the IgM repertoire, nature has apparently taken advantage of the broadening of detectable specificity due to the multi-dentate effect. As shown here, the avidity effect does not eliminate specificity; it simply broadens it.

For all of these kinds of reagents, there is a trade-off between achieving specificity by adding unique detectors compared to using some kind of pattern matching. These trade-offs have profound impact on the economics of assay development, manufacturing, and sales. In previous work, we demonstrated that even subtle molecular features can be discriminated using a panel of antibodies that individually do not provide clean differentiation among the related analytes [31].

These empirical results are consistent with available theoretical treatments of binding repertoires [32]. Binding pockets are generally modelled as a patchwork of subsites, so that total affinity is proportional to the number of subsites occupied by a ligand. At a moderate affinity threshold, only modest numbers of binding sites are required in such models to provide recognition elements for most small molecules. Likewise, macromolecules are predicted to have a higher probability of matching a given binding agent than simple molecules. From a fundamental perspective, all proteins can be viewed as related, with specificity a quantitative parameter, not a qualitative one [33].

Broad specificity is not unique to primary repertoire antibodies, but is also found in the olfactory system [6], which can recognize immense numbers of compounds with a modest panel of receptors. Further, nature has developed several other families of proteins that each recognize large numbers of compounds [34]. Many of these proteins function in defending cells against toxic chemicals, of either exogenous origin or generated endogenously from free radicals leaking out of the mitochondria. Families such as cytochrome P-450 and glutathione S-transferase contain dozens of proteins with broad and overlapping but distinctive specificities.

The theoretical and practical tools are thus available for constructing a universal chemical monitoring device, or artificial nose, based on cross-reactivity fingerprinting using a small panel of functionally diverse binding proteins [35,36]. In work to be presented elsewhere [37], we have shown that fingerprinting can also be used to cluster compounds into groups, based on their binding properties to a fixed reference panel of proteins. Such clustering has predictive utility for selecting compounds likely to bind to new proteins, for example in the context of drug development [38]. Finally, once the issue of specificity is firmly embedded in the context of cross-reactivity pattern matching, it becomes possible to build novel assay formats, utilizing structurally diverse ligands as surrogates of analytes.

Acknowledgments. We thank Bill Usinger for hybridoma advice, and Tracy Orvik, Lura Wilhelm and Drew Lau for technical assistance. We also thank Patricia Gearhart and Michael Buchmeier for comments on the manuscript.

Literature Cited.

1. Marks, J.D., H.R. Hoogenboom, A.D. Griffiths and G. Winter. *J. Biol. Chem.* 1992, **267**:16007.
2. Eigen, M., W. Gardiner, P. Schuster, and R. Winkler-Oswatitsch. *Sci.Am.* 1981, **244**(4):88.
3. Wu, T.T., E.A. Kabat. *J.Exp.Med.* 1970, **132**:211.
4. Jerne, N.K. *Basel Institute for Immunology Annual Report*, 1977, p.1.
5. Buck, L. and R. Axel. *Cell* 1991, **65**:175.
6. Lancet, D., E. Sadovsky and E. Seidemann. *Proc.Nat.Acad.Sci. USA*, 1993, **90**:3715.
7. Kauvar, L.M. *Am. Lab.* 1993, **25**(9):25.
8. Hooijkass, H., A.A. Preesman, A. van Oudenaren, R. Benner, and J.J. Haaijman. *J.Immunol.* 1983, **131**:1629.
9. Tonegawa, S. *Nature*, 1983, **302**: 575.
10. Malynn, B.A., G.D. Yancopoulos, J.E. Barth, C.A. Bona, and F.W. Alt. *J.Exp. Med.* 1990, **171**:843.
11. Striebich, C.C., R.M. Miceli, D.H. Schulze, G. Kelsoe, and J. Cerny. *J.Immunol.* 1990, **144**:1857.
12. Goldsby, R.A., B.A. Osborne, D. Suri, J. Williams, and A.D. Mandel. *Curr.Micro.* 1979, **2**:157.
13. Andersson, J., and F. Melchers. *Curr. Topics in Micro. and Immunol.* 1978, **81**:130.
14. Rousseau, P.G., C.P. Mallett, and S.J. Smith-Gill. *Mol. Immunol.* 1989, **26**:993.
15. Fischbach, M. and N. Talal in *Idiotypy in Biology and Medicine*, H. Kohler, J. Urbain, P-A. Cazenave, Eds. (NY: Academic Press, 1984), pp. 417-428.
16. Gearhart, P.J. in 3rd ed. *Fundamental Immunology*, W.E. Paul, Ed. (NY: Raven Press, 1993), pp. 865-885.
17. Wang, K., J.R. Feramisco, and J.F. Ash. *Meth. Enz.* 1982, **85**:514.
18. Harrison, R.O., M.H. Goodrow, S.J. Gee, and B.D. Hammock. in *Immunoassays for Monitoring Human Exposure to Toxic Chemicals in food and the Environment* M. Vanderlaan, L. Stanker, B. Watkins, R. Roberts, Eds. (Washington, D.C.: American Chemical Society, 1990).
19. Harlow, E., and D. Lane. *Antibodies: a laboratory manual* (NY: Cold Spring Harbor Laboratory, 1988).
20. Kauvar, L.M., and P.Y.K. Cheung. 1992. U.S. Patent No. 5,155,049.
21. Crothers, D.M., and H. Metzger. *Immunochem.* 1972, **9**:341.
22. Markl, J., A. Markl, W. Schartau, and B. Linzen. *J.Comp.Physiol.* 1979, **130**:283.
23. Varga, J.M., S. Lande, and F.F. Richards. *J.Immunol.* 1974, **112**:1565.
24. Getzoff, E.D., J.A. Tainer, R.L. Lerner, and H.M. Geysen. *Adv. Immunol.* 1988, **43**:1.
25. Huse, W., L. Sastry, S.A. Iverson, A.S. Kang, M. Alting-Mees, D.R. Burton, S.J. Benkovic, and R.A. Lerner. *Science* 1989, **246**:1275.
26. McCafferty, J., A.D. Griffiths, G. Winter, and D. J. Chiswell. *Nature* 1990, **348**:552.
27. Huston, J.S., D. Levinson, M. Mudgett-Hunter, M.S. Tai, J. Novotny, M.N. Margolies, R.J. Ridge, R.E. Bruccoleri, E. Haber, R. Crea, and H. Oppermann. *Proc.Nat.Acad.Sci. USA* 1988, **85**:5879.

28. Bird, R.E., K.D. Hardman, J.W. Jacobson, S. Johnson, B.M. Kaufman, S.M. Lee, T. Lee, S.H. Pope, G.S. Riordan, and M. Whitlow. *Science* 1988, **242**:423.
29. Pessi, A., E. Bianchi, A. Crameri, S. Venturini, A. Tramontano, and M. Sollazzo. *Nature* 1993, **362**:367.
30. Lowman, H.B., S.H. Bass, N. Simpson, and J.A. Wells. *Biochem.* 1991, **30**:10832.
31. Cheung, P.Y.K., L.M. Kauvar, A.E. Engqvist-Goldstein, S.M. Ambler, A.E. Karu and L.S. Ramos. *Anal. Chim. Acta* 1993, **282**:181.
32. Perelson, A. S. and G. F. Oster. *J. Theor. Biol.* 1979, **81**:645.
33. Villar, H.O., and L.M. Kauvar. *FEBS Letters* 1994, **349**:125.
34. Jakoby, W. B. and D. M. Ziegler. *J. Biol. Chem.* 1990, **265**:20715.
35. Kauvar, L.M. in *New Frontiers in Agrochemical Immunoanalysis*, D. Kurtz, L. Stanker, and J.H. Skerritt, Eds. (Washington, DC: AOAC, 1995), in press.
36. Kauvar, L.M. 1994. U.S. Patent No. 5,217,869.
37. Kauvar, L.M. in 8th ed. *Pharm. Mfr. Int'l.*, P.A. Barnacal, Ed. (London: Sterling Publications, 1995) in press.
38. Kauvar, L.M. in *Structure and Function of Glutathione S-Transferase*, K.D. Tew, C.B. Pickett, T.J. Mantle, B. Mannervik, and J.D. Hayes, Eds. (Boca Raton, FL: CRC Press, 1993); pp. 257-268

RECEIVED October 31, 1994

Chapter 8

Analytical Representation and Prediction of Macroscopic Properties

A General Interaction Properties Function

Peter Politzer, Jane S. Murray, Tore Brinck, and Pat Lane

Department of Chemistry, University of New Orleans,
New Orleans, LA 70148

We present a procedure whereby quantities computed for an isolated molecule can be used to represent and predict macroscopic properties that reflect molecular interactions. Such representations are all special cases of a General Interaction Properties Function (GIPF). The molecular quantities are evaluated on the surface of the molecule, defined as the 0.001 au contour of its electronic density; most of them are related to its electrostatic potential. Among the macroscopic properties for which GIPF expressions have been developed are aqueous acidities, boiling points, critical constants, partition coefficients, heats of vaporization, solubilities in supercritical fluids and hydrogen bonding parameters. The GIPF approach is expected to facilitate the design of molecules for specific purposes such as haptens in immunochemistry, since it identifies the key factors that determine particular properties and provides a means for evaluating proposed compounds prior to their syntheses.

We have developed a unified approach to correlating and predicting macroscopic condensed-phase properties that reflect molecular interactions (*1,2*). It involves representing each property in terms of some subset of a group of computed molecular quantities. Each resulting relationship is viewed as a special case of the General Interaction Properties Function (GIPF) described by equation 1:

$$\text{Property} = f\left[\text{area}, \bar{I}_{S,min}, V_{S,max}, V_{S,min}, \Pi, \sigma_{tot}^2, \nu\right] \tag{1}$$

The quantities in brackets in equation 1 are measures of various aspects of a molecule's interactive behavior; they will be defined and discussed in the next section.

The macroscopic properties that have been represented successfully by forms of equation 1 include aqueous acidities (pK$_a$ values) (*3-6*), gas phase protonation enthalpies (*4-6*), boiling points (*7*), critical constants (temperatures, pressures and volumes) (*7*), partition coefficients (*8,9*), solubilities in supercritical fluids (*10-12*), heats of vaporization (*13*) and hydrogen-bonding parameters (*14,15*). In no instance are all of the molecular quantities in equation 1 used in the expression for a particular property; typically two or three of them are involved. Since each of these quantities has a well-defined physical meaning, it is accordingly possible to achieve insight into the key factors that determine the macroscopic property of interest. This facilitates efforts to design molecules having improved performance in that respect. In the same context,

0097–6156/95/0586–0109$12.00/0

it is important to note that all of the quantities in equation 1 are evaluated computationally; no experimental data are needed. This permits predictions to be made for compounds that have not yet been prepared or isolated.

Finally, a very significant and somewhat surprising feature of the GIPF approach is that it allows the correlation and prediction of solution and liquid-phase properties solely from quantities calculated for isolated molecules. The effects of the medium need not be explicitly taken into account.

Molecular Quantities in the GIPF

The quantities in equation 1 are all evaluated on the surface of the molecule, which we define, following Bader $et\ al$ (16), as the 0.001 electrons/bohr^3 contour of its electronic density. A surface defined in this manner directly reflects features specific to the particular molecule, such as lone pairs. We obtain the area, to use in equation 1, by means of a grid of equidistant points converted to units of Å^2 ($7,10$-12).

All of the remaining quantities in equation 1 except $\bar{I}_{S,min}$ are related to the molecular electrostatic potential $V(\mathbf{r})$, which is defined rigorously by equation 2:

$$V(\mathbf{r}) = \sum_A \frac{Z_A}{|\mathbf{R}_A - \mathbf{r}|} - \int \frac{\rho(\mathbf{r}')\,d\mathbf{r}'}{|\mathbf{r}' - \mathbf{r}|} \qquad (2)$$

Z_A is the charge on nucleus A, located at $\mathbf{R}_A$, and $\rho(\mathbf{r})$ is the electronic density function of the molecule. The sign of $V(\mathbf{r})$ at any point $\mathbf{r}$ is the net result of the positive and negative contributions of the nuclei and electrons, respectively, as given by the two terms on the right side of equation 2.

The electrostatic potential is a real physical property, which can be determined experimentally by diffraction techniques (17), as well as computationally. It is well-established as an effective tool for interpreting and predicting molecular reactivity (17-22); sites reactive toward electrophiles can be identified and ranked by means of the locations and magnitudes of the most negative potentials on the molecular surface, $V_{S,min}$ (6), while the most positive surface potentials, $V_{S,max}$, play an analogous role for nucleophilic attack ($14,15,23$). $V_{S,min}$ and $V_{S,max}$ are accordingly site-specific quantities; they are measures of the tendencies for electrophilic and nucleophilic interactions, respectively, at particular points in the space of a molecule.

In contrast, Π, σ^2_{tot} and ν are statistically-based global quantities, which reflect the electrostatic potentials over the entire surfaces of molecules. They are defined by equations 3 - 5:

$$\Pi = \frac{1}{n}\sum_{i=1}^{n}\left|V(\mathbf{r}_i) - \bar{V}_S\right| \qquad (3)$$

$$\sigma^2_{tot} = \sigma^2_+ + \sigma^2_- = \frac{1}{m}\sum_{i=1}^{m}\left[V^+(\mathbf{r}_i) - \bar{V}^+_S\right]^2 + \frac{1}{n}\sum_{j=1}^{n}\left[V^-(\mathbf{r}_j) - \bar{V}^-_S\right]^2 \qquad (4)$$

$$\nu = \frac{\sigma^2_+ \sigma^2_-}{\left[\sigma^2_{tot}\right]^2} \qquad (5)$$

$V(\mathbf{r}_i)$ is the value of $V(\mathbf{r})$ at point $\mathbf{r}_i$ on the surface, and $\bar{V}_S$ is the average value of the potential on the surface: $\bar{V}_S = \frac{1}{n}\sum_{i=1}^{n}V(\mathbf{r}_i)$. In a similar fashion, $V^+(\mathbf{r}_i)$ and $V^-(\mathbf{r}_j)$

are the positive and negative values of $V(\mathbf{r})$ on the surface, and $\bar{V}_{S^+}$ and $\bar{V}_{S^-}$ are their averages: $\bar{V}_{S^+} = \frac{1}{m} \sum_{i=1}^{m} V^+(\mathbf{r}_i)$ and $\bar{V}_{S^-} = \frac{1}{n} \sum_{j=1}^{n} V^-(\mathbf{r}_j)$.

Π is equal to the average deviation of $V(\mathbf{r})$ on the molecular surface, which we take to be indicative of the local polarity, or charge separation, that is present even in molecules having zero dipole moments (24), e.g. BF_3 and p-dinitrobenzene. We have shown that Π correlates with dielectric constants (24).

σ^2_{tot} is the total variance of $V(\mathbf{r})$ on the molecular surface. It is a measure of the spread, or range of values, of the surface potential; because the terms are squared, σ^2_{tot} is particularly sensitive to the positive and negative extremes in $V(\mathbf{r})$. We have found it to be an effective indicator of a molecule's tendency for noncovalent electrostatic interactions (1,7-12). In some instances, it is preferable to use σ^2_+ or σ^2_- alone, instead of σ^2_{tot} (8,9).

The function of ν, the third global quantity, is to show the degree of balance between the positive and negative potentials on the surface (7,12). ν attains a maximum value of 0.250 when σ^2_+ and σ^2_- are equal; accordingly, the closer that ν is to 0.250, the better able is the molecule to interact to a similar extent (whether strongly or weakly) through both its positive and negative potentials. The product $\nu \sigma^2_{tot}$ has been found to be a key term in representing properties that reflect the electrostatic interactions of a molecule with others of its own kind, e.g. boiling points and critical temperatures (7).

For illustrative purposes, Table I gives Π, σ^2_{tot} and ν for a few selected molecules. They are listed in order of increasing Π. It should be noted that some of the larger Π values are for molecules having zero dipole moments, **5** and **7**. The data in the table clearly show that Π and σ^2_{tot} are quite different quantities, despite the superficial similarity in their definitions. σ^2_{tot} covers a much greater range of magnitudes than does Π; furthermore, they do not even necessarily vary in the same direction. A particularly interesting comparison in Table I is between **2** and **6**. The latter clearly has more local polarity, as measured by Π, and much stronger positive and negative regions; its σ^2_{tot} is 182.9 $(kcal/mole)^2$, vs. 15.9 for **2**. Nevertheless they both have $\nu \approx 0.25$, showing essential balance between positive and negative potentials in each molecule, notwithstanding the fact that they are far stronger in **6**.

The final quantity to be defined, $\bar{I}_{S,min}$, is again site-specific; it represents the minimum values(s), on the molecular surface, of the average local ionization energy, $\bar{I}(\mathbf{r})$, defined by equation 6 (25):

$$\bar{I}(\mathbf{r}) = \sum_i \frac{\rho_i(\mathbf{r})|\varepsilon_i|}{\rho(\mathbf{r})} \tag{6}$$

$\rho_i(\mathbf{r})$ is the electronic density of the ith molecular orbital, having orbital energy ε_i, and $\rho(\mathbf{r})$ is the total electronic density.

Table I. Values of global quantities for selected molecules [a]

	Molecule	Π (kcal/mole)	σ^2_{tot} (kcal/mole)^2	ν
1	(cyclohexane ring)	2.16	3.2	0.171
2	(naphthalene)	5.12	15.9	0.250
3	(indole)	8.39	96.6	0.169
4	(pyridine, N)	8.55	230.8	0.074
5	(hexafluorobenzene, F)	10.35	45.3	0.116
6	(indole-3-COOH)	11.09	182.9	0.248
7	(1,3,5-trinitrobenzene, NO_2, O_2N, NO_2)	18.70	152.7	0.214
8	CH_3NO_2	19.90	116.0	0.209

[a]Taken from reference 2.

SOURCE: Data taken from several tables in reference 2.

We interpret $\bar{I}(\mathbf{r})$ as the average energy required to remove an electron from the point $\mathbf{r}$; we are focusing upon a point in the space of the molecule rather than upon a particular molecular orbital. Accordingly the positions at which $\bar{I}(\mathbf{r})$ has its lowest values are the locations, on the average, of the least tightly bound, most readily transferred electrons. We have found it advantageous to evaluate $\bar{I}(\mathbf{r})$ on the molecular surface (3-6,25), where the minima are designated as $\bar{I}_{S,min}$.

One of our early uses of a calculated molecular quantity to correlate and predict a macroscopic property involved $\bar{I}_{S,min}$ and aqueous acidity, as measured by pK_a; this reflects the tendency to interact through charge transfer to the electrophile H^+. For a group of 27 acids of various types, we found a good relationship between pK_a and the $\bar{I}_{S,min}$ of the conjugate base (4,5,26); the correlation coefficient is 0.97. This provides a predictive capability which we used to estimate pK_a's for several compounds, including $HN(NO_2)_2$ (predicted $pK_a = -5.6$) and s-triazine (predicted $pK_a = -2.3$) (5). These computations were carried out at the *ab initio* HF/6-31G* level, using HF/3-21G optimized geometries (27).

Some Past Applications of the GIPF

General Approach. In order to establish a GIPF representation of some macroscopic property, it is first necessary to have known values for it for some group of compounds. We then evaluate the various quantities in eq. (1) for the corresponding molecules; this is done computationally, usually at an *ab initio* minimum-basis-set self-consistent field (SCF) level (27), e.g. HF/STO/5G*//HF/STO-3G*. (Polarization functions are included for second- and third-row atoms.) Minimum-basis SCF calculations are generally satisfactory for geometry optimizations (28) and for one-electron properties (20,21,29-31), which includes the electron density and the electrostatic potential. The SAS statistical analysis program (32) is then used to develop a relationship between the known values of the property and some subset of the computed quantities.

Relationships in Terms of Global Molecular Quantities. Table II presents GIPF relationships for some properties that can be represented in terms of the global molecular quantities: area, Π, σ_{tot}^2 (or σ_+^2 or σ_-^2) and ν. (In one instance, solubility in supercritical CO_2, molecular volume was used instead of area.) All of these properties can be viewed as involving noncovalent interactions.

One of our objectives is to develop as general relationships as possible, covering compounds of a variety of chemical types. This is reflected in the sizes of the data bases reported in Table II. The correlation coefficients and standard deviations would of course be expected to improve if we were to treat families of compounds separately, e.g. hydrocarbons, alcohols, acids, etc.

It has been convenient to use area as a measure of molecular size in GIPF equations, although it may be that volume is more appropriate. This would especially be so if, as seems likely, the size term is expressing the contribution of the molecular polarizability; this is directly related to molecular volume (33-35).

The equations in Table II illustrate our earlier statement that the product $\nu\sigma_{tot}^2$ is very important in representing properties that depend upon interactions of a molecule with others of its own kind. Another interesting observation is that the variables in the expression for T_c are the square roots of those in the T_{bp} equation. This may reflect the weaker interactions at the critical temperature, where the density is much less than at the normal boiling point.

Relationships Involving Site-Specific Molecular Quantities. We have already mentioned the correlation between aqueous acidity (pK_a) and $\bar{I}_{S,min}$, a site-

Table II. Some GIPF relationships in terms of global molecular quantities [a,b]

Relationship[c]	N	R	S. D.	Ref.
Normal boiling point, T_{bp}: $T_{bp} = \alpha(\text{area}) + \beta(v\sigma_{tot}^2)^{0.5} - \gamma$	100	0.948	37.0	6
Critical temperature, T_c: $T_c = \alpha(\text{area})^{0.5} + \beta(v\sigma_{tot}^2)^{0.25} - \gamma$	66	0.909	60.7	6
Critical volume, $\overline{V}_c$: $\overline{V}_c = \alpha(\text{area})^{1.5} + \beta$	58	0.986	15.2	6
Critical pressure, P_c: $P_c = -\alpha(\text{area}) + \beta(v\sigma_{tot}^2 / \text{area}) + \gamma$	57	0.910	4.8	6
Octanol/water partition coefficient, P_{ow}: $\log P_{ow} = \alpha(\text{area}) - \beta(\sigma_-^2) - \gamma(\text{area})\Pi - \varepsilon$	70	0.961	0.437	7
Heat of vaporization, $\Delta\overline{H}_v$: $\Delta\overline{H}_v = \alpha(\text{area})^{0.5} + \beta(v\sigma_{tot}^2)^{0.5} - \gamma$	40	0.971	2.03	d
Solubility in supercritical CO_2 at 14 MPa and 308 K: $\ln(\text{sol}) = \alpha(\text{vol})^{-1.5} - \beta(\sigma_{tot}^2)^2 - \gamma$	21	0.95		10
Enhancement factor, E, in supercritical CO_2 at 20 MPa and 308 K:[e] $E = -\alpha(\text{area})^{-1.5} + \beta(\sigma_{tot}^2) + \gamma(v) - \varepsilon(v\sigma_{tot}^2) - \eta$	12	0.921		11
Critical temperature of X/CO_2 mixture, mole fraction X = 0.10: $T_c = \alpha(\text{area})^2 - \beta(16.1 - \Pi)^2 - \gamma(\frac{\text{area}^3}{v}) + \varepsilon$	12	0.979	2.4	d

[a]N is number of systems in data base; R is correlation coefficient; S. D. is standard deviation.

[b]Units: T_{bp}, K; T_c, K; $\overline{V}_c$, cm^3/mole; P_c, bar; $\Delta\overline{H}_v$, kJ/mole.

[c]All coefficients (α, β, γ, ε, η) are positive numbers.

[d]Unpublished work.

[e]The enhancement factor is a measure of the role of solute-solvent interactions in supercritical solutions.

specific quantity (*4,5,26*). This involves only acids in which the hydrogen is bonded to a first-row atom (carbon, nitrogen or oxygen). We have also developed a representation of gas phase acidity, as measured by the enthalpy of protonation, for the first-, second- and third-row hydrides, and their anions, of Groups V - VII of the periodic table (*6*). All 18 of them can be described by equation 7, but it requires $V_{S,min}$ as well as $\bar{I}_{S,min}$:

$$\Delta H^{\circ}_{pr} = \alpha V_{S,min} + \beta \bar{I}_{S,min} - \gamma \qquad (7)$$

The correlation coefficient is 0.997 and the standard deviation is 7.5 kcal/mole.

The need for $V_{S,min}$ in equation 7 arises because of the inclusion of second- and third-row hydrides (*6*), and it seems likely, therefore, that our earlier pK_a correlation would also need a $V_{S,min}$ term if it included acids in which the hydrogen is bonded to a second- or third-row atom. Indeed, pK_a for the first-, second- and third-row hydrides of Groups V-VII was found to obey equation 8 (*6*),

$$pK_a = -\alpha V_{S,min} - \beta \bar{I}_{S,min} - \gamma \qquad (8)$$

with a correlation coefficient of 0.962 and standard deviation of 4.2 (for a range of values from (-9.5 to 34).

Since hydrogen bonding is often a key factor in molecular interactions, we have devoted considerable effort to relating it to our molecular quantities (*14,15,22,36*). We have developed GIPF expressions for solvatochromic parameters that are well-established quantitative measures of solute/solvent hydrogen-bond-donating and -accepting tendencies, and are used for this purpose in linear solvation energy relationships (*2,37-41*). These parameters are designated as α and α_2^H for solvent and solute hydrogen-bond acidity, and β and β_2^H for the respective hydrogen bond basicities. For 24 compounds of various types, we showed that,

$$\beta_2^H \text{ (or } \beta) = -\alpha V_{S,min} - \beta \bar{I}_{S,min} - \gamma \Pi + \varepsilon \qquad (9)$$

with correlation coefficients of 0.977 for β_2^H and 0.973 for β. For an assortment of 20 compounds,

$$\alpha_2^H = \alpha(V_{S,max})\sigma_+^2 + \beta(v\sigma_{tot}^2) + \gamma(area) + \varepsilon \qquad (10)$$

Equations 9 and 10 involve both site-specific and global molecular quantities.

Equations 7 - 9 show that there is a complementarity between $V_{S,min}$ and $\bar{I}_{S,min}$, despite an apparent similarity in that both are related to interactions with electrophiles. We view $V_{S,min}$ as being particularly relevant to the approach of an electrophile, and $\bar{I}_{S,min}$ to subsequent polarization/charge transfer.

Molecular Design

The relevance of the GIPF approach to molecular design is at least three-fold: First, GIPF representations of the desired properties help to identify the electronic and/or structural factors that are of prime importance in determining these properties. Second, the predictive capabilities of GIPF equations make it possible to evaluate proposed compounds prior to attempting their syntheses. Third, the quantities that appear in the GIPF, equation 1, provide a meaningful basis for analyzing and interpreting similiarities and differences in the interactive behavior of molecules.

Table III. Values of global molecular quantities for some benzene derivatives

Molecule	Surface Area (Å^2)	Π (kcal/mole)	σ_+^2	σ_-^2	σ_{tot}^2	ν
benzene	115	4.8	7	9	16	0.25
—CH$_3$	136	4.6	7	11	18	0.24
—F	118	5.6	12	33	45	0.20
—Cl	132	6.2	14	23	37	0.24
—Br	137	5.9	13	19	32	0.24
—O,H (Cl ortho)	138	6.8	24	70	94	0.19
—O,H (Cl meta)	140	8.6	80	54	134	0.24
—O,H	125	8.6	64	74	137	0.25
—O,CH$_3$	144	7.4	16	61	77	0.16

For example, we have used our GIPF representation of solubility in supercritical CO_2 to predict which of a group of suggested compounds best simulate certain highly toxic chemical defense agents in this respect (42). This is very relevant to the current active interest in disposing of hazardous materials by supercritical oxidation (43-45).

In the area of immunology, we anticipate that GIPF representations of antigen-antibody binding constants could be developed. Such relationships would facilitate the design of new haptens, and subsequently the introduction of new antigens. However even without binding constant correlations, useful comparisons of existing and potential haptens can be made in terms of the GIPF molecular quantities. As an example, selected because many haptens are benzene derivatives, Table III gives the calculated GIPF global quantities for some substituted benzenes, arranged in four groups. Within each group, similar values are enclosed in boxes. These reveal some interesting patterns. Benzene and toluene differ significantly in surface area, but are very much alike in all interactive aspects. Chlorobenzene and bromobenzene resemble each other in all respects, but fluorobenzene in none! The isomers *ortho* and *meta* chlorophenol have nearly the same areas, but differ in all other ways. Finally, phenol and anisole show no similarities.

Summary

The GIPF procedure permits macroscopic properties that reflect molecular interactions to be represented analytically in terms of a subset of computed molecular quantities. This allows the prediction of such properties from calculations for isolated molecules, without explicitly accounting for macroscopic and medium effects. The GIPF expression also helps to identify the key electronic and structural effects upon which the properties depend. Interaction involving charge transfer and polarization can be treated as well as noncovalent ones. The GIPF approach is continually evolving; as the range of applications is expanded, additional molecular quantities may have to be introduced. An important use is anticipated to be in the design and evaluation of molecules for specific purposes.

Acknowledgement

We greatly appreciate the support provided by ARPA/ONR Contract No. N00014-91-J-1897, administered by ONR.

References

1. Murray, J. S.; Brinck, T.; Lane, P.; Paulsen, K.; Politzer, P. *J. Mol Struct. (Theochem)* in press.
2. Murray, J. S.; Politzer, P. In *Quantitative Approaches to Solute/Solvent Interactions*; Murray, J. S. and Politzer, P., Ed.; Elsevier: Amsterdam, 1994,
3. Brinck, T.; Murray, J. S.; Politzer, P.; Carter, R. E. *J. Org. Chem.* **1991**, *56*, 2934.
4. Brinck, T.; Murray, J. S.; Politzer, P. *J. Org. Chem.* **1991**, *56*, 5012.
5. Murray, J. S.; Brinck, T.; Politzer, P. *J. Mol Struct. (Theochem)* **1992**, *255*, 271.
6. Brinck, T.; Murray, J. S.; Politzer, P. *Int. J. Quant. Chem.* **1993**, *48*, 73.
7. Murray, J. S.; Lane, P.; Brinck, T.; Paulsen, K.; Grice, M. E.; Politzer, P. *J. Phys. Chem.* **1993**, *97*, 9369.
8. Brinck, T.; Murray, J. S.; Politzer, P. *J. Org. Chem.* **1993**, *58*, 7070.
9. Murray, J. S.; Brinck, T.; Politzer, P. *J. Phys. Chem.* **1994**, *97*, 13807.
10. Politzer, P.; Lane, P.; Murray, J. S.; Brinck, T. *J. Phys. Chem.* **1992**, *96*, 7938.
11. Politzer, P.; Murray, J. S.; Lane, P.; Brinck, T. *J. Phys. Chem.* **1993**, *97*, 729.
12. Murray, J. S.; Lane, P.; Brinck, T.; Politzer, P. *J. Phys. Chem.* **1993**, *97*, 5144.
13. Murray, J. S.; Lane, P.; Politzer, P. submitted for publication.
14. Murray, J. S.; Politzer, P. *J. Org. Chem.* **1991**, *56*, 6715.
15. Murray, J. S.; Politzer, P. *J. Chem. Res.* **1992**, *S*, 110.
16. Bader, R. F. W.; Carroll, M. T.; Cheeseman, J. R.; Chang, C. *J. Am. Chem. Soc.* **1987**, *109*, 7968.

17. *Chemical Applications of Atomic and Molecular Electrostatic Potentials*; Politzer, P.; Truhlar, D. G., Ed.; Plenum Press: New York, 1981.
18. Scrocco, E.; Tomasi, J. In *Topics in Current Chemistry,* Springer-Verlag: Berlin, 1973; Vol. 42, 95.
19. Scrocco, E.; Tomasi, J. *Advances Quantum Chemistry,* **1978**, *11*, 115.
20. Politzer, P.; Daiker, K. C. In *The Force Concept in Chemistry*; Deb, B. M., Ed.; Van Nostrand Reinhold Company: New York, 1981, Chapter 6.
21. Politzer, P.; Murray, J. S. In *Theoretical Biochemistry and Molecular Biophysics: Vol. 2, Proteins*; Beveridge, D. L. and Labery, R., Ed.; Adenine Press: Schenectady, NY, 1991, ch. 13.
22. Politzer, P.; Murray, J. S. In *Reviews in Computational Chemistry*; Lipkowitz, K. B. and Boyd, D. B., Ed.; VCH Publishers: New York, 1991; Vol. 2, ch 7.
23. Murray, J. S.; Lane, P.; Brinck, T.; Politzer, P. *J. Phys. Chem.* **1990**, *95*, 844.
24. Brinck, T.; Murray, J. S.; Politzer, P. *Mol. Phys.* **1992**, *76*, 609.
25. Sjoberg, P.; Murray, J. S.; Brinck, T.; Politzer, P. *Can. J. Chem.* **1990**, *68*, 1440.
26. Murray, J. S.; Brinck, T.; Politzer, P. *Int. J. of Quantum Chem., Quantum Biol. Symp.* **1991**, *18*, 91.
27. Frisch, M. J.; Head-Gordon, M.; Schlegel, H. B.; Raghavachari, K.; Binkley, J. S.; Gonzalez, C.; Defrees, D. J.; Fox, D. J.; Whiteside, R. A.; Seeger, R.; Melius, C. F.; Baker, J.; Martin, R.; Kahn, L. R.; Stewart, J. J. P.; Fluder, E. M.; Topiol, S.; Pople, J. A. *GAUSSIAN 88*; Gaussian Inc.: Pittsburgh, PA, 1988.
28. Hehre, W. J.; Radom, L.; Schleyer, P. v. R.; Pople, J. A. *Ab Initio Molecular Orbital Theory*; Wiley-Interscience: New York, 1986.
29. Gatti, C.; MacDougall, P. J.; Bader, R. F. W. *J. Chem. Phys.* **1988**, *88*, 3792.
30. Boyd, R. J.; Wang, L.-C. *J. Comp. Chem.* **1989**, *1*, 367.
31. Seminario, J. M.; Murray, J. S.; Politzer, P. In *The Application of Charge Density Research to Chemistry and Drug Design,* Plenum Press: New York, 1991, p. 371.
32. SAS, SAS Institute Inc.: Cary, NC 27511.
33. Gough, K. M. *J. Chem. Phys.* **1989**, *91*, 2424.
34. Laidig, K. E.; Bader, R. F. W. *J. Chem. Phys.* **1990**, *93*, 7213.
35. Brinck, T.; Murray, J. S.; Politzer, P. *J. Chem. Phys.* **1993**, *98*, 4305.
36. Murray, J. S.; Ranganathan, S.; Politzer, P. *J. Org. Chem.* **1991**, *56*, 3734.
37. Kamlet, M. J.; Abboud, J.-L. M.; Abraham, M. H.; Taft, R. W. *J. Org. Chem.* **1983**, *48*, 2877.
38. Kamlet, M. J.; Doherty, R. M.; Abraham, M. H.; Marcus, Y.; Taft, R. W. *J. Phys. Chem.* **1988**, *92*, 5244.
39. Abraham, M. H.; Grellier, P. L.; Prior, D. V.; Duce, P. P.; Morris, J. J.; Taylor, P. J. *J. Chem. Soc. Perkin Trans. II* **1989**, 699.
40. Abraham, M. H.; Grellier, P. L.; Prior, D. V.; Morris, J. J.; Taylor, P. J. *J. Chem. Soc. Perkin Trans. II* **1990**, 521.
41. Abraham, M. H. *Chem. Soc. Rev.* **1993**, *22*, 73.
42. Politzer, P.; Murray, J. S.; Concha, M. C.; Brinck, T. *J. Mol Struct. (Theochem)* **1993**, *281*, 107.
43. Staszak, C. N.; Malinowski, K. C.; Killilea, W. R. *Environ. Prog.* **1987**, *6*, 39.
44. Webley, P. A.; Tester, J. W. In *Supercritical Fluid Science and Technology*; Johnston, K. P. and Penninger, J. M. L., Ed.; ACS Symp. Ser. 406; American Chemical Society: Washington, 1989, ch. 17.
45. Shaw, R. W.; Brill, T. R.; Clifford, A. A.; Eckert, C. A.; Franck, E. U. *Chem. Eng. News* **1991**, *69*, 26.

RECEIVED November 1, 1994

Chapter 9

Strategies for Immunoassay Hapten Design

Marvin H. Goodrow[1], James R. Sanborn[2], Donald W. Stoutamire[1],
Shirley J. Gee[1], and Bruce D. Hammock[1]

[1]Departments of Entomology and Environmental Toxicology,
University of California, Davis, CA 95616
[2]Department of Pesticide Regulation, California Environmental
Protection Agency, Sacramento, CA 95814

Immunoassay performance is a function of the affinity and
selectivity of the antibody. The immunizing hapten should represent
a near perfect mimic of the target molecule in structure, electronic
and hydrophobic properties. These haptens are tethered with an
antigenically inert handle distal to the determinant group(s) and do
not mask or alter any functional group. Optimal hapten design
criteria are based on extending an existing carbon chain, or replacing
a C-H moiety of the target molecule with a CH_2 chain terminated by
a functional group for conjugation to proteins. Careful selection of
immunizing hapten can lead to the production of compound or class
selective antibodies. A multiple hapten approach, based on handle
location, length, and composition, results in assays with sub-ppb
levels of detection and improved selectivity. Examination of cross-
reactivity data of the haptens led to the identification of the best
coating/enzyme-labeled haptens for improved heterologous assays.
Examples from research with triazine, arylurea,and chloracetanilide
pesticides illustrate these principles.

The rational design of haptens for the development of antibodies to small molecules
has been evolving since the pioneering work of Landsteiner (1). Our objective is to
maximize recognition of a single target molecule and/or class of molecules with the
greatest selectivity and lowest limit of detectability (LLD). Selection of the
immunizing hapten is the single most important factor in eliciting antibody
production for meeting these objectives. Immunizing haptens may also be used for
the development of a useful homologous assay (i.e. the same hapten is used for
immunizing and for assay purposes). Modification of the hapten for assay usage
(i.e. by altering handle composition, length or position; heterologous assay) has
resulted in sub-ppb levels of detection and improved selectivity. Enzyme-label

detection techniques and assay format evaluations are secondary strategies for altering and/or improving target molecule LLD. This report will concentrate on the more critical hapten design parameters of the immunizing hapten, drawing principally from our results and the literature since our last review on the subject (2). The design criteria are summarized in Table I.

Table I. Guidelines for Immunizing Hapten Synthesis

-1. Location of Handle on Target Molecule
   *Distal to important haptenic determinants
   *Avoid attachment to functional groups
-2. Handle Selection
   *Length of handle
   *Avoid functional groups in handle -- use alkyl or aralkyl handles
-3. Functional Group for Coupling
   *Type of reaction of coupling
   *Compatibility of reaction with target molecule functional groups
-4. Solubility of Hapten and/or Conjugates
-5. Stability of Hapten Under Coupling Conditions and Subsequent Use
-6. Ease of Synthesis
-7. Determination of Hapten:Protein Ratio

Design of Immunizing Haptens

An optimum immunizing hapten for a selected target analyte has to be a near-perfect mimic of that molecule. It should contain a handle terminated with a functional group capable of covalent bonding to a carrier protein. This hapten should be identical to the target analyte in structure, geometry, electronic (H-bonding capabilities), and hydrophobic properties. If possible, the number of synthetic steps and/or difficulties required for its synthesis should not be a major consideration.

Numerous studies demonstrate that the handle should be attached as far as possible from the unique determinant groups (1, 2). This maximizes presentation of the important structural features of the analyte to the immune system. This is particularly important for selectivity to a single chemical structure within a class of compounds. If a class-selective assay is desirable, the handle is best located at or near a position that differentiates members of the class and exposes features common to the class. The smaller the molecule, the more important is the retention of each determinant group's identity. Thus, one should resist the temptation to attach the handle to a determinant group as this alters the structure, geometry, and the electronic nature compared to the original target compound.

For the immunizing hapten, we concentrate on replacing a carbon-hydrogen bond with a carbon to carbon chain of methylene groups terminated by a function group for conjugation to proteins. Electronic properties thus undergo a minimal change though there may also be a slight alteration in geometry and decrease in water solubility. Since some change of the target molecule is unavoidable, any attachment regardless of location, will cause some geometry change. The most remote attachment site should have the least effect on the geometry at the distinctive electrostatic binding sites. Our approach is based on the quantitative treatment of hapten-antibody interactions of Kutter and Hansch (*3*) who concluded that steric repulsion by substituents is the most important variable with hydrophobic and electronic effects being less important.

For small molecules, it is important to retain the identity of all determinant groups. For example, we try to avoid the most commonly used approach when dealing with an amino group, i.e., the preparation of the hemisuccinate of the hapten. This technique completely alters the basic nature and the moderate H-bonding donor/acceptor properties of the amine group. This results in a neutral species displaying the strongest H-bonding donor/acceptor properties. Often the strong amide determinant group produces polyclonal antibodies that recognize only the amide handle at the expense of the target analyte. Similar reasons apply to the use of heterobifunctional conjugation reagents, commonly employed in the medicinal field. However, both approaches may be useful if there is a desire to mask the recognition of the amine, if the site of attachment is remote from the desired recognition site, or if one is preparing a coating antigen or enzyme labeled hapten rather than an immunizing hapten.

Design of Coating/Enzyme-labeled Haptens

There are many immunoassay formats, each of which has distinct advantages such as speed, cost, selectivity and/or sensitivity. Each format shares three common components: a specific antibody, a hapten conjugated to a protein/enzyme, and a target analyte. Reviews that describe many assay formats have been published (*4, 5*). The various formats all obey the Law of Mass Action, a reversible antibody-analyte equilibrium competition with an antibody-hapten-protein conjugate as illustrated Figure 1. This is the rationale for reducing detection limits through use of heterologous haptens (*2*). As K_H for coating hapten-protein (H) is decreased (relative to a fixed K_A for analyte A), by selecting a different H-(coating hapten-protein), the equilibrium shifts to the antibody-analyte complex providing a lower level of detection (LLD). Thus, for a fixed quantity of antibody, the lowest detection level is observed when the affinity of the antibody for the analyte is greater than the affinity of the antibody for the plate-coating hapten. That is, $K_A \gg K_H$. Although a near perfect mimic of the target molecule will provide the best immunizing hapten, the LLD of an assay may be further improved by using heterologous coating/enzyme-labeled haptens. The guidelines for obtaining this heterology are outlined in Table II.

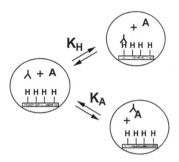

Figure 1. Schematic representation of the quasi-equilibria using heterologous haptens in immunoassay formats occurring on a hapten-protein coated plate. For a fixed K_A, the overall equilibrium shifts as K_H is decreased with a reduced affinity of antibody for H. For a fixed [A], the highest sensitivity is observed when $K_A \gg K_H$. K_A is the equilibrium constant for the binding of antibody (**Y**) to analyte (A). K_H is the equilibrium constant for the binding of antibody to hapten protein conjugate (H-) immobilized on a solid phase.

Table II. Guidelines for Coating / Enzyme Labeled Haptens

-1. Heterology of Hapten Structure
    *Position of handle
    *Composition of handle
    *Length of handle
    *Conjugation chemistry

-2. Alterations in Target Molecule Structure
    *Use of partial structure
    *Change of key determinants, i.e. sulfur for chlorine

-3. Cross-Reactivity Data of Hapten Structures (or Derivatives)
-4. Determination of Hapten:Protein Ratio

Triazines

Parent Compounds. The most extensively explored pesticides for immunoassay development has been the triazine herbicides, simazine and atrazine. They are particularly suited for study because of their commercial importance, extensive use, low mammalian toxicity, persistence in the environment, and very strong hydrogen bonding acceptor/donor properties (*6, 7*). In approaching the design of haptens for triazine immunoassays, three locations were identified for attachment of a handle (Figure 2).

Some of the earlier triazine assays used the sulfoxides of the herbicides ametryne (*8*) and terbutryne (*9*) as immunizing haptens (Figure 2). These modified triazines were conjugated directly to the protein carrier at the lysine amino and/or cysteine thio groups. This led to substitution at the 2-position of the triazine ring with a secondary nitrogen moiety (A, Figure 2), a non-chlorine atom, displaying hydrogen-bond donor/acceptor properties -- substantially dissimilar to a chlorine atom. This short handle also placed the hapten relatively close to the protein carrier surface. Assays with an acceptable IC_{50} (concentration that results in a 50% reduction in signal) value were obtained from ametryne sulfoxide-hemocyanin derived polyclonal antibodies (affinity purified) in a homologous assay using an alkaline phosphatase (AP) label (*8*). These antibodies, however, displayed substantial recognition of many closely related triazines making it a better candidate for a broader, class selective assay. Employing a longer handle for the enzyme labeled hapten, namely a simazine hexanoic acid derivative (B5, Figure 2 and Table III) conjugated to AP, resulted in a ten-fold reduction in the LLD for atrazine (IC_{50} = 3 ppb and an LLD of <1 ppb) accompanied by a substantial reduction in cross-reactivity towards related herbicides (*10*). Even lower amounts of atrazine (IC_{50} = 20 ppt and an LLD of 1 ppt) could be detected when the long handle (B10, Figure 2) was employed for both the immunizing and enzyme labeled hapten (*11*). Similarly,

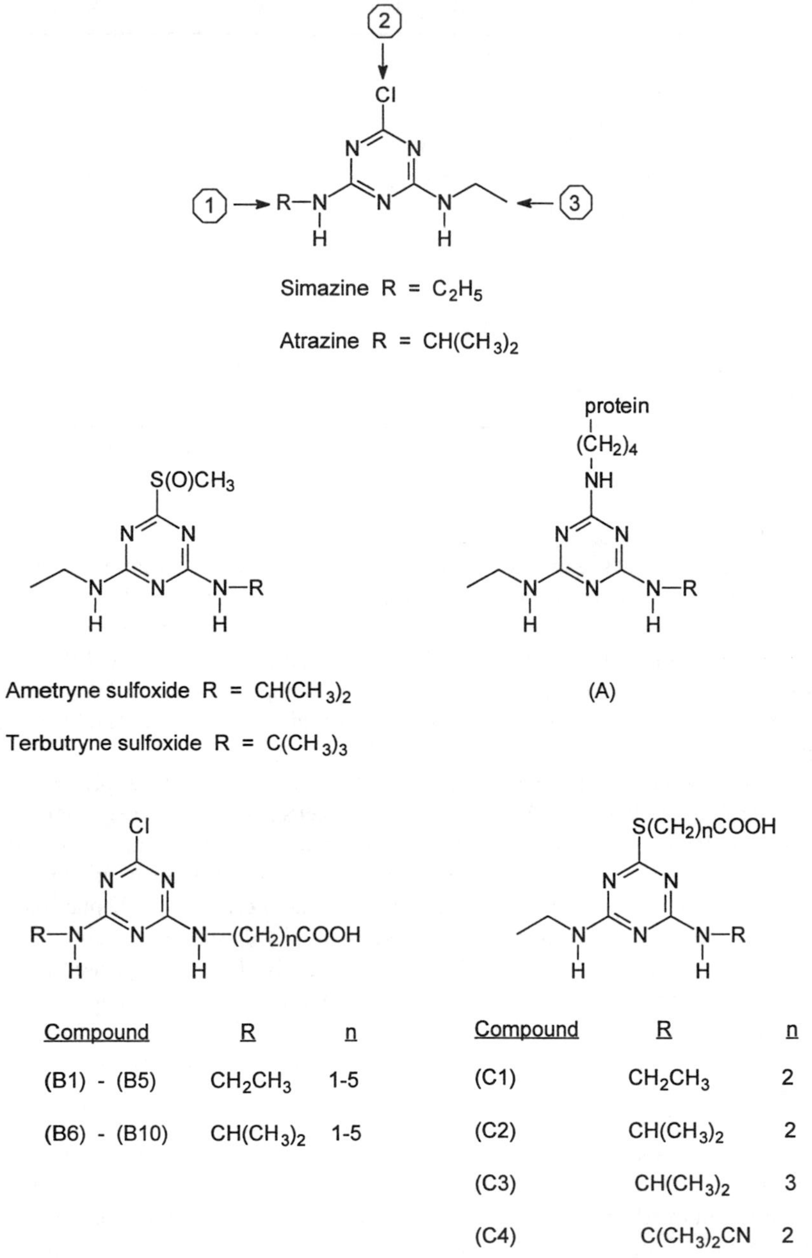

Simazine R = C_2H_5

Atrazine R = $CH(CH_3)_2$

Ametryne sulfoxide R = $CH(CH_3)_2$

Terbutryne sulfoxide R = $C(CH_3)_3$

(A)

Compound	R	n
(B1) - (B5)	CH_2CH_3	1-5
(B6) - (B10)	$CH(CH_3)_2$	1-5

Compound	R	n
(C1)	CH_2CH_3	2
(C2)	$CH(CH_3)_2$	2
(C3)	$CH(CH_3)_2$	3
(C4)	$C(CH_3)_2CN$	2

Figure 2. Structures of some triazine herbicides and haptens designed for their detection.

an homologous assay for terbutryne employing antibodies generated from terbutryne sulfoxide-BSA (bovine serum albumin) (Figure 2) and an AP label (*9*), had a LLD of 4.8 ng, but again with considerable cross-reactivity toward similar structures. This group later produced a very sensitive monoclonal assay (IC_{50} = 0.8 ppb) for terbuthylazine employing a similar non-tethered triazine immunogen and a hexanoic acid tethered triazine (Table III, *12*) that also displayed negligible responses to related triazines and metabolites.

Table III. 2-Chloro-4-alkylamino-6-ω-carboxyalkylamino-triazines Used as Immunizing (IH), Coating (CH) or Enzyme-labeled (TH) Haptens and Cross-Reactivity (CR) Evaluation.

Structure: R—NH— (on triazine ring with Cl at 2-position) —NH—$(CH_2)_nCOOH$

Compound	R	n	Usage	Reference
B1 - B5	Ethyl	1-5	IH, CH	(*4, 13*)
B5		5	TH	(*10, 14, 15*)
B5		5	IH, CH	(*16*)
B2, B3, B5		2, 3, 5	TH	(*17-20*)
B5		5	CRISP[a]	(*21*)
B5		5	CH	(*12*)
B5		5	CR	(*18, 22*)
B5	Isopropyl	5	IH, CH	(*11, 16, 17, 23-25*)
B10		5	TH	(*15, 25-27*)
B10		5	IH, TH	(*15, 28*)
B10		5	IH	(*18, 19*)
B5, B7, B10		1, 2, 5	IH, CH	(*4, 13*)
B9		4	IH, CH	(*29*)
B7, B10		2, 5	TH	(*20*)
B7, B10		2, 5	IH, CH, TH	(*30*)
B11	Cyclopropyl	5	TH	(*15, 12*)
B12	Cyclopentyl	5	TH	(*15*)
B11	t-Butyl	5	TH	(*15*)

[a]CRISP = cross reaction immuno-spectrum

In 1985 an atrazine homologous assay was reported that used the five-methylene group hapten (B10, Figure 2) conjugated to BSA and the same hapten on ovalbumin (OA) or rabbit serum albumin (RSA) as a coating antigen to yield an assay with an LLD of 0.1 ppb (*23, 24*). Selectivity was marginal as the assay also detected propazine and azidoatrazine (>50% CR); simazine and ametryne were detected to a lesser extent (<10% CR, *10*). These initial studies and the success of longer handles for atrazine (*11, 23*) and cyanazine (*31*), were the foundation for a multi-hapten approach that emphasized heterology by altering handle position and length to improve assay sensitivity and selectivity (*13, 32*).

The simazine molecule, because of its symmetry, contains only two possible handle attachments sites: extension of the N-ethyl side chain (site 1, Figure 2), or replacement of the chlorine (site 2, Figure 2) with a suitable mimic capable of accommodating two appendages. Atrazine, being unsymmetrical, permits an additional site (site 3, Figure 2) for handle attachment. Thus, to maximize antibody production utilizing the most unique features of the more widely used atrazine, and to improve selectivity among structurally similar triazines such as simazine and propazine, a series of N-alkyl carboxyalkyl acid haptens based on hapten structures B (Figure 2 and Table III) was prepared (*13, 16*). The longer chain haptens were selected for use as immunogens. These resulted in sera with very high titer, but of limited selectivity.

To enhance selectivity between atrazine and simazine their dissimilar structural features must be exposed. To this end, we synthesized haptens with a handle at site 2 (C, Figure 2). We used the sulfur in 3-mercaptopropanoic acid as a divalent connecting atom. The sulfur atom acted as a chlorine mimic that accommodated attachment to the triazine. It also contained a methylene handle terminated with a carboxylic acid for conjugation. The rationale for the selection of sulfur is reviewed by Goodrow et al. (*13*) and has been applied successfully by several others for conjugation of triazine (*22, 33, 34*), chloroacetanilide (see following section) and chlorpyrifos haptens (*35*).

Both homologous and heterologous immunoassay systems were evaluated using the multi-hapten approach based on target molecule handle position, chain length, and/or alkyl substituents. Heterology based on target molecule handle location provided the largest improvement in LLD. Both conjugation position and alkyl substitutions were important determinants for reducing the LLD (*16*). Although we achieved assays detecting atrazine at low ppb levels, this strategy, did not reach our objective of highly target-selective polyclonal antibodies. Simazine showed a cross-reactivity of about 2% with the best atrazine hapten B-10 (Table III) derived antibodies and atrazine displayed a cross-reactivity of about 10% with the best simazine hapten B5-derived antibodies (Table III). These compounds, however, were excellent as enzyme labeled haptens for assays based on the immunogenic hapten C2 antibodies (Table IV). An assay based on Mabs to C2 with a B2-AP label showed a five-fold improvement over the conventional antibody-coated plate format with an IC_{50} of 3 ppb and an LLD of 0.7 ppb for atrazine (*17*). Further improvements to assay sensitivity (IC_{50} of 0.25 ppb and a LLD of 0.03 ppb) with C2-derived Mabs were obtained using a B5-HRP label (Table III, *36*). Cross-reactivity patterns of these combinations have resulted in an improved reliability of analyte identification accompanied by an improvement in quantification accuracy for multianalyte samples (*21, 30*).

Table IV. 2-Carboxyalkylthio-4,6-bis-alkylaminotriazines Used as IH, CH or TH

S(CH₂)ₙCOOH structure: $S(CH_2)_nCOOH$ attached to triazine ring with N—H ethyl group and N—R group

Compound	R	n	Usage	Reference
C1	Ethyl	2	IH, CH	(*4, 13, 32*)
C1			IH, CH, CR	(*16*)
			CRISP	(*21*)
C2	Isopropyl	2	IH, CH	(*4, 13, 17, 34*)
			IH, CH, CR	(*2*)
			IH	(*14, 18, 19, 27*)
			CR	(*14, 18, 19, 22, 27*)
C3	Isopropyl	3	TH	(*37*)
C4	C(CH₃)₂CN	2	IH	(*31, 33*)

Triazine Metabolites. Recently, more emphasis has been placed on environmentally related concerns such as water quality assessment and exposure of field workers to toxic substances. This has generated an interest in developing sensitive, fast, convenient and low-cost means of monitoring drinking water sources and human body fluids. Since 1989 (*29*), several laboratories have developed immunoassays for the major metabolites of the triazines. The principal metabolites for atrazine are the 2-hydroxy, deethyl-, deisopropyl-, and didealkyl- compounds (Figure 3) found in soil and water. The mercapturate is the major triazine excretion product found in human urine.

Using the principles described previously for the parent triazines, Wittmann and Hock (*28, 38*) developed polyclonal antibodies to mono-N-dealkylated triazines from the hexanoic acid derivative located at site 3 (D, Figure 3). Employing a hapten homologous enzyme labeled hapten format, assays for de-ethyl atrazine and de-isopropyl atrazine were obtained with IC_{50}s of 0.2 ppb and 0.3 ppb respectively; LLD's were about 0.01 ppb.

Muldoon et al. (*34*) immunized mice with a hapten E whose handle was located at site 2. It was conjugated to keyhole limpet hemocyanin (KLH) to generate polyclonal antibodies that recognized didealkylated chlorotriazine. Interestingly, the assay using hapten D-HRP had an IC_{50} of about 600 ppb and an

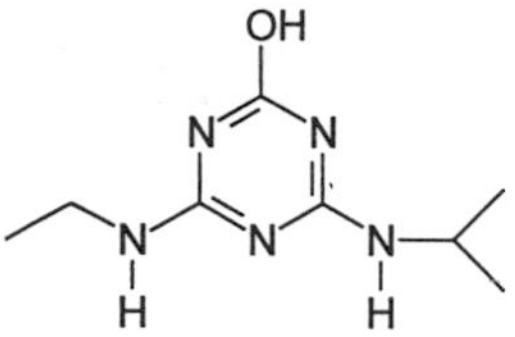

2-Hydroxyatrazine

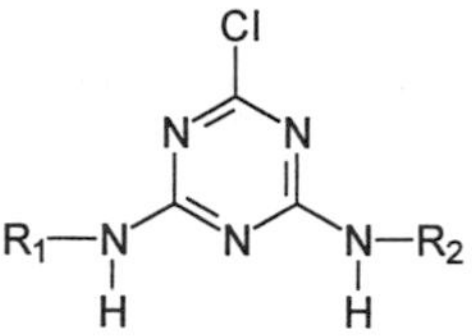

Compound	R_1	R_2
Deethyl	H	$CH(CH_3)_2$
Deisopropyl	C_2H_5	H
Didealkyl	H	H
(D)	H	$(CH_2)_5COOH$

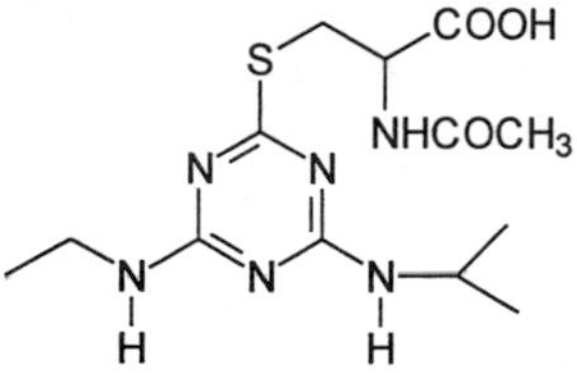

Atrazine mercapturate

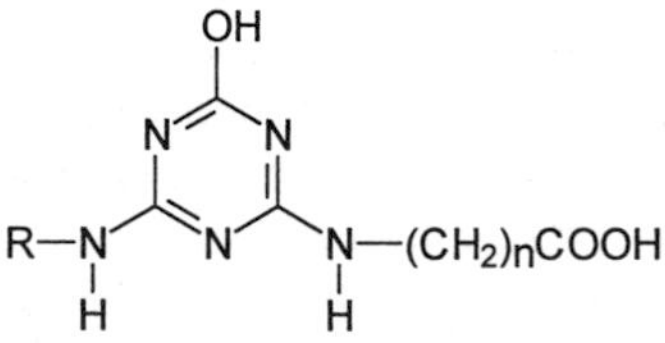

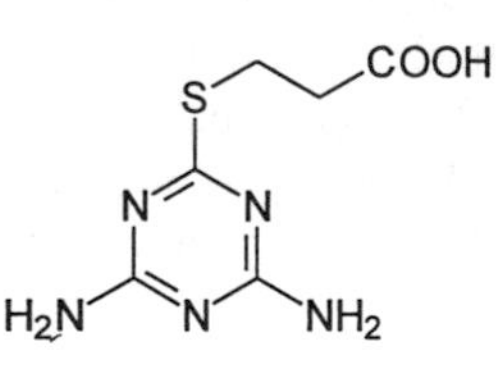

(E)

Compound	R	n
(F1)	$CH(CH_3)_2$	2
(F2)	$CH(CH_3)_2$	4
(F3)	CH_2CH_3	2

Figure 3. Structures of some metabolites of triazine herbicides and haptens designed for their detection.

LLD of about 300 ppb. Since these values were high compared to those obtained for other triazine assays, they concluded that antibody recognition for the triazines decreased as the number of alkyl side chains diminished. This observation is in agreement with conclusions reached by Kutter and Hansch (*3*).

The first hydroxytriazine immunoassay employed an immunizing hapten containing a pentanoic acid handle opposite the important hydroxy and isopropyl groups (*29*). Homologous assays with Mabs from an F2-KLH conjugate and an F2-BSA plate coating antigen provided IC_{50}s of about 0.5-1 ppb for hydroxyatrazine and hydroxysimazine. These Mabs were very selective and displayed cross reactivities only for triazines containing a hydroxyl and an ethyl (or isopropyl) appendage; the parent 2-chloro-, methoxy-, or methylthio-triazine compounds were not recognized (CR = <0.2%).

Lucas et al. (*18*) have derived polyclonal antibodies from a similar but shorter chain hydroxy structure (F1, Figure 3). The AP labeled hapten (B2, Figure 2) contained a chlorine instead of the hydroxy that reduced the hydrogen bond characteristics and an isopropyl group in place of an ethyl group that increased steric hindrance producing an overall decrease in affinity of the antibody for the coating/enzyme labeled hapten. This assay had a considerably higher LLD (IC_{50} of about 10 ppb) for hydroxyatrazine and hydroxysimazine than reported by Schlaeppi (*29*). The assay showed excellent selectivity for other hydroxytriazines, but replacement of the OH by other substituents reduced recognition of these analytes by the antibodies.

Atrazine metabolites in human urine have recently been examined by Lucas et al. (*22*), employing multiple immunoassays and an affinity extraction approach. Using atrazine, hydroxyatrazine and atrazine-mercapturate selective antibodies, generated from B10-KLH (Table 3), F3-KLH (Table 3), and C2-KLH (Figures 2 and 3), respectively and a B2-AP label, they evaluated potential metabolite components in urine samples from 18 field workers. The mercapturic acid conjugate of atrazine was found to be the major urinary metabolite, some 10 times more than any dealkyltriazine or the parent compound. Hydroxytriazines were not found. The highly sensitive Mab AM7B2.1 (from hapten C2; *17*) detected atrazine mercapturate down to 0.5 ppb in crude urine samples diluted one-to-four in buffer. As noted later, these thiotriazine-coupled haptens produce antibodies that recognize not only thiotriazines, but the corresponding chlorotriazines as well. This feature strongly supported our original selection of the two-appendage sulfur atom as a chloro-mimic for developing handles at this position. Similar observations have been made by others subsequently (*34, 35, 39, 40*).

Arylurea Herbicides

The arylurea herbicides, monuron and diuron (Figure 4), also have several potential sites for handle attachment. Some of these are especially appealing due to their synthetic simplicity and were utilized for developing a successful polyclonal assay

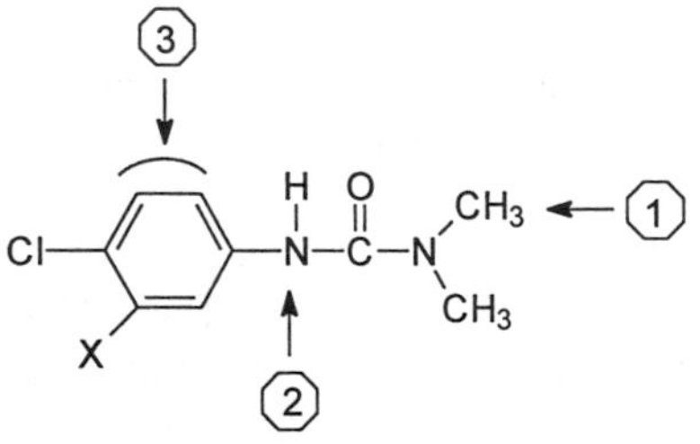

Monuron X = H

Diuron   X = Cl

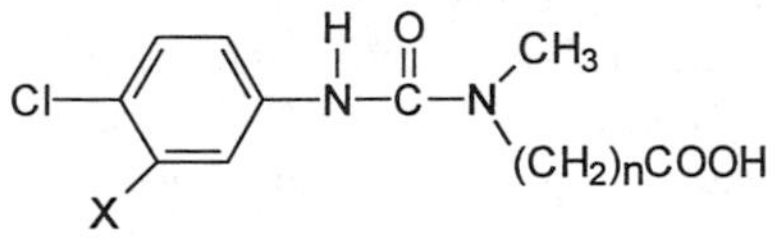

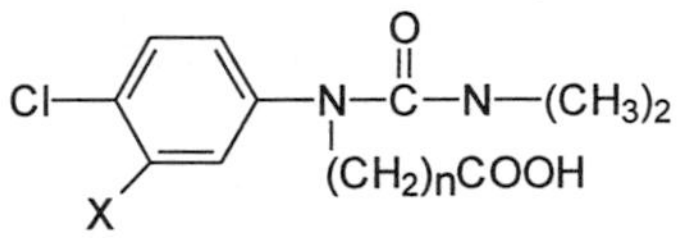

Compound	X	n	Compound	X	n
(G1) - (G5)	H	1-5	(I)	H	3
(H1) - (H3)	Cl	2, 3, 5	(J1) - (J2)	Cl	3, 5

Figure 4. Structures of some arylurea herbicides and haptens designed for their detection.

for monuron with an IC_{50} of 0.5 ppb and a LLD of 0.05 ppb (*36*) and a monoclonal assay for diuron with and IC_{50} of 2 ppb (*41*).

Fortuitously, of the three potential sites for handle attachment (see Figure 4) on the arylurea molecules, site 1 is synthetically easier and provides a near-perfect mimic of these molecules. This series was explored for both arylureas by the extension of the terminal N-methyl group at site 1 with innocuous methylene groups (G and H, Figure 4) terminated with a carboxylic acid group. The methylene groups alter the arylurea's structure, geometry, electronic and hydrophobic properties minimally. This attachment is far removed from the aromatic ring and imparts a negligible inductive effect change on the adjacent urea functional group.

In an initial study (*41*), mice immunized with a diuron hapten H2-KLH (Figure 4) produced antibodies substantially more selective for diuron than antibodies derived from an H3-KLH antigen. Eight IgG Mabs derived from the H2-KLH antigen gave IC_{50} values of 2-20 ppb for diuron in a heterologous assay using J1-BSA as a coating antigen. Selectivity was excellent as cross-reactivity measurements with monuron, linuron, other arylureas, and structurally similar carbamate herbicides varied from undetectable to 3%.

Polyclonal antibodies generated from the monuron hapten G5-BSA (Figure 4) in rabbits were highly sensitive for the three arylureas diuron, monuron, and linuron, demonstrating IC_{50} values of 0.5, 0.3 and 1.0 ppb respectively in a G3-HRP enzyme-labeled hapten format (*36*). These antibodies, however, failed to recognize the internal nitrogen hapten (I, Figure 4) and thus were not useful as enzyme labeled haptens.

Of the three haptens used for immunizing, Newsome and Collins (*42*) similarly found the terminal nitrogen handle (site 1, Figure 4) produced antibodies that resulted in assays with the lowest IC_{50} (monuron IC_{50} of 0.3 ppb), but they were not extremely selective. Satisfactory class- and compound-selective assays were reported by the judicious use of coating proteins with haptens heterologous to the immunogen haptens. Using a 4-succinamide hapten as the immunogen (site 3, Figure 4) they did achieve assays with excellent to average LLD's for monuron and diuron, but the polyclonal sera recognized many other structures containing the dimethylurea moiety. This hapten did prove to be the best coating antigen for site 1-type immunogens.

Site 2 hapten handles were considered mainly for coating/enzyme-labeled haptens and not for use as immunizing haptens. Whereas such hapten structures, with the replacement of the N-H by an N-alkyl chain, may not alter the geometry of the molecules significantly, converting a secondary amide nitrogen to a tertiary amide nitrogen does have drastic effects on the electronic and hydrophobic (H-bonding) characteristics of the molecule. Antibodies derived from such structures should not bind the target arylureas as well as the former. This reduced binding capability makes such structures good candidates for coating/enzyme labeled haptens (*43*). For example, the coating/enzyme labeled hapten, J1, that has weaker H-bond acceptor properties, resulted in a highly sensitive assay when used with antibodies generated from hapten H2 (Figure 4, *41*).

Using a multi-hapten approach we have screened all hapten molecules synthesized for cross-reactivity using an established assay. In the triazine series, we found a correlation between the ability of the antibody to recognize the hapten and the efficiency of this hapten as an enzyme-labeled hapten (16, 32). This observation was further verified in the arylurea assay development studies (36). In the latter, we also found that the esters of our haptens more closely mimicked the hapten-protein linkage found in enzyme-labeled haptens and thus some carboxylic acids that did not cross-react as free acids, when tested as esters, were recognized These were useful as enzyme labeled haptens. Furthermore, the monuron hapten with an internal nitrogen handle (I, Figure 4) tested as the carboxylic acid, was not recognized by the antibody (36). However, in the diuron assay (41), this internal nitrogen substituted hapten (J1, Figure 4) proved to be an excellent coating antigen. We are currently refining this theory of comparing ester cross-reactivities and utility as coating/enzyme labeled antigens for these and other target molecules.

Useful coating/enzyme labeled haptens have also been synthesized by making slight changes of the immunizing hapten structure, such as substituting a sulfur atom for an oxygen in an arylurea hapten. This led to reduced recognition of the coating/enzyme labeled hapten by the antibodies made to the oxygen-containing arylureas. The handle position and composition remain the same. Such a change reduced the affinity of the antibody for the thiourea hapten by decreased H-bonding. This thiourea hapten (i.e., oxygen-substituted G5), when used as a coating/enzyme labeled hapten provided an assay among the most sensitive for the arylurea targets (36). This strategy should also be applicable to development of assays for the phosphate pesticides.

Chloroacetanilide Herbicides

For alachlor, the most important member of this group, there are four apparent handle locations; the obvious and synthetically easiest (sites 1 and 2, Figure 5) have been explored the most. Although synthetically more difficult, an inviting challenge would be the 4-position handle consisting of a methylene chain terminated by a carboxylic acid (N1, Figure 6). Presumably antibodies derived therefrom would recognize alachlor significantly better than its analogues. This was shown to be true with an alkyl ether appendage at this site (44).

Feng et al. (39, 40) utilized the reactive chlorine (site 2) by a procedure involving thiolation of sheep immunoglobulin (IgG) with S-acetylmercaptosuccinic anhydride (SAMSA; Figure 5), or of BSA with acetyl homocysteine thiolactone (AHT; Figure 5), followed by treatment with alachlor. Rabbit polyclonal antibodies from immunization with the alachlor-IgG and using the BSA conjugate as the coating antigen led to an optimized alachlor immunoassay with an IC_{50} of 1 ppb. It was most effective in the range of 0.2-8.0 ppb. This substitution of S for Cl to facilitate an appendage attachment at this site supports the results obtained with the 2-thio-triazine haptens (13, 32). These antibodies also displayed high cross-reactivities for the N-acetylcysteine and glutathione derivatives formed by

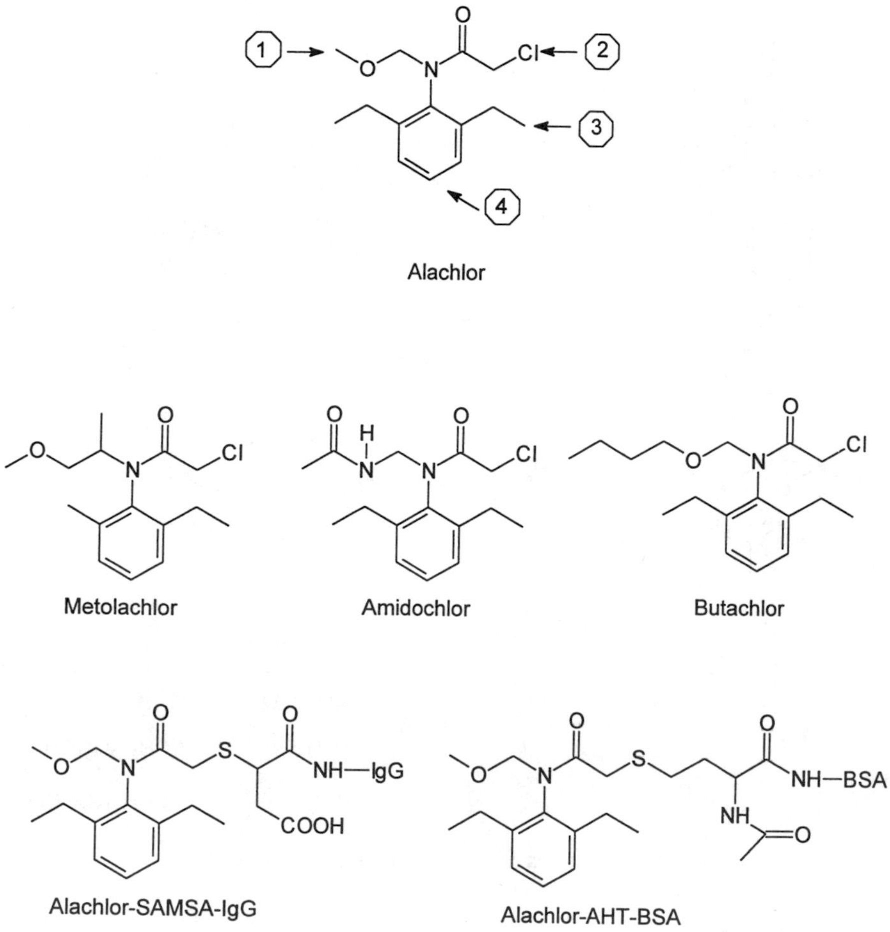

Alachlor

Metolachlor

Amidochlor

Butachlor

Alachlor-SAMSA-IgG

Alachlor-AHT-BSA

Figure 5. Structures of some chloroacetanilide herbicides and alachlor conjugates.

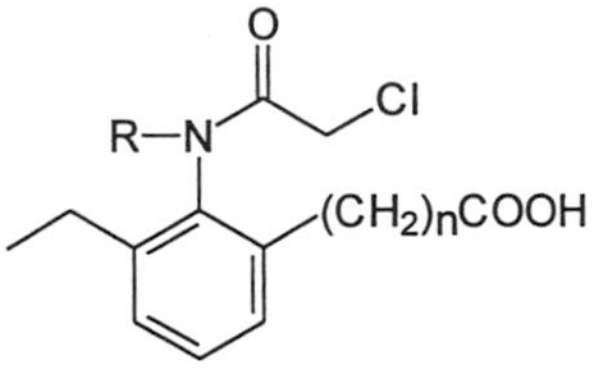

(K)

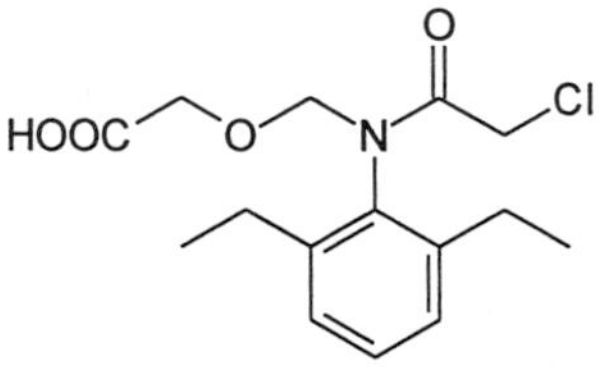

(L)

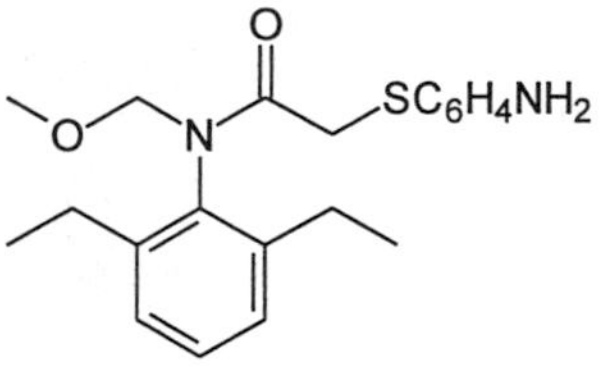

(M)

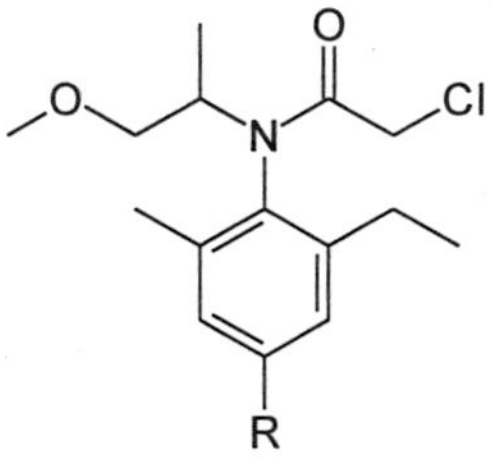

Compound	R
(N1)	$(CH_2)_nCOOH$
(N2)	$O(CH_2)_4COOH$

Figure 6. Structures of some chloroacetanilide haptens.

displacement of the active chlorine in alachlor (*45*). This confirms the same type of recognition pattern as observed for the triazines (*22*). Cross-reactivity studies reflected that the antibodies were particularly sensitive to modification in the methoxymethyl side chain and that sulfur (or chlorine) was necessary. Variations in the aromatic ring or appendages were not examined for cross reactivity.

This general hapten strategy has more recently (*46*) been applied to metolachlor, amidochlor and butachlor (Figure 5) with the development of immunoassays having low ppb levels of detection. The IC_{50}s for the three related compounds were 6, 10, and 7 ppb respectively with all three antibodies showing no significant cross-reactivity to other structurally related chloroacetanilides. From these results it was concluded that these antibodies distinguish one chloroacetanilide from another primarily by the N-alkyl side chain and this side chain varies dramatically in composition, size and polarity. Furthermore, they found that increasing the heterology for the coating antigens improved the sensitivity of the immunoassays. For example, the best assay for amidochlor from amidochlor-SAMSA-IgG conjugate as immunogen utilized the butachlor-AHT-human serum albumin (HSA) as a coating antigen. This produced an assay with an IC_{50} of <0.2 ppb, a 10-fold improvement in the LLD. As with alachlor mercapturates, these antibodies presumably will also show high cross-reactivity for their corresponding thioether metabolites. Haptens with a totally innocuous methylene handle and without the thioether (K, Figure 6) would be an easy compounds to synthesize and would still expose those determinants important for discrimination between these acetanilides. These haptens might elicit antibodies that recognize the parent target and not their thioether metabolites.

Conjugation through site 1, via the methoxymethyl side chain using the carboxy-alachlor L (Figure 6) hapten conjugated to BSA and sheep IgG provided antibodies with low cross-reactivities (0.1-9%) with other chloroacetanilides. Cross reactivities were generally lower than with the thioether derived antibodies, and were especially low (0.1-0.2%) for the thio analogues. Both the L-type and the thioether handle derived antibodies were equally useful for the analysis of alachlor in well water samples down to the 1 ppb range. However since the thioether derived antibodies recognized the ethane sulfonic acid metabolite, a step separating alachlor from the metabolite was necessary to obtain accurate results (*47, 48*).

The thioaromatic side chain compound M conjugated to BSA was used as an immunogen to produce antibodies for alachlor (*49*). Employing an alachlor-AHT-chicken albumin coating conjugate an assay was developed with an LLD of about 1 ppb. Cross reactivity for related compounds was 2-6%. These results suggested that the methoxymethyl appendage played a significant role in antibody selectivity as previously shown by Feng (*46*). A more sensitive assay with an LLD of about 0.1 ppb and an IC_{50} of about 2 ppb was obtained using an antibody coated plate format and alachlor labeled with HRP. Cross-reactivities were also improved, being less than half the former values.

These studies confirm the importance of the chloroacetamide, the alkylether, and the aryl appended alkyl groups. Applying the criteria presented here, the optimal hapten for immunization would likely have an alkyl handle of about 4-6 methylene groups in the 4-position of the aryl ring (N1, Figure 6). A close mimic of this hapten for metalochlor having a four carbon methylene chain in the form of an ether attached in the 4-position satisfies these criteria with the exception of a hetero ether oxygen (N2, Figure 6). Although unreactive, it does allow for strong H-bond attraction which may result in some handle recognition. Nevertheless, extremely high affinity Mabs were generated from mice using the N2-KLH conjugate. The assays had LLDs of 0.05 and 0.1 ppb with IC_{50}s of 1.0 and 0.6 ppb, respectively. The hydroxy metabolite (OH substituted for Cl) showed 1% cross-reactivity; the glutathione metabolite was 0.7%; and other related metabolites or similar analogs were <0.1% (44). This study points out the importance of immunizing hapten design to development of assays with low cross reactivities. In this case, the hapten derived from the more difficult six step synthesis resulted in the most sensitive and selective assays.

Conclusions

Development of a good immunoassay for small molecules depends greatly on the affinity and selectivity of the antibody for the analyte. The affinities and selectivities of the coating/enzyme labeled haptens for the antibody, the choice of protein carriers, and assay formats are also important for decreasing the LLDs. By first having a clear idea of the goals of the assay and then approaching the design of haptens in a systematic process, it usually is possible to synthesize only a few haptens and obtain a superior assay. Of course having a large library of antibodies raised to different haptens and a large library of haptens for coating/enzyme labels allows one to screen for the assay with the desired properties. For example, antibodies generated to recognize alachlor, when used with a different hapten labeled enzyme could be employed to analyze for the thioether metabolites of alachlor (45). Both polyclonal and monoclonal based immunoassays often can be improved in terms of both selectivity and sensitivity by careful hapten design. As demonstrated here, this is true even after the monoclonal line or serum pool is selected.

Acknowledgments

This work was supported in part by NIEHS Superfund 2P42-ES04699, U.S. Environmental Protection Agency (EPA) Cooperative Research Grant CR819047, NIEHS Center for Environmental Health Sciences (at U.C. Davis) 1P30 ES05707; U.S. EPA. Center for Ecological Health Research at U.C. Davis CR819658 and U.S.D.A./P.S.W. 5-93-25 (Forest Service NAPIAP). Although the information in this document has been funded in part by the United States Environmental Protection Agency, it may not necessarily reflect the views of the Agency and no official endorsement should be inferred.

Literature Cited

(*1*) Landsteiner, K. The Specificity of Serological Reactions; Harvard University Press: Cambridge, 1945; 293 pp.

(*2*) Harrison, R. O.; Goodrow, M. H.; Gee, S. J.; Hammock, B. D. In *Immunoassays for Trace Chemical Analysis: Monitoring Toxic Chemicals in Humans, Food, and the Environment.*; Vanderlaan, M.; Stanker, L. H.; Watkins, B. E.; Roberts, D. W., Ed.; American Chemical Society: Washington, D.C., 1991, 14-27.

(*3*) Kutter, E.; Hansch, C. *Arch. Biochem. Biophys.* **1969,** *135,* 126-135.

(*4*) Jung, F.; Gee, S. J.; Harrison, R. O.; Goodrow, M. H.; Karu, A. E.; Braun, A. L.; Li, Q. X.; Hammock, B. D. *Pesticide Science* **1989,** *26,* 303-317.

(*5*) Tijssen, P. Practice and Theory of Enzyme Immunoassays; Laboratory Techniques in Biochemistry and Molecular Biology; Elsevier: Amsterdam, 1985; 15; 549 pp.

(*6*) Welhouse, G. J.; Bleam, W. F. *Environ. Sci. Tech.* **1993,** *27,* 500-505.

(*7*) Welhouse, G. J.; Bleam, W. F. *Environ. Sci. Tech.* **1993,** *27,* 494-500.

(*8*) Huber, S. J. *Chemosphere* **1985,** *14,* 1795-1803.

(*9*) Huber, S. J.; Hock, B. *Z. Pflanzenkrankh. Pflanzenschutz* **1985,** *92,* 147-156.

(*10*) Giersch, T.; Hock, B. *Food Agric. Immunol.* **1990,** *2,* 85-97.

(*11*) Wittmann, C.; Hock, B. *Food Agric. Immunol.* **1989,** *1,* 211-224.

(*12*) Giersch, T.; Kramer, K.; Hock, B. *Sci. Total Env.* **1993,** *132,* 435-448.

(*13*) Goodrow, M. H.; Harrison, R. O.; Hammock, B. D. *J. Agric. Food Chem.* **1990,** *38,* 990-996.

(*14*) Minunni, M.; Mascini, M. *Anal. Lett.* **1993,** *26,* 1441-1460.

(*15*) Ulrich, P.; Weller, M.; Weil, L.; Niessner, R. *Vom Wasser* **1991,** *76,* 251-266.

(*16*) Harrison, R. O.; Goodrow, M. H.; Hammock, B. D. *J. Agric. Food Chem.* **1991,** *39,* 122-128.

(*17*) Karu, A. E.; Harrison, R. O.; Schmidt, D. J.; Clarkson, C. E.; Grassman, J.; Goodrow, M. H.; Lucas, A.; Hammock, B. D.; Emon, J. M. V.; White, R. J. In *Immunoassays for Trace Chemical Analysis: Monitoring Toxic Chemicals in Humans, Food, and the Environment.*; Vanderlaan, M.; Stanker, L. H.; Watkins, B. E.; Roberts, D. W., Ed.; American Chemical Society: Washington, D.C., 1991, 59-77.

(*18*) Lucas, A. D.; Bekheit, H. K. M.; Goodrow, M. H.; Jones, A. D.; Kullman, S.; Matsumura, F.; Woodrow, J. E.; Seiber, J. N.; Hammock, B. D. *J. Agric. Food Chem.* **1993,** *41,* 1523-1529.

(*19*) Lucas, A. D.; Schneider, P.; Harrison, R. O.; Seiber, J. N.; Hammock, B. D.; Biggar, J. W.; Rolston, D. E. *Food Agric. Immunol.* **1992,** *3,* 155-167.

(*20*) Schneider, P.; Hammock, B. D. *J. Agric. Food Chem.* **1992,** *40,* 525-530.

(*21*) Cheung, P. Y. K.; Kauvar, L. M.; Engqvist-Goldstein, A. E.; Ambler, S. M. *Anal. Chim. Acta* **1993,** *282,* 181-192.

(*22*) Lucas, A. D.; Jones, A. D.; Goodrow, M. H.; Saiz, S. G.; Blewett, C.; Seiber, J. N.; Hammock, B. D. *Chem. Res. Toxicol.* **1993,** *6,* 107-116.

(23) Dunbar, B. D.; Niswender, G. D.; Hudson, J. M. United States Patent, July 23, 1985, Patent Number 4,530,786.
(24) Dunbar, B.; Riggle, B.; Nisender, G. *J. Agric. Food Chem.* **1990,** *38,* 433-437.
(25) Wust, S.; Hock, B. *Anal. Lett.* **1992,** *25,* 1025-1037.
(26) Wortberg, M.; Cammann, K. *Fresenius J. Anal. Chem.* **1993,** *346,* 757-760.
(27) Kim, B. B.; Vlasov, E. V.; Miethe, P.; Egorov, A. M. *Anal. Chim. Acta* **1993,** *280,* 191-196.
(28) Wittmann, C.; Hock, B. *J. Agric. Food Chem.* **1993,** *41,* 1795-1799.
(29) Schlaeppi, J.; Fory, W.; Ramsteiner, K. *J. Agric. Food Chem.* **1989,** *37,* 1532-1538.
(30) Muldoon, M. T.; Fries, G. F.; Nelson, J. O. *J. Agric. Food Chem.* **1993,** *41,* 322-328.
(31) Robotti, K. M.; Sharp, J. K.; Ehrmann, P. R.; Brown, L. J.; Hermann, B. W., 192nd American Chemical Society Meeting, Anaheim, CA, September 7-12, 1986, Abstract #42 of the Agrochemicals Division.
(32) Goodrow, M. H.; Harrison, R. O.; Gee, S. J.; Hammock, B. D., 102nd Annual International Meeting of the Association of Official Analytical Chemists, Palm Beach, FL, August 29-September 1, 1988,
(33) Lawruk, T. S.; Lachman, C. E.; Jourdan, S. W.; Fleeker, J. R.; Herzog, D. P.; Rubio, F. M. *J. Agric. Food Chem.* **1993,** *41,* 747-752.
(34) Muldoon, M. T.; Huang, R.-N.; Hapeman, C. J.; Fries, G. F.; Ma, M. C.; Nelson, J. O. *J. Agric. Food Chem.* **1994,** *42,* 747-755.
(35) Manclus, J. J.; Primo, J.; Montoya, A. *J. Agric. Food Chem.* **1994,** *42,* 1257-1260.
(36) Schneider, P.; Goodrow, M. H.; Gee, S. J.; Hammock, B. D. *J. Agric. Food Chem.* **1994,** *42,* 413-422.
(37) Rubio, F. M.; Itak, J. A.; Scutellaro, A. M.; Selisker, M. Y.; Herzog, D. P. *Food Agricul. Immunol.* **1991,** *3,* 113-125.
(38) Wittmann, C.; Hock, B. *J. Agric. Food Chem.* **1991,** *39,* 1194-1200.
(39) Feng, P. C. C.; Wratten, S. J.; Horton, S. R.; Sharp, C. R.; Logusch, E. W. *J. Agric. Food Chem.* **1990,** *38,* 159-163.
(40) Feng, P. C. C.; Wratten, S. J.; Logusch, E. W.; Horton, S. R.; Sharp, C. R. In *Immunochemical Methods for Environmental Analysis*; Van Emon, J. M.; Mumma, R. O., Ed.; American Chemical Society: Washington D.C., 1990, 442; 180-192.
(41) Karu, A. E.; Goodrow, M. H.; Schmidt, D. J.; Hammock, B. D.; Bigelow, M. W. *J. Agric. Food Chem.* **1994,** *42,* 301-309.
(42) Newsome, W. H.; Collins, P. G. *Food Agric. Immunol.* **1990,** *2,* 75-84.
(43) Wie, S. I.; Hammock, B. D. *J. Agric. Food Chem.* **1984,** *32(6),* 1294-1301.
(44) Schlaeppi, J.; Moser, H.; Ramsteiner, K. *J. Agric. Food Chem.* **1991,** *39,* 1533-1536.
(45) Feng, P. C. C.; Sharp, C. R.; Horton, S. R. *J. Agric. Food Chem.* **1994,** *42,* 316-319.

(*46*) Feng, P. C. C.; Horton, S. R.; Sharp, C. R. *J. Agric. Food Chem.* **1992,** *40,* 211-214.

(*47*) Aga, D. S.; Thurman, E. M.; Pomes, M. L. *Anal. Chem.* **1994,** *66,* 1495-1499.

(*48*) Sharp, C. R.; Feng, P. C. C.; Horton, S. R.; Logusch, E. W. In *Pesticide Residues and Food Safety*; Tweedy, B. G.; Dishburger, H. J.; Ballantine, L. G.; McCarthy, J., Ed.; American Chemical Society: Washington, DC, 1991, 446; 87-95.

(*49*) Rittenburg, J. H.; Grothaus, G. D.; Fitzpatrick, D. A.; Lankow, R. K. In *Immunoassays for Trace Chemical Analysis: Monitoring Toxic Chemicals in Humans, Food, and the Environment.*; Vanderlaan, M.; Stanker, L. H.; Watkins, B. E.; Roberts, D. W., Ed.; American Chemical Society: Washington, D.C., 1991, 28-39.

RECEIVED October 24, 1994

Chapter 10

Hapten Versus Competitor Design Strategies for Immunoassay Development

Robert E. Carlson

ECOCHEM Research, Inc., Suite 510, 1107 Hazeltine Boulevard, Chaska, MN 55318–1043

Separate design strategies for immunizing haptens and labeled competitor reagents are critical to the development of sensitive and specific immunoassays. Immunoassay development has generally focused on hapten and competitor designs which faithfully mimic the target analyte and vary mainly in the attachment site, chain length and functional groups of the spacer arm. We have investigated the concept that while the immunizing hapten defines assay specificity, it is the labeled competitor which determines sensitivity. Thus it is not an inherent assay development requirement that the labeled competitor substantially duplicate the analyte. The development of a polychlorinated biphenyl (PCB) immunoassay is described which uses a "core" heterology approach, in which the competitor is a structural fragment of the analyte, to maximize analyte sensitivity while retaining assay specificity.

The basis of immunoassay is the competitive interaction of an antibody with the analyte and a labeled or conjugated derivative of the analyte (1). In the absence of analyte, the labeled or conjugated competition reagent is fully bound to the antibody. When analyte is present, the degree of competitor to antibody binding is reduced in proportion to the analyte concentration. This differential interaction results in a measurable colorimetric, fluorescent, etc. signal which is proportional to analyte concentration.

HAPTEN DERIVED ANTIBODY
ANALYTE CONCENTRATION -interaction produces->PROPORTIONAL SIGNAL
LABELED COMPETITOR

Thus, the properties of the assay depend on its key components: the hapten derived antibody and the labeled or conjugated competition reagent. To differentiate the labeled competitor reagent from the immunizing hapten and to reflect its role in the competition step of the assay, we refer to these materials as "competitor reagents" or "competitor conjugates". The simplest approach to assay development is to use the

0097–6156/95/0586–0140$12.00/0

immunizing hapten as the competitor in a homologous (competitor = hapten) assay format. However, it has long been recognized (2) that the preference of the antibody for the hapten usually results in significantly less than optimum analyte sensitivity. Consequently, heterologous assays utilize competitors which are closely related but non-identical to the hapten (2).

The basic principle upon which our assay development program is based is that the hapten defines specificity while the competitor determines sensitivity. This approach suggests that each component plays a key role in the assay. The role of the hapten is to develop a binding pocket in the antibody which closely reflects the structure of the analyte. However, the sensitivity of the assay depends on the affinity of the antibody for the analyte in relation to the affinity of the antibody for the competitor. Thus hapten and competitor performance criteria are different. Consequently, hapten and competitor design should be approached as independent processes.

METHODS

Experimental details on the development of the Aroclor directed PCB immunoassay will be described in detail elsewhere (Carlson, R.E. et al., in preparation).

Synthesis and Conjugation. Hapten synthesis utilized Cadogen (3) coupling of a trichloroaniline with a chloroanisole to produce a methoxytetrachlorobiphenyl intermediate which was converted to the hapten. Competitor synthesis was based on the addition of the anion of di- or trichlorotoluene to a 4-bromobutyrate synthon. Bovine serum albumin (BSA) and keyhole limpet hemocyanin (KLH) conjugate preparation was based on 1-ethyl-3-(3-dimethylaminopropyl)carbodiimide / n-hydroxysuccinimide activation of the pendant carboxyl group.

Antisera Development. Antisera were prepared in female New Zealand White rabbits using a KLH conjugate of the hapten. The immunization schedule used 300 ug of conjugate on day 1 in complete Fruend's adjuvant, 200 ug of conjugate in incomplete adjuvant on day 14, followed by 100 ug of conjugate in incomplete adjuvant on days 21, 28, 42, 56, 84, 112 etc. Rabbits were bled at 14 day intervals beginning on day 26. Serum samples were stored at -20° C until analyzed.

Immunoassay. Standard enzyme linked immunosorbent assay (ELISA) procedures were followed (4). Microtiter plates (Dynatech Immulon-2) were coated with BSA hapten or competitor conjugates. Titer determinations were performed by incubating the coated wells with antisera diluted in a phosphate buffered saline buffer which contained BSA / dimethylformamide / Triton X-100. Bound antibody was measured using alkaline phosphatase labeled anti-rabbit IgG second antibody. The same format, with the addition of methanol solutions of the Aroclors or specific PCB congeners, was used to determine the analyte response of the assay. Reported part-per-million values reflect concentration of the analyte in the assay solution.

RESULTS / PCB IMMUNOASSAY DEVELOPMENT

An example of our approach to hapten and competitor design is the development of a polychlorinated biphenyl immunoassay. The PCBs are a class of chloroaromatics produced by the chlorination of biphenyl to give complex mixtures that can theoretically contain as many as 209 mono- to decachlorobiphenyl congeners (5).

$X + Y = 1 - 10$

The commercial PCB formulations (e.g., Aroclor 1221, 1248, 1260) each contain fewer than 100 congeners at a concentration greater than 0.1 mole% (5-8). Consequently, for maximum sensitivity, a successful assay must be responsive to those congeners which are most prevalent in the common, target Aroclor formulations (i.e., Aroclors 1248, 1254 and 1260). Previous studies on the development of PCB immunoassays have been reported by Luster, et al. (9), Newsome and Shields (10), Franek, et al. (11) and Mattingly, et al. (12). None of these studies led to the development of a commercially successful assay. In particular, these assays exhibited marginally useful sensitivity and a strong preference for the immunizing hapten. The goal of this assay development program was to achieve useful Aroclor sensitivity. Moreover, the assay must be specific for PCBs and non-responsive to all other potential environmental co-contaminants.

Hapten Design. Based on the notion that the hapten defines specificity, the design of the immunizing hapten should closely duplicate the structure of the analyte within the limits set by the incorporation of a linker moiety for immunizing conjugate preparation. Three factors were considered to be critical to the design of the PCB hapten:
    1. Congener specific analysis (6-8) has shown that although 81 congeners are present at >0.1 mole% concentration in Aroclor 1254, only 13 of the 81 congeners are greater than 3.0 mole%. However, these 13 congeners comprise 66 mole% of this Aroclor. This suggests that a hapten can be designed which will represent the most prevalent congeners in the Aroclor mixtures.
    2. The 2,4,5- substitution pattern and its 2,4- and 2,5- subsets represent a significant portion of the mole% composition of these complex Aroclor mixtures (6-8). For example, the six congeners with a common 2,4,5- substitution pattern (e.g., 2,2',4,4',5-pentachloro-, 2,2',4,4',5,5'-hexachloro-, 2,2',3,4,4',5,5'-heptachloro-) account for 32 mole% of Aroclor 1254. Addition of the three congeners which are 2,4- or 2,5- substituted (e.g., 2,2',3,4,5'-pentachloro-) brings this total to 46 mole%. Furthermore, Figure 1 illustrates that the 6 congeners which are 2,4,5- substituted in one ring are dominated by 2,4- substitution in the second ring. Thus incorporation of the 2,4- / 2,5- / 2,4,5- substitution patterns would be expected to maximize mole% congener binding and assay response to the Aroclor analytes.
    3. The dihedral angle between the phenyl moieties of a biphenyl derivative is dependent on the steric effect of the substituents at the 2, 2', 6 and 6' (ortho) positions (13). PCB's which are 2, 2', 6, and/or 6' substituted are not coplanar. The size of the dihedral angle between the phenyl groups is dependent on the degree of substitution. The 2,2'-chlorine substitution pattern that predominates in the target Aroclors results in a dihedral angle of about 70° (13). Clearly, this angle should be preserved in the design of the hapten.
    Incorporation of both the dominant patterns of chlorine substitution and the requirement for specific chlorine substitution at the critical 2 and 2' (ortho) positions of the biphenyl nucleus, directs the design of the immunizing hapten to incorporate the 2,4,5-/2',4'- substitution pattern. However, incorporation of a linker moiety for hapten conjugation requires modification of this pattern. Three factors were considered in this step of the design process:
    1. The common 2,4,5- substitution pattern should be incorporated in unmodified form into one of the two phenyl moieties.

2. As discussed above, the presence of 2 chlorines in the 2,4,5-/2',4'-substitution pattern which are <u>ortho</u> to the biphenyl bond establishes a particular value (ca. 70°) for the dihedral angle between the two phenyl rings (*13,14*). Incorporation of the linker moiety into the 2,2',6 or 6'- positions would be expected to have a detrimental effect on analyte binding to the antibody and assay performance because of the potential for torsional angle and steric differences between the analyte and the hapten derived antibody binding pocket.

3. The functional group which serves as the attachment point between the linker and the phenyl group should be a good "chlorine mimic" (*9*) to increase the potential for recognition of congeners with chlorine substitution in the same position as the linker attachment site.

These criteria led to the design of PCB/Hapten I (Figure 2). PCB/Hapten I incorporates both 2,4,5- and 2,2'- chlorine substitution with an ether moiety based linker as the chlorine mimic. The linker has been placed in the 4'- attachment site to minimize the impact of the linker on the biphenyl directed antibody binding pocket. In addition, the 4'- linker site is attractive because any flexibility in the antibody binding pocket due to linker steric requirements would be expected to improve cross-reaction within the target congener group (e.g., 2,4,5-/2',4'- versus 2,4,5-/2',5'- versus 2,4,5-/2',4',5'-).

Figure 3 allows comparison of a molecular model of PCB/Hapten-I with its analogous PCB congener, 2,2',4,4',5-pentachlorobiphenyl (*14*).   The hapten duplicates the analyte, both in substitution pattern and spatially, except for the distal chlorine-to-ether linker moiety exchange. Figure 3 also compares the pentachloro congener with a representative, 2'-amide based hapten of the type used in prior PCB immunoassay development studies (*9-11*). The dihedral angle difference between the analyte, PCB/Hapten I and the amide hapten is relatively small (dihedral angles of 70°, 68° and 67° respectively (*14*)). However, comparison of the molecular model of the amide hapten and the model of the PCB congener clearly illustrates that the <u>ortho</u> positioned, sterically bulky amide linker would be expected to lead to the development of an antibody binding pocket which is different from the spatial requirements of the PCB congener.

Homologous Assay Evaluation. A KLH conjugate of PCB/Hapten I produced a strong anti-hapten titer in rabbits. However, inhibition ELISA using a BSA conjugate of PCB/Hapten I as the coating conjugate gave a poor PCB response using Aroclor 1248 as the analyte. As shown in Table I, the minimum sensitivity of the assay for Aroclor 1248 was >10 ppm. Thus, this homologous (hapten as competitor) assay would not be sufficiently sensitive for the development of a useful environmental matrix PCB immunoassay.   More importantly, Table I compares the sensitivity of the assay to PCB/Hapten I versus Aroclor 1248. The assay was significantly more responsive to the hapten than to the analyte. This result indicates that the hapten-KLH immunizing conjugate has successfully elicited a PCB <u>hapten</u> directed antibody and that the homologous assay is strongly biased toward the hapten's functional groups.

These observations are consistent with: 1) prior PCB immunoassay development experience based on amide and ether haptens as described by Luster, et.al. (amide (*9*)), Newsome and Shields (amide (*10*)), Franek, et.al. (amide (*11*)) and Mattingly, et.al. (ether (*12*)) which all demonstrated strong anti-hapten assay bias, 2) the expectation, based on the subjective notion that an ether is a stronger epitope than the chloroaryl moiety, that the relative affinity of the antisera for the hapten ether moiety would be greater than its affinity for the chloroaryl moiety and 3) the accepted notion (*2*) that hapten designs which are optimum for the generation of an anti-hapten / anti-analyte antibody often are not optimum as competitors for assay development.

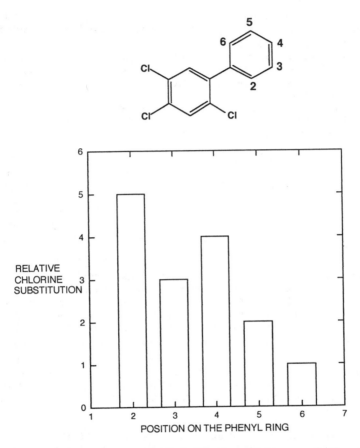

Figure 1.   Relative Phenyl Substitution Pattern of the 2,4,5-Substituted PCB Congeners. Bar height equals relative chlorine frequency among the 6 selected congeners. The 2-position is chlorinated in 5 of the 6 congeners, 3- in 3 of 6, 4- in 4 of 6, 5- in 2 of 6 and 6- in 1 of 6.

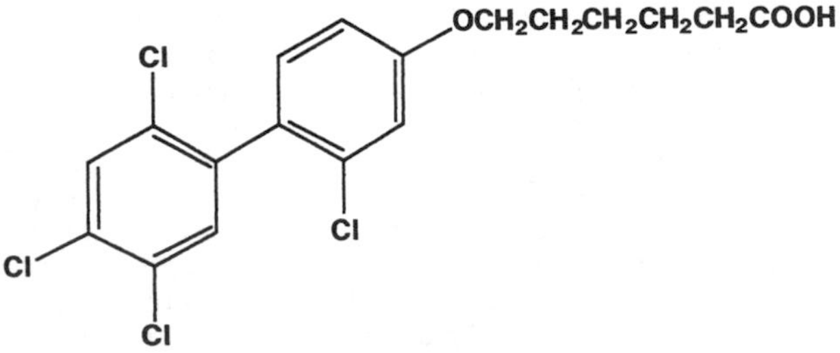

Figure 2.   Structure of PCB.Hapten-I.

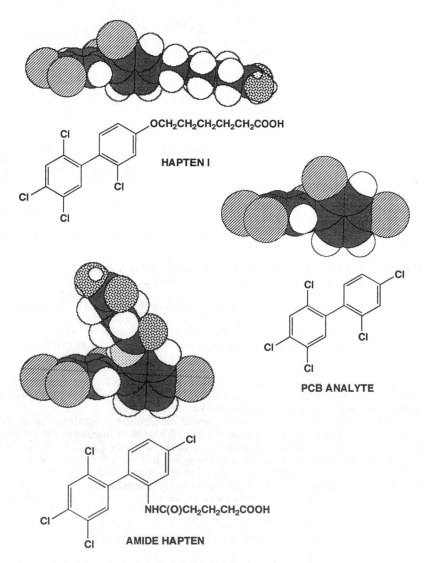

Figure 3. Molecular Model Comparison of PCB Haptens and a PCB Congener.

Table I.   Response of the Homologous ELISA to Hapten and PCB Analytes.

ANALYTE	I_{10} (ppm)	I_{50} (ppm)
PCB/HAPTEN I (R = -O-linker)	0.18	6.3
AROCLOR 1248 (R = Cl_n)	11	200 (estimated)

Heterologous Assay Development. The definitions of VanWeeman and Shuurs (2) summarize the principles of heterologous assay design:

    1. Hapten Heterology - the competitor is structurally related, but not identical to the hapten.
    2. Bridge Heterology - only the linking chain of the hapten is varied.
    3. Site Heterology - only the attachment site of the linking chain to the hapten is varied.

These definitions, particularly as systematically evaluated by Jung, et al. (15), Harrison, et al. (16), Jockers, et al (17) and Karu, et al. (18) form the current basis for the rational development of sensitive and specific immunoassays.

    Clearly, the optimum competitor should incorporate a combination of the standard hapten, linker and/or site heterologies (2,15-18) to maximize the relative difference in competitor versus analyte binding to the antibody (17). However, the strongly hapten biased response of this assay suggests that modest enhancement in the interaction of the analyte with the antibody relative to a hapten derived heterologous competitor would not significantly improve on analyte sensitivity. For example, in Luster, et.al. (9), the assay was 8-fold more sensitive to a hapten analog than to a comparable PCB congener and >>10-fold more sensitive to a hapten analog than to Aroclor 1248. It seemed probable that relatively minor variations in the competitor would not be sufficient to achieve the required improvement in sensitivity.

"Core" Heterology Design. The development of heterologous competitors has traditionally focused on preserving the structure of the analyte (2,15-18). We hypothesized that although the design of the hapten must faithfully represent the analyte in order to produce an anti-analyte antibody binding pocket, the competitor need only bind to the antibody sufficiently to produce an adequate, analyte competitive assay signal. As long as the assay has an adequate signal, the most sensitive assays would be expected to result from a competitor which has the lowest antibody affinity relative to the analyte. To achieve this goal, the competitor may only have to be a fragment of the analyte. We have termed this concept "Core Heterology".

"Core" heterologous competitor design for the PCB assay was based on five factors:

1.   A polychlorobenzene would be the logical core fragment for a polychlorinated biphenyl. (core heterology)

2.   The functional group that serves as the attachment point between the linker and the phenyl group should not be an ether. (bridge heterology)

3.   Linker chain length heterology should be incorporated through the use of a shorter spacer. (bridge heterology)

4.   Various chlorine substitution patterns that mimic the "2,4,5-" pattern should be evaluated. (site, hapten heterology)

5.   The site of linker attachment should be consistent with or in addition to the "2,4,5-" pattern. (site, hapten heterology)

A variety of polychlorobenzene based competitors could be designed to satisfy these criteria. For example, the attachment functional group could be an amide, carbonyl, amine or methylene moiety. Additionally, the linker could incorporate amide or polyether groups and the aromatic nucleus could incorporate various 2,4-, 2,5- or 2,4,5- substitution patterns. However, we initially focused on alkyl based linkers and "2,4,5-" substitution as the most direct approach to incorporation of the above factors. This focus led the the synthesis of Competitors 1, 2 and 3 (Figure 4) and their evaluation in the ELISA format with Aroclor 1248 as the analyte. Table II compares antiserum titer and ELISA sensitivity for assays based on these competitors to the homologous hapten based assay. The titer results, which may be an indirect measure of antibody to competitor affinity, indicate the relatively weak antiserum binding of these competitors compared to the hapten. The improvement in assay sensitivity over the homologous system using these three competitors was 6- to 150-fold at the minimum detection limit (I_{10}) and 70 to 220-fold at the assay mid-point (I_{50}). These results are consistent with our expectation that lower competitor titers will be reflected in a significant improvement in relative competitor versus analyte binding and, most importantly, analyte sensitivity.

Assay Characterization. The results in Table II indicated that competitors 1, 2 or 3 could be used to develop a sensitive PCB immunoassay. However, the utility of the assay is dependent on both sensitivity and specificity. Three aspects of assay performance were evaluated, using the competitor 3 based ELISA, to define specificity:

1. The congener specificity data in Table IIIa shows that there is a general loss of cross-reactivity relative to 2,4,5-/2,4,5- substitution as the 4- and 5- chlorines are removed. Table IIIb demonstrates the loss of cross-reaction relative to 2,4,5-/2,4,5- substitution as ortho chlorination is altered through either the loss of 2- position chlorines or the addition of chlorines at the 6-position. These results indicate that the assay is responsive, as would be expected from the design of the immunizing hapten, to 2,4-, 2,5- and 2,4,5- substitution and ortho chlorination.

2. The Aroclor cross-reaction data in Table IV demonstrate that the assay is broadly responsive to the Aroclor 1242, 1248, 1254 and 1260 analytes. The response of the assay was within a factor of 2 for Aroclor 1242 to 1260, in agreement with the significance of 2,4- / 2,5- / 2,4,5- substitution in these congener mixtures. Cross-reaction to Aroclor 1221 was only 1/10th that of Aroclor 1248. This result is also in agreement with expectation because Aroclor 1221 is composed of congeners (e.g., 2-monochloro, 4-monochloro, 2,4'-dichloro) which are <10% cross-reactive when compared to the 2,4,5- substituted congeners (Carlson, R.E. et al., in preparation). These results confirm the 2,4-, 2,5- and 2,4,5- hapten design strategy.

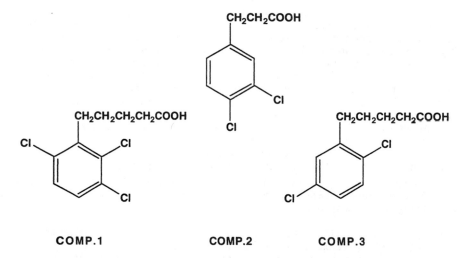

COMP.1   COMP.2   COMP.3

Figure 4. Structures of the PCB Assay Competitors 1, 2 and 3.

Table II. Evaluation of Competitors-1, 2 and 3 versus PCB/Hapten I in the ELISA Format using Aroclor 1248 as the Analyte.

COMPETITOR	COMPETITOR TITER	I_{50} (ppm)
1. 2,3,6-trichloro-	1/ 1,600	1.1
2. 3,4-dichloro-	<1/ 400	0.9
3. 2,5-dichloro-	1/ 1,000	2.8
PCB/Hapten I	1/28,000	200 (estimated)

Table III.   Selected Congener Cross-reaction Data.   Cross-reactions are versus 2,2',4,4',5,5'-hexachlorobiphenyl (BZ #153) and are determined at the minimum detection limit (mdl, I_{10}) of the assay.

A. 4- AND 5- CHLORINE SUBSTITUTION SERIES.

CONGENER (BZ#)	RING A	RING B	CROSS-REACTION VERSUS BZ #153
4	2-	2'-	0.00
18	2,5-	2'-	0.12
52	2,5-	2',5'-	0.31
101	2,4,5-	2',5'-	0.24
153	2,4,5-	2',4',5'-	1.0

B. ORTHO CHLORINATION SERIES.

CONGENER (BZ#)	RING A	RING B	ORTHO NUMBER	CROSS-REACTION VERSUS BZ #153
77	3,4-	3',4'-	0	0.00
118	2,4,5-	3',4'-	1	0.15
153	2,4,5-	2',4',5'-	2	1.0
154	2,4,5-	2',4',6'-	3	0.36
155	2,4,6-	2',4',6'-	4	0.00

Table IV.   Aroclor 1221-1260 Cross-Reaction Relative to Aroclor 1248.

AROCLOR	CROSS-REACTION
1221	0.08
1242	1.00
1248	1.00
1254	0.51
1260	0.47

3. Assay discrimination requires minimal cross-reaction to non-target analytes. No competition was observed at up to 100 ppm (cross-reaction << 0.2% versus Aroclor 1248) for the following commonly observed hazardous waste analytes: chlorobenzenes [1,2-dichlorobenzene, 1,3-dichlorobenzene, 1,4-dichlorobenzene, 1,2,4-trichlorobenzene]; chlorophenols [3-chlorophenol, 2,5-dichlorophenol]; aryl ethers [2,4-dichlorophenoxyacetic acid, 3-chloroanisole].

CONCLUSIONS

The development of this assay was guided by three key factors:
1. The chlorine substitution pattern of the hapten was chosen to incorporate both the 2,4-, 2,5- and 2,4,5- substitution patterns and the degree of <u>ortho</u> chlorination which are dominant in the target Aroclor analytes.
2. Hapten design incorporated a linker that is a chlorine mimic, positioned so that it does not interfere with the critical <u>ortho</u> substitution pattern of the target analyte.
3. Competitor design does not require substantial duplication of the analyte.

Our success in applying these factors to the development of a PCB/Aroclor immunoassay suggests that hapten design which is based on the structure of the analyte can be combined with an expanded view of competitor heterology to develop sensitive and specific immunoassays for a variety of analytes.

ACKNOWLEDGMENTS

I would like to thank Mary Cooper, Andrew Buirge and Todd Swanson for their technical assistance on and Viorica Lopez-Avila (Acurex Environmental Systems, Inc.) and Jeanette VanEmon (EPA) for their partial support of the PCB immunoassay development project.

LITERATURE CITED

1. Sherry, J.P. *Crit. Reviews Anal. Chem.* **1992**, *64*, 217-300.
2. Van Weeman, B.K. et al. First International Symposium on Immunoenzymatic Techniques: INSERM Symposium No. 2, Feldmann, M. et al. **1976**, 125-133.
3. Cadogan, J.I.G. et al *J. Chem. Soc. (C)* **1966**, 1249-1250.
4. Kurstak, E. Enzyme Immunodiagnostics, Academic Press, Inc. New York **1986**, 23-54.
5. Aldford-Stevens, A.L. *Environ. Sci. Technol.* **1986**, *20*, 1194-1199.
6. Albro, P.W. et al. *J. Chromatogr.* **1977**, *136*, 147-153.
7. Albro, P.W. et al. *J. Chromatogr.* **1979**, *169*, 161-166.
8. Albro, P.W. et al. *J. Chromatogr.* **1981**, *205*, 103-111.
9. Luster, M.I. et al. *Toxicol. Appl. Pharmacol.* **1979**, *50*, 147-155.
10. Newsome, W.H. et al. *Intern. J. Environ. Anal. Chem.* **1981**, *10*, 295-304.
11. Franek, M. et al. *J Agric. Food Chem.* **1992**, *40*, 1559-1565.
12. Mattingly, P. et al. United States Patent #5,145,790 **1992**.
13. Egolf, D.S. et al. *Anal Chem.* **1990**, *62*, 1746-1754.
14. Cambridge Scientific Computing, Inc. CSC Chem3D Molecular Modeling and Analysis Software, Cambridge, MA.
15. Jung, F. et al. *Pestic. Sci.* **1989**, *26*, 303-317.
16. Harrison, R.O. et al. *J. Agric. Food Chem.* **1991**, *39*, 122-128.
17. Jockers, R. et al. *J. Immunol. Methods* **1993**, *163*, 161-167.
18. Karu, A.E. et al. *J. Agric. Food Chem.* **1994**, *42*, 301-309.

RECEIVED October 13, 1994

Chapter 11

Miniaturized Microspot Multianalyte Immunoassay Systems

Roger P. Ekins and Frederick W. Chu

Department of Molecular Endocrinology, University College London
Medical School, University College London, London W1N 8AA, England

The next major development in the immunoassay field is likely to
take the form of miniaturized systems permitting the simultaneous
determination of many analytes in the same sample. Preliminary
studies suggest that assays based on antibody microspots are of
greater sensitivity, and are more rapidly performed, than those based
on conventional formats.

Current Trends in the Immunoassay of Substances of Biological Importance.

In the past three decades, radioimmunoassay (RIA) and other analogous "binding
assays" based on radioisotope labels been developed for thousands of substances of
biological importance (disregarding their widespread use to identify individual
nucleotide sequences in DNA). Such assays were first developed to determine
hormone concentrations in blood and other body fluids (*1, 2*); however their use
rapidly extended to the measurement of vitamins, viral and tumor antigens, drugs
and other similar trace constituents of living matter. Subsequently they have
penetrated into many fields other than medicine, including agriculture, food analysis,
environmental studies, forensic investigation, etc.

The utility of these widely-used methods has derived from the high "structural
specificity" of antibodies and certain other binding agents of biological origin (i.e.
their ability to recognize three-dimensional molecular shapes or structures) and on
the high "specific activities" of radioisotopes, which permit the binding reactions
between exceedingly small amounts of analyte and binder to be observed. In
combination, these attributes underlie the high specificity and sensitivity of
isotopically-based binding assay methods, and guarantee their continued inclusion in
the micro-analytical armamentarium for many years to come. However, in the past
decade, non-isotopic labels (e.g. enzymes, chemiluminescent compounds and
fluorophors) have been increasingly employed in this context, in part to circumvent
the environmental, legal, economic and logistic problems associated with the use of
radioactive materials, but also - and perhaps more importantly - to exploit a number
of important analytical advantages offered by non-isotopic methods.

0097–6156/95/0586–0153$12.50/0
© 1995 American Chemical Society

Recent activities reflecting this trend have primarily centered on the development of:
1. methodologies for home or doctor's office use (e.g. pregnancy test kits);
2. "ultrasensitive" assays (i.e. assays of greater sensitivity than isotopically-based methods);
3. transducer-based "immunosensors";
4. "multifunctional" and random access analyzers.

This presentation discusses concepts which are relevant to each of these topics, but relates primarily to the development of a miniaturized multianalyte "array" technology permitting the simultaneous, ultrasensitive, measurement of a virtually unlimited number of analytes in a small biological sample (e.g. a single drop of blood). Technologies of this kind (if successfully industrialized) can be anticipated to prove of particular importance to clinical diagnosis - for example, in genetic testing - and to bring about a revolution in medical microanalysis comparable to that which has occurred in the past decade in other areas, such as computing, home entertainment, etc. However we also foresee them replacing current analytical methods in fields other than medicine, extending the use of binding assay techniques into areas from which they are presently precluded by virtue of their inconvenience, complexity and/or cost.

Why Develop Miniaturized, Multianalyte, Binding Assays?

The general grounds for developing "miniaturized" analytical techniques are largely self evident, and similar to those underlying similar trends in other areas, for example computer design. Interest has therefore recently focused on the possibility offered by techniques such as have been developed in the micro-electronics industry (e.g. microlithography) to construct centimeter-sized instruments capable of performing the functions of conventional large-scale laboratory equipment. Certain particular benefits are likely to arise from these developments in the medical diagnostics field, e.g. that of conducting analyses on finger-tip blood samples, thereby obviating the discomfort and distress caused to patients by conventional methods of taking blood. Likewise the problems and overhead costs associated with the transport of samples to centralized hospital laboratories are likely to be obviated by assay miniaturization, permitting the "devolution" to doctors' surgeries, clinics, etc., of sophisticated microanalytical equipment, or even portable "diagnostic smart-cards", yielding immediate diagnostic information. Miniaturization can thus be anticipated to make accessible, to doctors and others, assay techniques whose use is presently restricted to central laboratories, with the attendant delays, administrative and other costs, and liability to error, that this implies.

Equally importantly, immunoassay miniaturization offers the possibility of multianalyte determination in very small samples. The need to determine many analytes in blood and other biological fluids has become increasingly apparent in many branches of medicine such as endocrinology, where knowledge of the plasma concentrations of a number of different hormones is often required to resolve a diagnostic problem. An even more pressing need is evident in other areas, such as allergy testing, the screening of transfusion blood for viral contamination, genetic analysis, etc.

A more subtle requirement arises from the finding that many hormones and other "substances" of biological importance are of heterogeneous molecular composition, i.e. they comprise a number of "isoforms", differing in molecular structure, biological potency, and (questionably) in physiological function. The measurement by present methods of their "amounts" or "concentrations" is, in consequence, entirely illusory (3), leading to results which differ depending on the method used and the particular reagents on which the method relies (e.g., the antibody employed in an immunoassay). Thus, aside from the observation that *in vitro* assay results frequently fail to correlate with the "substance's" *in vivo* biological effects, assays in different laboratories cannot be standardized by the distribution of international standards or reference preparations. In short, such conventional stratagems do not, and indeed cannot, ensure that assay results obtained in different laboratories (or by different methods) will agree, or can be validly expressed in terms of "international units" supposedly constituting a "common currency". This creates major difficulties in diagnostic medicine which will never be satisfactorily overcome other than by the development of "ultrasensitive" multianalyte techniques capable of accurately quantifying each of the (major) individual constituents of complex substances of this kind.

More generally, multianalyte technologies clearly permit the analysis of analyte mixtures, a requirement which is obviously likely to arise in many fields other than clinical medicine *per se*, such as that of pesticide analysis.

Conceptual Blocks to Miniaturized Multianalyte Immunoassay Development.

The creation of assay technologies of the kind discussed in this presentation depends largely on the availability of non-isotopic labels exhibiting much higher specific activities than radioisotopes, i.e. labels generating larger numbers of "detectable events" per unit mass in an acceptable signal-measurement time. However such labels' existence has been known for many years, and - given the potential advantages of assay miniaturization - it is necessary to seek and clarify the reasons for a past failure to perceive the possibility of such a development.

These relate, in our view, to certain concepts relating to assay design which, though specious, have become widely accepted in the immunodiagnostics field. Immunoassays have generally been developed to measure substances present in biological fluids at low concentrations, and have therefore been intended to be of high sensitivity. Unfortunately, the concept of "assay sensitivity" has itself been both widely misunderstood and occasionally controversial, leading to dispute and confusion regarding immunoassay design. This issue should therefore be briefly addressed before turning to the other basic concepts on which our current activities in this area are based.

Differing Perceptions of Assay Sensitivity. The need to establish conditions yielding maximal assay sensitivity (and precision) underlay the construction of mathematical theories of immunoassay design by both Berson and Yalow and Ekins *et al* in the course of these workers' independent development of "competitive", isotopically-based, binding assay methods in the late 50's and early 60's (1, 2). However the concepts of "sensitivity" and "precision" adopted by the two groups differed (Figure 1). Briefly, Berson and Yalow, in their many publications relating to immunoassay design (e.g. 4), defined "sensitivity" as the slope of the response curve relating the fraction or percentage of labeled antigen bound (b) to analyte concentration ([H]). In contrast, Ekins *et al* (e.g. 5) defined "sensitivity" as the (im)precision of measurement of a zero analyte concentration (this quantity being indicative of, and essentially equivalent to, the lower limit of detection of <u>any</u> measuring system).

The definition of "sensitivity" as the slope of the dose response curve (or the quotient {response/stimulus}) has been common (and continues to be adhered to) in many areas of science. Indeed it represented the formal definition originally adopted by a number of bodies, including the American Chemical Society (6). Nevertheless the concept has often been criticized as unworkable (7), and leads to various absurdities. For example, since the response variables characterizing different analytical methods used to determine the same quantity generally differ, the slopes of the response curves they yield are likely to be dimensionally different, making it impossible - on the basis of the "slope" definition - to compare their relative "sensitivities". Furthermore, plotting conventional RIA data in terms (for example) of the response variable B/F (i.e. the "bound to free ratio") suggests that assay "sensitivity" is increased by increasing the antibody concentration in the system; however, the converse conclusion is reached if identical data are plotted in terms of F/B. Plotting assay results in terms of another well known and widely used response variable ("fraction-" or "%-bound" (b)) leads to yet a third conclusion (see below).

In contrast, the lower limit of detection is independent of the measurement method used, or the manner in which response curves are plotted, and permits the "sensitivities" of different techniques to be compared without ambiguity.

The key distinction between the two concepts clearly lies in the dependence of the detection limit on the random error (i.e. imprecision) in the measurement of the response variable. By neglecting this crucial factor, the "slope" definition leads to many absurdities of the kind illustrated above. For these and other reasons, the lower detection limit is now widely accepted as indicative of the sensitivity of an analytical system. Nevertheless the slope definition has governed many of the studies conducted in the immunoassay field during the past thirty years, and is the source of much of the mythology that has become embedded within it.

Current beliefs regarding the optimal amount of antibody required to maximize assay sensitivity provide an example of such mythology that is especially germane to the topic of this presentation. Elementary consideration of the mass action laws reveals that, when theoretical response curves corresponding to different antibody concentrations are plotted in terms of (b) versus [H] (where [H] represents the analyte concentration), maximal slope at zero dose ($\{db/d[H]\}_0$) is observed using a concentration of $0.5/K$ (where K is the affinity constant of the binding reaction), in which circumstance the zero dose response (b_0) is 33% (see Figure 2). This theoretical observation led to Berson and Yalow's well-known dictum that - to maximize RIA "sensitivity" - an antibody concentration of $0.5/K$ (i.e. a concentration binding 33% of labeled antigen in the absence of the unlabeled analyte) should be used (4, 8).

However this strategy would minimize the lower limit of detection of the assay system if (and only if) the standard deviation of b (σ_b) were to remain constant regardless of the value of b (a phenomenon known as "homoscesdacity"). Otherwise changes in the slope $\{db/d[H]\}_0$ do not imply corresponding (reciprocal) changes in the detection limit. Thus the notion that "sensitivity" (using this term in its now generally-accepted sense) is maximized using antibody concentrations equal to $0.5/K$ has no theoretical or experimental basis, although most immunoassay practitioners and kit manufacturers have nevertheless continued to design assays in broad accordance with Berson and Yalow's ideas.

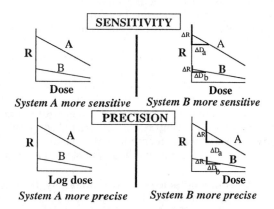

Figure 1. Slope definitions of sensitivity and precision (left) relied on by Berson and Yalow, eg (*4*); error-related definitions (right) on which Ekins et al's theoretical analyses were based, eg (*5*). Reproduced with permission from ref 16. Copyright 1991 American Association for Clinical Chemistry Inc.

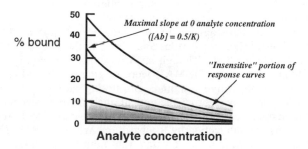

Figure 2. Maximal slope of the RIA response curve relating % bound to analyte concentration is obtained using an amount of antibody binding 33% of an analyte concentration approaching zero (ie 0.5/K). Note the common description of portions of response curves of lesser slope as "insensitive".

Similar incorrect conclusions govern current concepts regarding the design of so-called "non-competitive" immunoassay methods, such as certain forms of immunoradiometric assay (IRMA). "Sandwich" (or "two-site") assays belong to this class, and typically rely on reaction of the analyte with a "capture" antibody situated on a solid support, followed by exposure of the captured analyte to a second (labeled) antibody. It is generally believed that the concentrations of both capture and labeled antibodies should exceed the analyte concentrations the system is designed to measure, ensuring, *inter alia*, that all (or the majority) of the analyte present is captured (9), thereby (supposedly) maximizing the sensitivity of systems of this type (Figure 3).

As discussed below, these beliefs - which also implicitly rely on the response-curve-slope definition of sensitivity - are equally erroneous. Thus, in summary, it can be shown that fallacious concepts have governed immunoassay design for many decades, leading (*inter alia*) to the belief that antibody concentrations in the order of $0.5/K$ -$1/K$ (or greater) are optimal in the case of "competitive" immunoassays such as RIA, and much higher concentrations are required in the case of "non-competitive" assays. These false perceptions have constituted a major conceptual obstacle to the development of the miniaturized systems described in this paper.

"Competitive" and "Non-Competitive" Immunoassay Designs. It is likewise important in the present context to clarify the concepts underlying the basic immunoassay formats often referred to as "competitive" and "non-competitive" (although, as indicated below, this terminology is ambiguous and misleading), and the constraints on their respective sensitivities. Conventional RIA and analogous non-isotopic methods rely on a labeled analyte marker to reveal the products of the binding reactions between analyte and binder. This approach is frequently portrayed as relying on "competition" between labeled and unlabeled analyte molecules for a limited number of binding sites, this perception underlying the popular description of assays conforming to this approach as "competitive".

Subsequent to the development of the "labeled analyte" methods, Wide in Sweden (*10*), followed shortly by Miles and Hales in the UK (*11, 12*), developed radiolabeled antibody methods. Though such methods (commonly referred to as "immunoradiometric assay", or "IRMA") were originally claimed (*12*) to be more sensitive than the "competitive" methods employing radiolabeled analyte, these claims were unsupported by rigorous theoretical analysis or experimental evidence, and for some time remained controversial. In particular, major doubt regarding their validity was cast by the publication by Rodbard and Weiss in 1973 (*13*) of detailed theoretical studies purporting to demonstrate that both labeled analyte and labeled antibody methods possess essentially equal sensitivities (i.e. lower detection limits). (Note: these authors suggested that IRMA methods were *more* sensitive for the assay of small polypeptides in which radioiodine incorporation into the analyte molecule was restricted; conversely they concluded such methods were *less* sensitive for the measurement of high molecular weight analytes. These conclusions were likewise implicitly based on the notion that the magnitude of the "signal" emitted by the label constitutes the principal determinant of immunoassay sensitivity.) Nevertheless the erroneous belief that labeled antibody methods *per se* are intrinsically more sensitive (as a corollary of their reliance on labeled antibody) has persisted in some quarters.

Continuing confusion on this issue stems from the widespread perception that the distinction between labeled analyte and labeled antibody methods coincides with the distinction between "competitive" and "non-competitive" methods. This view is mistaken. For example, certain labeled antibody methods (e.g. those relying on solid-phase analyte, which can be regarded as competing with analyte in the sample for a small number of labeled antibody binding sites) can also be classed as "competitive", and can be theoretically shown to be essentially of equal sensitivity to conventional labeled analyte methods. As shown below, the greater potential sensitivity of certain "non-competitive" immunoassay formats is not a consequence of the labeling of antibody, nor are labeled antibody methods *per se* necessarily more sensitive. The true explanation lies in the differences in magnitude of the random errors incurred in the quantitative determination of the products of the binding reaction, as discussed in the next section.

The "Antibody Occupancy Principle" of Immunoassay.

When a "sensor" antibody is introduced into an analyte-containing medium, sensor antibody binding sites are occupied by analyte molecules to a fractional extent which reflects both the equilibrium constant governing the binding reaction, and the final unbound (i.e. free) analyte concentration in the mixture. This conclusion follows from the Mass Action Laws, which can be written as:

$$[AbAn]/[fAb] = K[fAn] \tag{1}$$

i.e. final fractional occupancy of antibody binding sites is given by:

$$[AbAn]/[Ab] = K[fAn]/(1 + K[fAn]) \tag{2}$$

where [AbAn], [Ab], [fAb] and [fAn] represent the concentrations (at equilibrium) of bound and total antibody, and free antibody and antigen (analyte) respectively, and K represents the equilibrium constant. The final free analyte concentration is generally dependent on both total analyte and antibody concentrations; however when the total antibody concentration [Ab] approximates 0.05/K or less, free ([fAn]) and total analyte ([An]) concentrations do not differ significantly, and fractional occupancy of antibody is given by:

$$[AbAn]/[Ab] = K[An]/(1 + K[An]) \tag{3}$$

(As further discussed below, assays utilizing this concept may be termed "ambient analyte immunoassays" (*14*), antibody fractional occupancy being independent of both sample volume and antibody concentration.)

All immunoassays essentially depend upon measurement of the "fractional occupancy" of the sensor antibody following its reaction with analyte (see Figures 4, 5). Techniques relying on the measurement (by whatever method) of *unoccupied* antibody binding sites (from which sensor antibody occupancy is implicitly deduced by subtraction) generally necessitate - for the attainment of maximal sensitivity - the use of sensor-antibody concentrations tending to zero, and constitute the class of techniques generally categorized as "competitive". Conversely techniques in which *occupied* sites are directly measured often (but not invariably) permit the use of relatively high sensor-antibody concentrations tending to infinity and may be described as "non-competitive". This difference in the amounts of sensor antibody that may be used is a corollary of the rule (well known in physics) that - to minimize

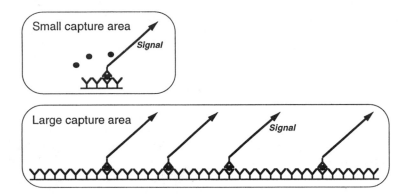

Figure 3. Widespread perception of "non-competitive" immunoassay design. A larger amount of "capture" antibody binds a greater fraction of the antigen present, implying higher assay "sensitivity".

"NON-COMPETITIVE"   "COMPETITIVE "

strategies of determination of antibody occupancy

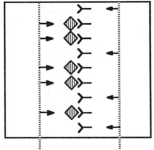

Measure occupied sites   Measure unoccupied sites

Figure 4. "Non-competitive" and "competitive" assay strategies. Reproduced with permission from ref 21. Copyright 1992 International Atomic Energy Agency.

error in the measurement of a small quantity - a direct measurement of the quantity is generally preferable to an estimation of the difference between two large quantities, each subject to error.

Conventional RIA and other similar "labeled analyte" techniques invariably rely on measurement of unoccupied binding sites, this being generally effected by back titration (either simultaneous or sequential) using labeled analyte, though labeled anti-idiotypic antibody (reactive only with unoccupied sites on the sensor antibody) may likewise be employed.

In labeled antibody assays of the type originally developed by Miles and Hales (*12, 13*), the labeled antibody itself constitutes the "sensor "antibody which, following reaction with analyte, may be separated into occupied and unoccupied fractions using, for example, an immunosorbent (comprising analyte, analyte analog or anti-idiotypic antibody linked to a solid support). If - following separation of bound and free labeled antibody - the "signal" emitted by labeled antibody bound to analyte (i.e. the "occupied" antibody fraction remaining in solution) is measured *directly*, the assay can be classed as "non-competitive". Conversely, if labeled antibody unbound to analyte (i.e. that attached to the immunosorbent) is measured, then the assay is "competitive".

Two-site "sandwich" assays rely on two antibodies directed against different epitopes on the analyte molecule, only one of which is generally labeled, the other - the "capture" antibody - being linked to a solid support. Either of these can be regarded as the sensor antibody in the present context. However, whichever view is taken of the sandwich approach, it is evident that - as normally performed - assays falling into this category can be classed as "non-competitive".

Sensitivities of "Competitive" and "Non-Competitive" Immunoassays.

Competitive and non-competitive measurement strategies can be shown to differ significantly in regard to their potential sensitivities. Clearly (other factors being equal) an approach relying on *direct* measurement of occupied sites is preferable to an *indirect* determination relying on subtraction of measurements of total and unoccupied sites when the analyte concentration is small, and antibody occupancy is therefore low. Moreover other broad conclusions regarding assay design are readily apparent from these considerations. For example, assuming the use of a competitive strategy, reduction of the amount of sensor antibody used maximizes its fractional occupancy at low analyte concentrations, thus minimizing the proportion of unoccupied sites. It is therefore advantageous to minimize the amount of antibody used in a competitive system, thereby also minimizing the error in the determination of occupied sites. Similarly the higher the affinity constant, the greater the fractional occupancy, implying that the use of a high affinity antibody is likewise advantageous with regard to assay sensitivity.

Conversely, increasing the amount of sensor antibody used in a non-competitive assay is, in principle, desirable, since this increases the total number of occupied binding sites until a point is reached beyond which essentially all analyte molecules in the test sample are "captured". Nevertheless certain constraints prevent the use of excessive amounts of sensor antibody in a non-competitive system. For example, in a typical sandwich assay, "non-specific" binding of labeled antibody to the solid support to which capture antibody is attached generates "noise" or "background" against which the specific signal must be determined. In short, increase in the amounts of either capture or labeled antibody beyond a certain point leads to a reduction in the signal/noise ratio, and hence to a loss of assay sensitivity. Thus although non-competitive assay systems are often perceived as relying on large amounts of antibody, this is not an inevitable or universal feature of non-competitive methods. Indeed as shown below, the optimal amount of antibody used in a non-competitive system may, in certain circumstances, be closely comparable to that used in a competitive assay.

Theoretical models have been developed to express these concepts in more rigorous form, and to provide guidance on optimal assay design. These models confirm that in both competitive and non competitive assays, the affinity constant (K) of the antibody and the specific activity of the label are invariably of importance, albeit, in practice, the sensitivity of competitive assays is primarily limited by the affinity constant of the antibody, whereas the specific activity of the label is often of greater significance in the case of non-competitive systems.

In both cases, the error (σ_{R_0}) in the measurement of the zero-dose response (R_0) is a crucial determinant of assay sensitivity. Errors in the response measurement are divisible into two components: i. manipulation, instrumental and other "experimental" errors arising from pipetting and other operations, instrument instabilities, etc., and ii. statistical "signal-measurement" errors arising from the counting of a limited number of the discrete "observable events" that constitute the "signal" (e.g. counts accumulated in a radioisotope counter, photons detected in a fluorometer, etc.). If the label used is assumed to be of infinite specific activity, or the signal measurement time is assumed infinite, the statistical error in signal measurement *per se* falls to zero, and assay sensitivity is limited, in these hypothetical circumstances, solely by "experimental" or "manipulation" errors. The sensitivity achieved in these circumstances may be termed the "potential" or "limiting" sensitivity of the methodology. Thus the potential sensitivity of a competitive assay can be shown to be σ_{R_0}/KR_0, where σ_{R_0}/R_0 is the relative (manipulation) error in R_0 (*15*). For example, if the relative error (i.e. coefficient of variation (cv)) in the measurement of the assay response (e.g. the fraction of labeled analyte bound) is 1% (i.e. $\sigma_{R_0}/KR_0 = 0.01$), and $K = 10^{12}$ l/mol, the maximal sensitivity achievable is 10^{-14} mol/l (i.e. *ca* 6×10^6 molecules/ml). Likewise the potential sensitivity of a non-competitive assay is given by $[bAb]_0\sigma_{R_0}/[Ab]KR_0$, where $[bAb]_0$ represents the labeled antibody misclassified as bound, i.e. the "non-specifically bound" antibody (*16*). Thus $[bAb]_0/[Ab = f_0$, the non-specifically bound labeled antibody fraction, and $[bAb]_0\sigma_{R_0}/[Ab]KR_0 = f_0\sigma_{R_0}/KR_0$. Furthermore, assuming the relative experimental error (σ_{R_0}/R_0) in the measurement of the zero dose response is similar in the case of both competitive and non-competitive assays, it is evident that the potential sensitivity of non-competitive methods is greater than that of competitive methods by a factor approximating f_0, i.e. the "non-specifically bound" labeled antibody fraction. For example, if f_0 is 0.01%, a non-competitive strategy is potentially capable of a sensitivity some 10,000-fold greater than that of a competitive approach, other factors being equal.

These findings are summarized in Figure 6. Figure 6a shows the relationship between sensitivity (expressed in terms of molecules/ml) and antibody affinity in an optimized competitive assay assuming (a) use of a label of infinite specific activity, and (b) use of ^{125}I as a label, samples being counted for one minute. (Computations of the theoretically-optimal reagent concentrations (on which the calculations represented in this figure rely) were based on the further assumptions (c) that the antibody-bound labeled analyte fraction was counted, and (d) that the "manipulation error" component in the measurement of the bound fraction (σ_b/b) was 1%.)

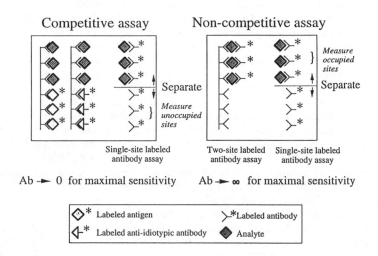

Figure 5. Typical "non-competitive" and "competitive" assay designs. Reproduced with permission from ref 21. Copyright 1992 International Atomic Energy Agency.

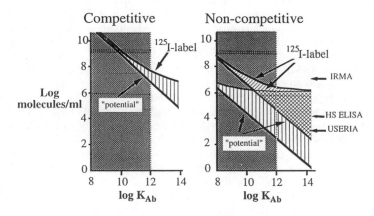

Figure 6. Potential and actual (using ^{125}I labeled reagents) immunoassay sensitivities as a function of antibody affinity. Adapted from ref 22. Copyright 1986 Elsevier Science Ltd.

However the additional "signal-measurement error" arising in consequence of counting radioactive samples for a finite time implies a loss of assay sensitivity, as shown by the upper curve in the figure, though the resulting sensitivity loss is relatively small for antibodies of affinities less than ca 10^{12} l/mol, and is negligible for antibodies with affinities less than ca 10^{11} l/mol. In other words, provided the assayist is prepared to accept individual sample counting times of a few minutes, little is gained in regard to sensitivity improvement by using non-isotopic labels displaying higher specific activities than ^{125}I. Nevertheless similar considerations suggest that radioisotopic labels (such as ^3H) of much lower specific activity than ^{125}I may significantly limit the sensitivities of assays (such as steroid assays) in which they are employed, notwithstanding the use of relatively long sample counting times. An important conclusion emerging from the analysis is the near-impossibility, in practice, of achieving immunoassay sensitivities higher than ca 10^7 molecules/ml using a competitive approach, irrespective of the nature of the label used, assuming an upper limit to antibody binding affinities in the order of 10^{12} l/mol.

The results of a similar analysis relating to non-competitive (two-site) assays are illustrated in Figure 6b. Two sets of curves are portrayed in this Figure, corresponding to assumptions of 1% and 0.01% non-specific binding (nsb) of labeled antibody to the capture-antibody substrate. The crucial importance in such assays of minimizing nsb of labeled antibody is evident, the sensitivities potentially achievable being respectively some 100-fold and 10,000-fold greater than those attainable (using antibodies of identical affinities) in a competitive system. Furthermore, assuming nsb is reduced to ca 0.01%, it emerges that a sensitivity as high as that achievable using an antibody of 10^{12} l/mol in a competitive method can be attained using an antibody of affinity 10^8 l/mol in an optimized non-competitive assay design.

Another important conclusion deriving from Figure 6 is that the sensitivities *potentially* attainable when using high affinity antibodies (K > ca 10^{10} l/mol) are beyond the reach of isotopically-based methods, which are restricted, in practice, to sensitivities of the order of 10^6 - 10^7 molecules/ml and above due to the limited specific activities of isotopes such as ^{125}I. In short, although - assuming nsb is low - non-competitive IRMA methods can be theoretically predicted to offer slightly greater sensitivity than corresponding RIA techniques (assuming the use of the same antibody), the <u>potential</u> superiority of the non-competitive approach can only be fully realized using non-isotopic labels of much higher specific activity than ^{125}I, the advantages of such labels being most apparent when they are combined with antibodies (or other binding agents) of very high affinity.

These theoretical conclusions (together with the publication by Köhler and Milstein of methods of in vitro monoclonal antibody production (*17*), which greatly facilitated the production - on an industrial scale - of relatively pure labeled antibodies) constituted the basis of our own laboratory's collaborative development (initiated $ca.$ 1976) with the instrument manufacturer LKB/Wallac of the time-resolved fluorometric immunoassay methodology now known as DELFIA (*18,19*) - the first commercially-available "ultra-sensitive" non-isotopic immunoassay technique to be developed. The same approach has subsequently been adopted by many other manufacturers of immunoassay kits using other high specific activity non-radioisotopic labels (principally chemiluminescent and enzyme labels), the use of such labels in non-competitive assay designs characterizing all present attempts to further improve binding assay sensitivities.

Ambient Analyte Immunoassay.

The recognition that all immunoassays essentially rely on measurement of sensor antibody occupancy leads to the concept of "ambient analyte immunoassay" (*14*). This term describes assay systems which, unlike conventional methods, measure the analyte concentration in the medium to which an antibody is exposed, being independent both of sample volume, and of the amount of antibody present. The possibility of developing such assays follows from the Mass Action Laws which lead to the following equation, representing the fractional occupancy (F) of antibody binding sites by analyte (at equilibrium):

$$F^2 - F(1/K[Ab] + [An]/[Ab] + 1) + [An]/[Ab] = 0 \tag{4}$$

where [An] = analyte concentration, [Ab] = antibody concentration.

From this equation it is evident that, for antibody concentrations tending to 0, $F \approx K[An]/(1 + K[An])$. This conclusion is illustrated in Figure 7 in which the fractional occupancy of sensor antibody binding sites in the presence of varying analyte concentrations, plotted against antibody concentration, is portrayed. This figure shows that, when antibody at a concentration of less than *ca.* 0.01/K - 0.05/K is exposed to an analyte-containing medium, the resulting (fractional) occupancy of antibody binding-sites solely reflects the ambient analyte concentration and is independent of the total amount of antibody in the system. Analyte binding by antibody causes analyte depletion in the medium, but because the amount bound is small, reduction in the ambient analyte concentration is insignificant. For example, if the antibody binding site concentration is less than 0.01/K, analyte depletion in the medium is less than 1% irrespective of analyte concentration, the system therefore being effectively sample volume independent.

Ambient Analyte Microspot Immunoassay. These conclusions lead to two further concepts. First, the sensor antibody may be confined to a "microspot" located on a solid support, the total number of antibody binding sites within the microspot being less than $v/K \times 10^{-5} \times N$ (where v = the sample volume to which the microspot is exposed (mls) and N = Avogadro's number (6×10^{23})).

Dual Label Ambient Analyte Microspot Immunoassay. The second concept deriving from the ambient analyte principle is that of dual label, "ratiometric", microspot immunoassay, this representing a simple approach to the measurement of antibody binding site occupancy.

This term embodies the idea that fractional occupancy may be determined by labeling the antibody itself with one label, and the second "developing" reagent (i.e. the customarily-labeled antibody or antigen) with a second label, and observing the ratio of signals emitted by the two labels. Thus, following exposure of the (labeled) capture antibody microspot described above to an analyte-containing medium, the solid probe bearing the microspot may be removed and exposed to a solution containing an appropriate concentration of a "developing" antibody (labeled with a second label) directed either against a second epitope on the analyte molecule if this is large (i.e. the occupied site), or against unoccupied antibody binding sites in the case of analytes of small molecular size (Figure 8).

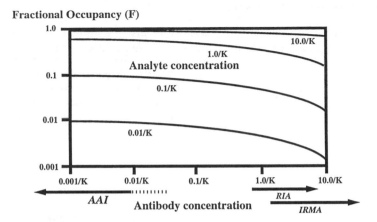

Figure 7. Fractional occupancy of sensor antibody as a function of antibody concentration. (All concentrations expressed in units of 1/K.) Reproduced with permission from ref 23. Copyright 1989 John Wiley & Sons Ltd.

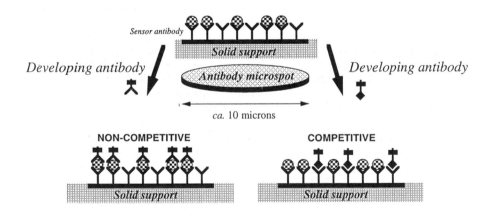

Figure 8. Non-competitive and competitive approaches to the determination of occupancy of antibody located in microspot. Reproduced with permission from ref 24. Copyright 1993 VCH Verlagsgesellschaft mBH.

An alternative approach is to label two developing antibodies with different labels, one directed against occupied sites, the other against unoccupied sites on the unlabeled sensor antibody. (This approach implies that the assay can be described as both competitive and non-competitive, and combines some of the advantages of both formats.) The ratio of signals emitted is, in each case, a measure of the analyte concentration to which the microspot has been exposed.

"Sensor" and "developing" antibodies may be labeled, for example, with a pair of radioisotopic, enzyme or chemiluminescent markers. However fluorescent labels (Figure 9) are especially useful because they enable arrays of "microspots" distributed over a surface (each microspot directed against a different analyte) to be readily scanned, thereby permitting multianalyte assays to be performed on the same sample (Figure 10). Several advantages stem from adopting a dual fluorescence-measurement approach of this kind. For example, neither the amount nor the distribution of the sensor-antibody within the detector's field of view are of importance, since the fluorescent signal ratio remains unaffected. Likewise fluctuations in the exciting light beam's intensity are of little significance.

"Microspot" immunoassay sensitivity and speed.

The suggestion that microspot immunoassays may be more sensitive and rapid than conventional systems clearly challenges established precepts regarding immunoassay design. Detailed discussion of this proposition's theoretical justification is beyond the scope of this presentation; however its basis is simply illustrated in Figure 11. This portrays areas of differing diameter, each area being assumed antibody-coated at the same surface density. When exposed to equal volumes of an analyte-containing solution, antibody concentrations will differ in proportion to the total areas as shown. Though, as the coated area and the amount of antibody decrease, the total amount of analyte bound to antibody also decreases, the fractional occupancy of binding sites on the surface <u>increases</u>, reaching a plateau when the antibody concentration falls below $0.01/K$. In other words, the surface density of the analyte on the solid support reaches a maximum in this circumstance. Thus, assuming use of a non-competitive approach (i.e. observation of the signal generated by occupied sites), the signal/background ratio will be greatest when the antibody coated area is small, and the antibody concentration is below $0.01/K$. Further reduction in spot size reduces the signal, but does not further improve the "visibility" of the microspot against background.

These considerations - which assume equilibrium in the system - show in a simple manner that immunoassay sensitivities may be improved by using small amounts of antibody located at a high density on a microspot, notwithstanding the fact that only a small fraction of the analyte present (i.e. 1% or less) is antibody-bound. However, since (in accordance with the Mass Action Laws) the use of a large amount of antibody increases the velocity of the antigen-antibody reaction, it might be thought that reliance on a "vanishingly small" amount of antibody would protract assay performance times.

In reality the converse is true. Consideration of antigen-antibody reaction kinetics in a homogeneous fluid reveals that fractional occupancy of antibody by antigen is greatest at all times prior to the attainment of equilibrium when the antibody concentration is $0.01/K$ or less (Figure 12). This implies (disregarding, for the moment, the diffusion effects on reaction kinetics resulting from the location of

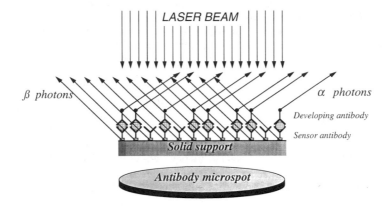

Figure 9. Dual (fluorescent) label approach to antibody occupancy measurement. Reproduced with permission from ref 21. Copyright 1992 International Atomic Energy Agency.

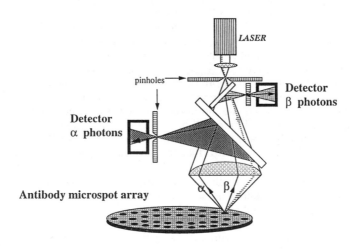

Figure 10. Laser-scanning confocal microscope used to interrogate microspot array. Reproduced with permission from ref 23. Copyright 1989 John Wiley & Sons Ltd.

Sensor antibody concentration
(constant antibody surface density)

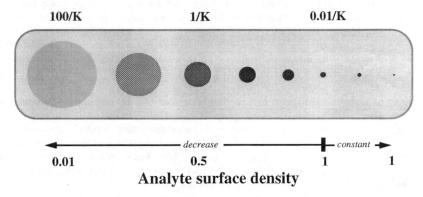

Figure 11. Fractional occupancy of antibody (and signal/background ratio) increase as area of antibody deposition is reduced.   Reproduced with permission from ref 25. Copyright 1994 Elsevier Trends Journals.

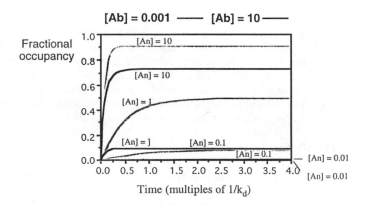

Figure 12. Fractional occupancy of antibody is greater at all times prior to the attainment of equilibrium using the lesser concentration of antibody. (All concentrations expressed in units of 1/K.) Reproduced with permission from ref 21. Copyright 1992 International Atomic Energy Agency.

antibody on a solid support) that the surface density of occupied sites is *at all times* greater when the antibody concentration is below 0.01/K.

It is, of course, well known that attachment of one of the participants of an antibody-antigen reaction on a solid support reduces (in consequence of diffusion constraints) the speed of the reaction, albeit (assuming no alteration in the shape or charge, etc., of solid-linked molecules) the equilibrium constant of the reaction is thereby unaltered. Simple consideration of the diffusion constraints limiting the rate at which analyte binds to antibody binding sites on a support demonstrates that the diffusion flux increases in proportion to the microspot radius. However the number of sites within the microspot increases in proportion to the square of the radius (see Figure 13), implying that the rate at which binding sites are occupied is proportional to the reciprocal of the microspot radius. Thus an even greater benefit stems from the use of very small spots than predicted by a consideration of the kinetics of homogeneous liquid-phase reactions.

Such analysis of diffusion effects suggests that the analyte-antibody reaction rate increases as the microspot area approaches zero. Clearly, however, a point is reached when either the signal generated from the "developing" antibody, or captured analyte molecules are so small in number, that statistical fluctuations cause unacceptable loss of precision in the determination. Such effects - which depend upon the specific activity of the label, the properties of the measuring instrument and other parameters - impose a lower limit on microspot size.

In summary, such theoretical considerations reveal that use of a "vanishingly small" amount of antibody located on a microspot yields an immunoassay system which is both more sensitive, and more rapid, than many conventional formats. Although this conclusion may appear surprising to those accustomed to conventional concepts of immunoassay design, a simple analogy may persuade doubters of its plausibility. A small antibody spot located within an analyte-containing solution can be compared to a thermometer immersed in a heat-containing medium. The thermometer determines the ambient temperature by absorbing heat from its surroundings until thermal equilibrium is reached, albeit the thermometer must be sufficiently small that the amount of heat extracted does not cause a significant change in the ambient temperature. The smaller the thermometer, the more rapid its response. Similar concepts apply to an antibody microspot, whose fractional occupancy by analyte reflects the ambient analyte concentration, and which thus effectively acts as an analyte concentration "sensor".

Experimental verification.

It is not proposed to provide here a detailed description of the extensive experimental studies we have performed to validate the concepts described above (some of which we have discussed elsewhere (*16*)), and which have been primarily designed as a prelude to the development of industrial methods of producing microspot arrays and appropriate instrumentation. Briefly, we have relied on the use of conventional fluorophors and a laser-scanning confocal microscope possessing facilities for dual fluorescence measurement, with an argon laser emitting two excitation lines at 488 and 514 nm. It is thus particularly efficient in exciting blue/green emitting fluorophores such as FITC (excitation maximum 492 nm), but less so in the case of fluorophores such as Texas Red (excitation maximum 596 nm). However, the ratiometric assay principle permits considerable variation in the detection efficiencies of two such labels since, *inter alia*, the specific activities of the labeled antibody species forming the antibody couplets can be chosen to yield signal ratios approximating unity. Inefficiency of the argon laser in exciting Texas Red is thus not a major handicap in the present context. Though the instrument relies on a conventional microscope, it permits quantitation of fluorescence signals generated from an entire microspot or from an area within a microspot of any selected size. It nevertheless possesses disadvantages in the present context, including a relatively high background deriving from fluorescent and scattered light originating in the optical system.

Likewise the microspotting techniques we have employed have been relatively primitive, generally relying on exposure (for periods of *ca* 1 sec) of a variety of solid support materials to minute droplets of antibody-containing solutions, followed by conventional washing and protein-blocking procedures. The solid supports used in this context should display a capacity to adsorb (in the form of a monolayer) - or the ability to covalently link - a high surface density of antibody combined with low intrinsic signal-generating properties (e.g. low intrinsic fluorescence), thus minimizing the fluorescent background. We have examined a number of candidate materials, such as polypropylene, Teflon®, cellulose and nitrocellulose membranes, microtitre plates (black, white and clear polystyrene plates), glass slides and quartz optical fibres coated with 3-(amino propyl) triethoxy silane, etc., and a number of alternative protocols for the achievement of the high monolayer coating densities which are crucial to the achievement of high sensitivity. White Dynatech Microfluor microtitre plates - formulated for the detection of low fluorescence signals, and yielding high signal/noise ratios and high coating densities of functional antibodies (*ca.* 5×10^4 IgG molecules/μm^2) - have generally proved satisfactory and have been mostly used for assay development, albeit occasional manufacturing changes have caused problems (in consequence of non-uniform background fluorescence) and such plates are not in any case ideal. Meanwhile developing antibodies have been coated onto a range of different fluorescent microspheres (*ca.* 2.0-0.5 μm in diameter) using a variety of coupling and washing procedures.

Notwithstanding these limitations, we have verified the concepts outlined above (eg, the constancy of antibody fractional occupancy irrespective of antibody amount and of sample volume when operating under ambient analyte assay conditions), and compared the performance of several assays in microspot format and when conventionally designed. As an example of the latter, the results of a thyroid stimulating hormone (TSH) assay are shown in Figure 14. Such experiments have confirmed that sensitivities comparable with, or indeed superior to, those of conventional methods are achievable using the microspot approach.

Future Prospects.

That ultrasensitive microspot immunoassays can be developed using far smaller amounts of antibody than are conventional, and do not merely constitute an attractive but unattainable theoretical goal, is likely to open a new era in immunoassay development. The finding permits, in principle, the construction of microspot arrays enabling the simultaneous measurement of hundreds of analytes in samples of volume *ca.* 1 ml or less. Our preliminary studies have been limited to the simultaneous assay of 3-4 different analytes, and difficulties may become apparent with increase in analyte number, due, for example, to non-specific binding of labeled antibodies in the developing antibody mixture and cross-reaction effects. Nevertheless, in consequence of the availability of improved solid supports and

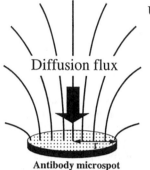

Under diffusion controlled
conditions:

- antigen diffusion flux
  proportional to r
- antibody on spot
  proportional to r^2
- rate of antibody binding
  site occupation by antigen
  proportional to $1/r$

Figure 13. Rate of occupation of antibody binding sites by antigen is inversely proportional to microspot radius. Reproduced with permission from ref 25. Copyright 1994 Elsevier Trends Journals

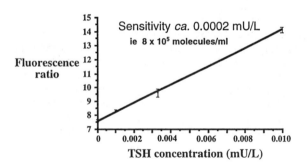

Figure 14. TSH microspot immunoassay dose response curve. Reproduced with permission from ref 21. Copyright 1992 International Atomic Energy Agency.

antibodies or antibody fragments (produced, for example, by recombinant DNA techniques (*20*), and by the development of improved instrumentation (relying, for example, on time-resolution of fluorescent signals to eliminate background generated by the instrument and solid supports), the capacity of workable arrays is likely to progressively increase. The possibility also exists of relying on labels other than fluorophors, such as chemiluminescent labels, in microspot assays of this type.

Clearly the same concepts are applicable to other "binding assays", including those relying on the use of hormone receptors, oligonucleotide probes, etc. For example, participants in the Genosensor Consortium project (launched in the US in 1992) are investigating the localization of tens of thousands of oligonucleotide strands (each comprising eight bases) on the surface of a chip, binding of complementary DNA strands to individual nucleotides being signaled by corresponding optical or electronic detectors located on the chip. Though, in principle, transduction-based systems of detecting microspot "occupancy" are ‑ attractive (permitting, *inter alia*, such chips to be linked directly to computers for data analysis), the attainment of sufficient sensitivity and specificity to permit their use in an immunoassay context presents a major challenge.

Given the financial resources now being devoted to assay miniaturization in many countries, the development within the foreseeable future of technologies comparable to that here described seems virtually certain. Amongst other likely consequences, it is possible to foresee a major increase in population screening for early biochemical signs of, or genetic propensity to, disease, and indeed a considerable transformation in the way that diagnostic medicine is practiced, accompanied by a significant reduction in costs. Clearly such technology can also be anticipated to prove of major importance within the field of pesticide analysis.

Literature cited

1. Yalow, R. S.; Berson, S. A. *J. Clin. Invest.* **1960,** *39,* 1157.
2. Ekins, R. P. *Clin. Chim. Acta.* **1960,** *5,* 453.
3. Ekins, R. *Scand J Clin Lab Invest.* **1991,** *Suppl 205,* 33.
4. Berson, S. A.; Yalow, R. S. *Peptide Hormones;* Methods in Investigative and Diagnostic Endocrinology 2A; American Elsevier: New York, 1973; Chapter 4, Section 2.1, pp. 84-120.
5. Ekins, R. P.; Newman, B.; O'Riordan, J. L. H. *Statistics in Endocrinology;* MIT Press: Cambridge, Mass., 1970; Chapter 19, pp. 345-378.
6. Macurdy, L. B.; Alber, H. K.; Beneditti-Pilcher, A. A.; Carmichael, H.; Corwin, H. A.; Fowler, R. M.; Huffman, E. W. D.; Kirk, P. L.; Lashof, T. W. *Anal Chem.* **1954,** *26,* 1190.
7. Jones, R. C. *Proc. Inst. Radio Engineers* **1959,** *47,* 1495.
8. Yalow, R. S.; Berson, S. A. *Statistics in Endocrinology;* MIT Press: Cambridge, Mass., 1970; Chapter 18, pp. 327-344.
9. Hay, I. D.; Bayer, M. F.; Kaplan, M. M.; Klee, G. G.; Larsen, P. R.; Spencer, C. A. *Clin. Chem.* **1991,** *37,* 2002.
10. Wide, L.; Bennich, H.; Johansson, S. G. O. *Lancet* **1967,** *ii,* 1105.
11. Miles, L. E. H.; Hales, C. N. *Nature* **1968,** *219,* 186.
12. Miles, L. E. H.; Hales, C. N. *Protein and Polypeptide Hormones Part 1,* Excerpta Medica: Amsterdam, Netherlands, 1968; pp. 61-70.
13. Rodbard, D.; Weiss, G. H. *Anal. Biochem.* **1973,** *52,* 10.

14. Ekins, R. P. *British Patent no. 8224600* **1983**.
15. Ekins, R.; Newman, B. *Steroid Assay by Protein Binding;* Karolinska Symposia on Research Methods in Reproductive Endocrinology 2; Karolinska sjukhuset, Stockholm, Sweden 1970; pp. 11-30.
16. Ekins, R. P.; Chu, F. W. *Clin. Chem.* **1991,** *37,* 1955.
17. Köhler, G.; Milstein, C. *Nature* **1975,** *256,* 495.
18. Marshall, N. J.; Dakubu, S.; Jackson, T.; Ekins, R. P. *Monoclonal antibodies and developments in immunoassay,* Elsevier/North Holland, Amsterdam, Netherlands, 1981; pp. 101-108.
19. Soini, E.; Lövgren, T. *Critical Reviews in Analytical Chemistry 18,* CRC press, Boca Raton, Florida, 1987; pp. 105-154.
20. Winter, G.; Milstein, C. *Nature* **1991,** *349,* 293.
21. Ekins, R. *Developments in Radioimmunoassay and Related Procedures,* International Atomic Energy Agency, Vienna, 1992; pp. 3-51.
22. Jackson, T.; Ekins, R. P. *Journal of Immunological Methods* **1986,** *87,* 13.
23. Ekins, R. P.; Chu, F. W.; Micallef, J. *Journal of Bioluminescence and Chemiluminescence* **1989,** *4,* 59.
24. Ekins, R. P. *Methods of Immunological Analysis,* VCH Verlagsgesellschaft mbH, Weinheim, Germany, 1993; pp. 227-257.
25. Ekins, R. P.; Chu, F. W. *Trends in Biotechnology* **1994,** *12,* 89.

RECEIVED November 4, 1994

Chapter 12

Very Sensitive Antigen Detection by Immuno-Polymerase Chain Reaction

Takeshi Sano, Cassandra L. Smith, and Charles R. Cantor

Center for Advanced Biotechnology and Departments of Biomedical
Engineering and Pharmacology, Boston University, Boston, MA 02215

Immuno-PCR (immuno-polymerase chain reaction) is a new antigen
detection system, in which PCR is used to amplify a segment of
marker DNA that has been attached specifically to antigen-antibody
complexes. Because of the enormous amplification capability and
specificity of PCR, immuno-PCR offers a greater sensitivity for
specific detection of antigens than any existing antibody-based antigen
detection system. Here, we describe the concept and general methods
of immuno-PCR, along with its potential applications to the detection
of relatively small antigens.

Immuno-PCR (immuno-polymerase chain reaction) (1-4) was developed to enhance
the capability of antibody-based detection systems for specific antigens by combining
two very powerful tools, antibodies and PCR. The power of antibodies originates
from their considerable specificity for the particular antigens (epitopes). The specific
binding affinity of antibodies for their antigens has made antibody-based antigen
detection systems one of the most powerful and versatile tools used in various
molecular and cellular analysis including clinical diagnostics. PCR has become a
general, yet very powerful tool in molecular biology and genetic engineering. The
efficacy of PCR is based on its ability to amplify specifically a DNA segment flanked
by a set of short oligonucleotides (primers). The immuno-PCR technology, derived
from the combination of these two powerful tools, offers an enhanced sensitivity over
existing antigen detection systems and, in principle, could be applied to the detection
of single antigen molecules, for which no method is currently available.

Concept of Immuno-PCR

The concept of immuno-PCR (Figure 1) (1-4) is quite simple, and is similar to those
of conventional antibody-based antigen detection systems, such as enzyme-linked
immunosorbent assays (ELISA) and radioimmunoassays (RIA).

In immuno-PCR, a specific molecular linker which has bispecific binding affinity
for antibody and DNA is used to attach an arbitrary marker DNA specifically to an
antigen-antibody complex. A segment of the attached marker DNA is amplified by
PCR with appropriate primers, and the resulting PCR products are analyzed by an
appropriate method. The production of specific PCR products demonstrates that the
marker DNA molecules are attached specifically to antigen-antibody complexes,
indicating the presence of antigen.

0097–6156/95/0586–0175$12.00/0

Reagents

Molecular Linkers. In conventional antibody-based antigen detection systems, such as ELISA and RIA, secondary antibodies, directed against primary antibody, are generally used to attach labels, such as enzymes in ELISA and radioisotopes in RIA, to antigen-primary antibody complexes. In a similar way, molecular linkers are used in immuno-PCR to attach marker DNA specifically to antigen-antibody complexes. Thus, such a molecular linker is a key component in immuno-PCR.

One unique, particularly versatile molecular linker used in immuno-PCR is a chimera consisting of streptavidin and staphylococcal protein A, which was designed and produced by genetic engineering (5). Streptavidin, a protein produced by *Streptomyces avidinii*, binds a small water-soluble vitamin, D-biotin (vitamin H), with an extremely high affinity ($K_d \sim 10^{-15}$ M) (6-9). Biotin can be incorporated relatively easily into various biological materials, including nucleic acids and proteins (10). These characteristics have made the streptavidin-biotin system a very powerful, versatile biological tool for detection, purification, and characterization of various biological materials (10, 11). Protein A is a cell-wall constituent of *Staphylococcus aureus*, and binds specifically the Fc domain of an immunoglobulin G (IgG) molecule without disturbing its antigen-binding ability (12-14). The specific IgG-binding ability of protein A offers a variety of immunological applications, including purification of antibodies and immunodetection of biological targets.

An expression vector for a streptavidin-protein A chimera (5) was constructed by fusing a truncated protein A gene, encoding two IgG-binding domains (15, 16), to a truncated streptavidin gene (Sano, T., Pandori, M. W., Cantor, C. R., Boston University, unpublished data). The encoded streptavidin-protein A chimera can be produced efficiently in *Escherichia coli* using the T7 expression system (17, 18). This streptavidin-protein A chimera has two independent binding abilities; one is to biotin, derived from the streptavidin moiety, and the other is to the Fc domain of an IgG molecule, derived from the protein A moiety. The chimera forms a subunit tetramer, which binds four biotins and four IgG molecules independently. This bispecific multivalent binding ability for biotin and immunoglobulin G allows the specific conjugation of any biotinylated DNA molecule to antigen-antibody complexes.

Other simple molecular linkers usable in immuno-PCR are mono-specific multivalent binder molecules. Particularly useful molecular linkers in this group are streptavidin and avidin, each of which has four biotin-binding sites per molecule (8, 9). Although these proteins are unable to bind antibody and marker DNA independently, biotinylated marker DNA can be attached tightly to antigen-antibody complexes by the use of biotinylated antibodies (19, 20).

Specific chemical conjugates of antibody with DNA can also be used in immuno-PCR without the use of molecular linkers. Such conjugates can be produced by chemically attaching derivatized DNA molecules covalently to antibody molecules. However, because of laboriousness of many chemical conjugation procedures and the need for purification of functional conjugates, the overall versatility and performance of immuno-PCR could be significantly reduced.

Antibodies. Both monoclonal and polyclonal antibodies can be used in immuno-PCR. In addition, genetically engineered and recombinantly produced antibodies can also be used. However, when the streptavidin-protein A chimera is used as the molecular linker, such antibodies must have the Fc domain, to which the protein A moiety of the chimera binds. The Fc domain of some IgG subclasses does not have sufficient binding affinity for protein A (12-14). For example, the Fc domains of sheep immunoglobulins have only very low binding affinity for protein A. These antibodies cannot be used directly in basic immuno-PCR protocols. However, these antibodies can be used in immuno-PCR with the use of secondary antibodies, which have high binding affinity for protein A, directed against the primary antibodies.

When streptavidin or avidin is used as the molecular linker, antibody molecules must be biotinylated before used in immuno-PCR. Many biotinylation reagents and methods, particularly for IgGs, are available (*10, 11*). It is important that the binding ability of antibodies for their particular antigens is investigated after biotinylation, because the antigen-binding ability of antibodies is sometimes lost or considerably reduced by biotinylation, particularly when many biotins are incorporated into an antibody molecule.

Marker DNA. One particularly unique feature of immuno-PCR is that marker DNA sequences are purely arbitrary. Thus, any DNA molecules can be used, unless their amplified segments are present also in sample-derived nucleic acid molecules.

The marker DNAs used most frequently in immuno-PCR are end-biotinylated double-stranded linear DNA molecules. Generally, such marker DNA contains one biotin per molecule. Incorporation of multiple biotins into a marker DNA molecule should be avoided, because conjugation of marker DNA containing multiple biotins to multi-valent molecular linkers leads to the formation of aggregates.

Two general methods are available for the preparation of biotinylated marker DNA. The first method uses a template-dependent DNA polymerase to incorporate a biotinylated deoxynucleotide, such as biotin-dATP, by a filling-in reaction to one of the termini of a linear DNA molecule, such as a linearized plasmid. Another method is the enzymatic amplification (PCR) of a template DNA with a set of primers, one of which contains one biotin per molecule. The resulting PCR product contains one biotin at one of its termini.

Methods

There are several formats of immuno-PCR. One of the most versatile formats uses microtiter plates as a solid support (microtiter plate format). In this format, antigen is immobilized on the surface of microtiter place wells.

Microtiter Plate Format. The basic microtiter plate format of immuno-PCR for the detection of antigen immobilized on microtiter plate wells using the streptavidin-protein A chimera as the molecular linker consists of six steps, illustrated schematically in Figure 2: Steps 1, immobilization of antigen on the surface of microtiter plate wells; 2, blocking of remaining reactable sites on microtiter plate wells; 3, binding of antibody to antigen (formation of antigen-antibody complexes); 4, binding of the streptavidin-protein A chimera, bound to biotinylated marker DNA, to antigen-antibody complexes (formation of antigen-antibody-chimera-marker DNA complexes); 5, PCR amplification of a segment of the attached marker DNA with appropriate primers; 6, analysis of PCR products by appropriate methods. The key factor in the protocol is efficient and complete removal of unbound and non-specifically bound antibodies and marker DNA, which contribute to background signals. Thus, extensive washing is carried out after each step. One of the advantages of the microtiter plate format is that washing steps can be performed quite easily even with large numbers of samples.

An example of the detection of antigen using the microtiter plate format (*1*) is shown in Figure 3. In this system, bovine serum albumin (BSA) was used as antigen, and mouse monoclonal antibody (subclass, IgG_{2a}) against BSA was directed to the antigen immobilized on microtiter plate wells. By using thirty amplification cycles in the PCR step and agarose gel electrophoresis as the detection method for PCR products, as few as 580 molecules (9.6×10^{-22} mol) of antigen were specifically and reproducibly detected. Direct comparison with a conventional antigen detection system, ELISA, indicated that approximately 10^5 times enhancement in detection sensitivity was obtained by using immuno-PCR.

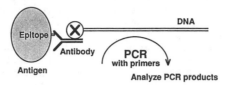

Figure 1. Concept of immuno-PCR. See text for explanations. (Reproduced with permission from ref. 3. Copyright 1994 Shujunsha Co., Ltd.)

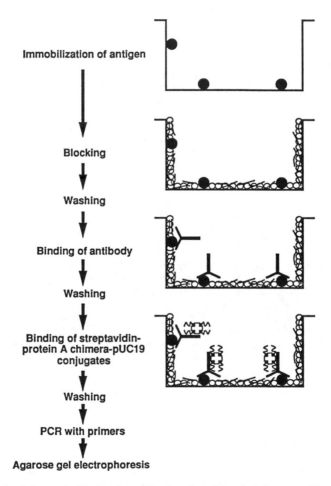

Figure 2. Schematic illustration of basic microtiter plate format of immuno-PCR. (Reproduced with permission from ref. 3. Copyright 1994 Shujunsha Co., Ltd.)

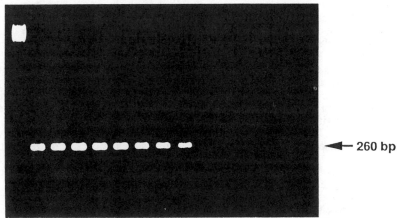

Figure 3. Detection of bovine serum albumin (BSA) by immuno-PCR using the microtiter plate format. BSA was immobilized on microtiter plate wells, and the immobilized BSA was detected by immuno-PCR using mouse monoclonal anti-BSA (subclass IgG_{2a}) and the streptavidin-protein A chimera containing a 2.67-kb end-biotinylated linear plasmid as a marker DNA. A 260-bp segment of the marker DNA was amplified by PCR, and the resulting PCR products were analyzed by agarose gel electrophoresis, stained with ethidium bromide. Lanes 1 - 9 contain PCR amplification mixtures with immobilized antigen: Lane 1, 95 fmol (5.8×10^{10} molecules); 2, 9.6 fmol (5.8×10^9 molecules); 3, 960 amol (5.8×10^8 molecules); 4, 96 amol (5.8×10^7 molecules); 5, 9.6 amol (5.8×10^6 molecules); 6, 0.96 amol (5.8×10^5 molecules); 7, 9.6×10^{-20} mol (5.8×10^4 molecules); 8, 9.6×10^{-21} mol (5.8×10^3 molecules); 9, 9.6×10^{-22} mol (5.8×10^2 molecules). Lanes 10 - 12 are derived from control wells, where no antigen was immobilized. (Reproduced with permission from ref. 1. Copyright 1992 American Association of Advancements in Science).

Other Formats. With simple modifications of the original microtiter plate format described above, several other immuno-PCR formats are available without loss in detection sensitivity. The availability of many different formats should expand the application of immuno-PCR to a wider range of biological and non-biological systems.

The basic microtiter plate format is likely to generate high background signals when antigen in physiological fluid containing endogenous immunoglobulins, such as blood, is to be detected (21). This occurs because of the binding of the protein A moiety of the streptavidin-protein A chimera to sample-derived immunoglobulins. One format that can be used for the detection of antigen in immunoglobulin-containing samples is sandwich assays (2, 4, 21). Antigen is captured by immobilized binding molecules, which have high affinity for the antigen. For example, immobilized antibody fragments, such as Fab and F(ab')$_2$, can be used to capture specific antigen, thereby removing sample-derived immunoglobulins. Subsequently, the basic immuno-PCR protocol can be used to detect the captured antigen. Another group of molecule to capture specific antigen in the sandwich assay format is mouse monoclonal antibodies of subclass IgG$_1$. Mouse IgG$_1$ has only very low affinity for protein A (12-14). Thus, the binding of the protein A moiety of the streptavidin-protein A chimera to mouse IgG$_1$ monoclonal antibodies immobilized on a solid support is negligible. Another advantage of the sandwich assay format is that variations of the PCR amplification efficiency from sample to sample may be minimized, because PCR inhibitors present in samples are removed at the first antigen capture step.

Another attractive format that is particularly useful for the detection of antigen in immunoglobulin-containing samples is pre-conjugation (2, 4, 21). In the pre-conjugation format, antibody and biotinylated marker DNA are conjugated to the streptavidin-protein A chimera prior to the application to samples. Because the streptavidin-protein A chimera binds both biotin and IgG stoichiometrically (5), one can easily produce specific conjugates, consisting of the streptavidin-protein A chimera, antibody, and biotinylated marker DNA, in which the IgG-binding domains of the protein A moiety of the chimera are saturated with the antibody. Since no free IgG-binding sites are available, such conjugates should not bind to immunoglobulins present in samples. Another advantage of the pre-conjugation format is that the number of steps in an protocol can be reduced. Protocols with fewer steps are particularly useful to analyze a large number of samples simultaneously, such as in clinical diagnostics.

Another simple, yet attractive format would be the detection of antigen molecules located on cell surfaces by immuno-PCR (cell surface immuno-PCR). Centrifugation or filtration steps can be incorporated into the protocol to separate unbound antibody and marker DNA from sample cells. Because of the extremely high sensitivity of immuno-PCR, this format should allow the detection and analysis of a variety of cell surface molecules with much smaller numbers of cells than conventional antigen detection methods. It should be possible to analyze such cell surface molecules even at single cell levels.

An attractive solid support that can be used in immuno-PCR is magnetic microbeads (22, 23). Magnetic microbeads are available with surfaces coated covalently with various binding molecules, such as antibodies and streptavidin. Because magnetic microbeads can be separated and transferred by the application of magnetic fields, most simply using paramagnetic bars, the use of magnetic microbeads in immuno-PCR should greatly facilitate the manipulation of samples during analysis.

General Precautions

Immuno-PCR has extremely high sensitivity, derived primarily from the enormous amplification capability of PCR. Thus, any non-specific binding of antibody and marker DNA will cause serious background problems. Extensive washing after the application of antibody and marker DNA is indispensable. Even though some fraction of specifically bound antibody and marker DNA is removed by washing, the overall sensitivity can be recovered easily by using, for example, additional amplification cycles in the PCR step.

The use of effective blocking reagents is of great importance in avoiding non-specific binding. Both protein blockers, such as BSA and non-fat dried milk, and nucleic acid blockers, such as sheared sperm DNA, are used. Because of the high specificity of PCR to target DNA segments defined by primers, the presence of other nucleic acid molecules does not cause the generation of background signals or false positive signals.

One most important factor in avoiding background signals and false positive signals is control of contamination, which is a problem common to all sensitive detection systems. Even though all procedures are conducted very carefully, repeated use of the same marker DNA and primers may generate false positive signals. One of the major advantage of immuno-PCR is that marker DNA is purely arbitrary. Thus, marker DNA molecules and their primers can be changed frequently, as needed, to avoid the generation of false positive signals caused by contamination. This characteristic offers easier control of false positive signals than other PCR-based detection methods, in which specific sample-derived nucleic acids are directly amplified.

Applications

Potentially, there are many areas where immuno-PCR can be applied practically and proficiently. In biological and biomedical sciences, specific detection of biological molecules of interest is one of the most important steps in analysis. The extremely high sensitivity of immuno-PCR should allow the specific detection of antigens that cannot be detected by conventional antigen detection systems. Thus, the use of immuno-PCR will allow the analysis of specific antigens at the microscopic scale, for example, at single cell levels.

One of the most practical applications of immuno-PCR is to clinical diagnostics. The extremely high sensitivity of immuno-PCR will enable the specific detection of rare antigens, which are present only in very small numbers. This characteristic should allow the diagnosis of diseases and infections at much earlier stages of disease or infection development. Another important characteristic of immuno-PCR is its simplicity, which should allow the development of fully automated immuno-PCR systems. Such automated systems are extraordinarily useful in clinical diagnostics, in which large numbers of samples are analyzed repeatedly.

Applications of Immuno-PCR to Sensitive Detection of Relatively Small Antigens. Although immuno-PCR has not, to our knowledge, yet been used practically for the detection of relatively small antigen molecules, there is considerable interest in applying this technology to such targets. Among small antigens of general interest are peptides, steroids, nucleotide derivatives, drugs, bacterial toxins, mycotoxins, haptens, pesticides, and other environmental organic and inorganic contaminants. Because these small target antigens are unlikely to be immobilized stably on solid supports by non-covalent interaction, the original microtiter plate format may not be applicable. Here we describe a few potential immuno-PCR formats usable in the detection of relatively small antigen molecules.

Covalent Immobilization of Antigen on Solid Supports. Because unstable, inefficient immobilization of antigen to a solid support hinders the use of the basic microtiter plate format of immuno-PCR for the detection of such small antigens, covalent chemistry could be used to attach antigen stably to a solid support. For example, antigen containing a primary amino group can be immobilized covalently to microtiter plate wells containing activated tresyl or tosyl groups on their surface. A variety of chemical methods are available to coat the surface of solid supports with activated groups, and some of such activated solid surfaces are commercially available. Once target antigen is covalently immobilized, the original microtiter plate format of immuno-PCR can be used to detect the immobilized antigen. If non-antigen molecules that have the same conjugation group are present in samples, purification or enrichment of antigen will be required prior to covalent immobilization to maximize the immobilization efficiency of antigen. This can be performed efficiently by using magnetic microbeads (22, 23) containing specific antibody against target antigen.

This scheme using chemical immobilization of antigen covalently to solid supports would allow the use of immuno-PCR for the detection of small antigens. However, the potential problem in this protocol would be the efficiency of each step, particularly the chemical immobilization of antigen. For example, loss of a fraction of target antigen during chemical immobilization prevents accurate quantitation of the antigen. A more serious problem would be that covalently immobilized antigen may have limited accessibility to antibodies, because other macromolecules, used as blockers, may surround the immobilized antigen molecules which may be sterically hindered.

Sandwich Assays. If an antigen molecule can be bound simultaneously by two different antibodies regardless of its small size without possible steric hindrance, the sandwich assay format could be used to detect such small antigens by immuno-PCR. As described in a preceding section, antigen molecules are first captured by a primary antibody or another binding molecule immobilized non-covalently on a solid support, such as microtiter plates. Then, the captured antigen is detected by immuno-PCR with the use of secondary antibody. The applicability of this sandwich assay format to the detection of small antigens is determined by whether two different antibodies against a particular antigen can be generated and whether these two different antibodies can bind to single antigen molecules simultaneously. Recent antibody technologies, particularly genetic engineering of antibody molecules including phage display systems for efficient selection (24-32), will surely help to solve the first problem. However, if the second issue is a problem that is a very likely case for very small antigens, such as many environmental organic and inorganic contaminants, the sandwich format is not applicable.

Competitive Immuno-PCR. Competitive assays are frequently used for the immunological detection of small antigens. The principle of competitive assays is based on the competition for the binding to antibodies between an antigen and its competitor, which is a modified form of the antigen. The modification of competitor molecules includes labeling with radioisotopes which allows the quantitation of competitor molecules bound to antibodies. By calibration using a series of antigen-competitor mixtures, the antigen can be detected and quantitated from the quantitation of the competitor. When antigen exists at a very low concentration in a sample, the concentration of its competitor must also be very low to be able to observe the competition between the antigen and the competitor. Thus, very sensitive detection of a label attached to competitors is required for accurate quantitation of antigen present at low concentrations.

Here we propose the competitive assay format of immuno-PCR (Figure 4), in which DNA is used as the marker, as in other immuno-PCR formats, that allows the extremely efficient amplification of its segments by PCR. Competitor molecules used in this format are labeled with biotin. Because of its small size (molecular mass = 243

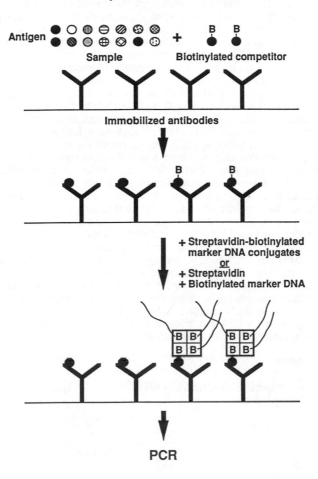

Figure 4. Schematic illustration of competitive assay format of immuno-PCR. Antigen molecules are indicated as black circles. Other circles represent non-antigen molecules present in a sample. See text for explanations.

Da) and high stability, a variety of chemical biotinylation methods are available that allow the stable attachment of biotin to a specific site of a competitor molecule (*10, 11*). In fact, biotinylation of haptens, followed by conjugation to streptavidin or avidin, is a common method to generate specific antibodies. In addition, the relatively small size of biotin should allow subsequent conjugation of streptavidin and biotinylated marker DNA without serious steric hindrance problems.

The basic competitive assay format of immuno-PCR (Figure 4) uses microtiter plates as the solid support and consists of six steps: 1, mixing of an antigen-containing sample and a competitor labeled with biotin at various ratios; 2, application of each sample-competitor mixture to a microtiter plate well on which specific antibody has been immobilized (formation of antibody-antigen and antibody-biotinylated competitor complexes); 3, binding of streptavidin to biotinylated competitors bound to immobilized antibodies (formation of antibody-biotinylated competitor-streptavidin complexes); 4, binding of biotinylated marker DNA to streptavidin bound to biotinylated competitor (formation of antibody-biotinylated competitor-streptavidin-biotinylated marker DNA complexes); 5, PCR amplification of a segment of the bound marker DNA with appropriate primers; 6, analysis of PCR products by appropriate methods. Steps 4 and 5 could potentially be combined by using streptavidin-biotinylated marker DNA complexes, each of which has a free biotin-binding site. However, such conjugates are likely to have limited accessibility to biotin attached to a competitor because of steric hindrance caused by marker DNA molecules. One characteristic of this format is that streptavidin is used as a cross-linker between biotinylated competitors and biotinylated marker DNA. Similar formats have also been used, in which avidin or streptavidin is used as a cross-linker between biotinylated antibody and biotinylated marker DNA (*19, 20*). Because a segment of marker DNA attached to biotinylated competitors via streptavidin can be amplified efficiently by PCR, bound competitor molecules should be quantitated with extremely high sensitivity by analyzing the resulting PCR products. This should allow the considerable enhancement in detection and quantitation ranges of antigen, and thus this competitive assay format should be particularly useful for sensitive detection of small antigens, for which the microtiter plate and sandwich assay formats of immuno-PCR are inapplicable or impractical.

Acknowledgments.

This work was supported by Grant CA39782 from the National Cancer Institute, National Institutes of Health.

Literature Cited.

1   Sano, T.; Smith, C. L.; Cantor, C. R. *Science* **1992**, *258*, 120-122.
2   Sano, T. *Exp. Med.* **1993**, *11*, 1497-1499.
3   Sano, T. *Cell Technol.* **1994**, *13*, 77-80.
4   Sano, T.; Smith, C. L.; Cantor, C. R. In *Encyclopedia of Molecular Biology and Biotechnology;* Meyers, R. A., Ed.; VCH Publishers: New York, NY, 1994; in press.
5   Sano, T.; Cantor, C. R. *Bio/Technology* **1991**, *9*, 1378-1381.
6   Chaiet, L.; Miller, T. W.; Tausing, F.; Wolf, F. J. *Antimicrob. Agents Chemother.* **1963**, *3*, 28-32.
7   Chaiet, L.; Wolf, F. J. *Arch. Biochem. Biophys.* **1964**, *106*, 1-5.
8   Green, N. M. *Adv. Prot. Chem.* **1975**, *29*, 85-133.
9   Green, N. M. *Methods Enzymol.* **1990**, *184*, 51-67.
10 Wilchek, M.; Bayer, E. A. *Methods Enzymol.* **1990**, *184*, 5-13.
11 Wilchek, M.; Bayer, E. A. *Methods Enzymol.* **1990**, *184*, 14-45.
12 Sjoholm, I. *Eur. J. Biochem.* **1975**, *51*, 55-61.

13 Surolia, A.; Pain, D.; Khan, M. I. *Trends Biochem. Sci.* **1982**, *7*, 74-76.
14 Lindmark, R.; Thoren-Tolling, K.; Sjoquist, J. *J. Immunol. Methods* **1983**, *62*, 1-13.
15 Uhlen, M.; Guss, B.; Nilsson, B.; Gatenbeck, S.; Philipson, L.; Lindberg, M. *J. Biol. Chem.* **1984**, *259*, 1695-1702.
16 Lowenadler, B.; Nilsson, B.; Abrahmsen, L.; Moks, T.; Ljungqvist, L.; Holmgren, E.; Paleus, S.; Josephson, S.; Philipson, L.; Uhlen, M. *EMBO J.* **1986**, *5*, 2393-2398.
17 Studier, F. W.; Moffatt, B. A. *J. Mol. Biol.* **1986**, *189*, 113-130.
18 Studier, F. W.; Rosenberg, A. H.; Dunn, J. J.; Dubendorff, J. W. *Methods Enzymol.* **1990**, *185*, 60-89.
19 Zhou, H.; Fisher, R. J.; Papas, T. S. *Nucleic Acids Res.* **1994**, *21*, 6038-6039.
20 Ruzicka, V.; Marz, W.; Russ, A.; Gross, W. *Science* **1993**, *260*, 698-699.
21 Sano, T.; Smith, C. L.; Cantor, C. R. *Science* **1993**, *260*, 698-699.
22 Ugelstadt, J.; Soderberg, L.; Berge, A.; Bergstrom, J. *Nature* **1983**, *303*, 95-96.
23 *Magnetic Separation Techniques Applied to Cellular and Molecular Biology;* Kemshead, J. T.; Bristol, U. K., Eds.; Wordsmiths' Conference Publications: Somerset, UK 1991.
24 Malavasi, F.; Albertini, A. *Trends Biotechnol.* **1992**, *10*, 267-269.
25 Lehner, R. A.; Kang, A. S.; Bain, J. D.; Burton, D. R.; Barbas, III, C. F. *Science* **1992**, *258*, 1313-1314.
26 Ward, E. S. *FASEB J.* **1992**, *6*, 2422-2427.
27 Chiswell, D. J.; McCafferty, J. *Trends Biotechnol.* **1992**, *10*, 80-84.
28 Geisow, M. J. *Trends Biotechnol.* **1992**, *10*, 75-76.
29 Marks, J. D.; Hoogenboom, H. R.; Griffiths, A. D.; Winter, G. *J. Biol. Chem.* **1992**, *267*, 16007-16010.
30 Pluckthun, A. *Bio/Technology* **1991**, *9*, 545-551.
31 Wetzel, R. *Prot. Eng.* **1991**, *4*, 371-374.
32 Winter, G.; Milstein, C. *Nature* **1991**, *349*, 293-294.

RECEIVED November 7, 1994

Chapter 13

Self-Regenerating Fiber-Optic Sensors

David R. Walt, Venetka Agayn, and Brian Healey

Max Tishler Laboratory for Organic Chemistry, Chemistry Department, Tufts University, Medford, MA 02155

Monitoring the level of environmental contaminants demands continuous and accurate data sometimes on several analytes simultaneously. Conventionally, contaminant levels are determined using techniques such as gas chromatography, mass-spectroscopy, high performance liquid chromatography or a combination thereof. A more recent development in environmental sensing is the use of immunoanalysis. Despite recent developments, these techniques are discontinuous and labor intensive. In this paper we discuss the use of fiber-optics in combination with immunoanalysis using two alternative methods to provide continuous sensing: first, based on a combination of antibodies with high binding constants and passive delivery of reagents and second, based on fast off-rate antibodies. These techniques allow for continuous analysis of environmental contaminants. We further introduce the concept of using imaging fiber bundles to provide spatial resolution of multiple reagents. These fibers can be used in combination with degradable polymers for continuous delivery of reagents or as substrates for the preparation of immunosensor arrays with multiple cross reactive antibodies.

Industrial development has been accompanied by the release of many toxic substances into the environment. As a result, there has been an effort to monitor and control the impact of these substances on the environment. For example, the widespread use of agrochemicals in contemporary farming potentially can lead to large scale contamination of ground water and runoff. Sites around chemical plants or accidental spills are other areas where sensors are needed that are capable of providing continuous, sensitive and fast responses. Among the widely accepted methods of environmental monitoring in current use are GC/MS and HPLC analysis. These techniques are sensitive but are discontinuous, involve numerous manipulation procedures and are time consuming. In the case of water monitoring, use of these instrumental techniques often requires an extraction procedure prior to the analysis. Conventional immunoassays overcome some of these drawbacks because they can be conducted without extraction, but these methods are also discontinuous in nature. Moreover, because of the high binding

0097–6156/95/0586–0186$12.00/0

constants of the antibodies used, harsh reagents are required to dissociate the antigen-antibody complexes, and the immunoassay reagents usually are not reusable. An attractive alternative to overcome some of the drawbacks of conventional immunoassay methods is to use fiber-optic immunosensors. Optical sensors (Figure 1) can be used for remote monitoring, are amenable to miniaturization and multiplexing, are not susceptible to electro-magnetic interferences and can be designed with internal calibration (1). Fiber-optic immunosensors offer the specificity and sensitivity of immunomethods combined with the valuable characteristics of optical fibers. Recently, much progress has been achieved in the preparation of fiber-optic immunosensors (2). Examples include sensors for BSA (3), phenytoin (4), benzo[a]pyrene (5) and atrazine (6). Two different approaches for using fiber-optic sensors are available: evanescent wave and distal end sensing. In distal end sensors the signal is collected from a fluorophore attached at the end of the fiber while evanescent wave sensors collect signal from the clad surface and require larger surfaces of the exposed fiber. In our discussion we concentrate on sensors involving distal-end sensing using fiber-optic waveguides.

Use of Degradable Polymers for the Development of Optical Fiber Immunosensors

The use of optical fibers does not resolve the problem of immunoreaction irreversibility. In an effort to avoid harsh regeneration conditions which might result in a decrease of antibody affinity, we employ controlled release polymers for continuous release of reagents (Figure 2) (7,8). In this approach a passive supply of reagents is provided from the polymer matrix to the fiber surface that reacts with solution analyte. The reaction products are removed by diffusion away from the fiber surface. This approach can be used with antibodies with high binding constants because it provides fresh reagents without the need for a regeneration step. A second approach is to use antibodies with low binding constants. In this case a higher off-rate is achieved and a reversible sensor can be devised. The response time of sensors based on such antibodies has been reported to be on the order of 5 to 30 minutes (4).

Another characteristic of environmental analysis is the need to monitor the concentrations of several analytes simultaneously. This need can be addressed by "multianalyte sensors", a concept that has been suggested for immunoanalysis in conjunction with fluorescent microscopy (9). In our research, we have demonstrated the use of imaging bundles for multianalyte sensing (10). We now extend this research to the design of multianalyte sensors using antibodies with different specificities confined to separate areas on the distal end of an imaging bundle. We suggest the use of a combination of controlled release polymers in conjunction with imaging fiber bundles to achieve multiple analyte sensitivity. We also discuss the use of antibodies with low-binding constants to achieve reversibility in fiber-optic sensors. Applying this approach to imaging bundles should enable us to monitoring multiple analytes simultaneously. We will elaborate on the advantages and shortcomings of the approach in various configurations of fiber-optic immunosensors based on the principles described above.

Design of Sensors Based on Degradable Polymers

We have designed a regenerable immunosensor using controlled-release non-degradable polymers (11). There are two key components in this design: the

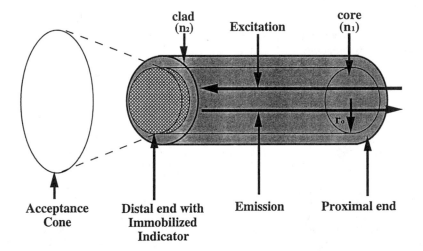

Figure 1.   Optical fiber as a waveguide in sensor design. Proximal end is
attached to the detector; the distal end has the immobilized sensing layer and is
put in contact with the sample. Light travels through the fiber to excite analyte-
sensitive fluorescent indicator immobilized at the distal end. The emitted light
is accepted by the distal end and carried back through the fiber to the detecting
device. The generated fluorescent signal is proportional to the analyte of
interest and can be used as a quantitative basis of analysis.

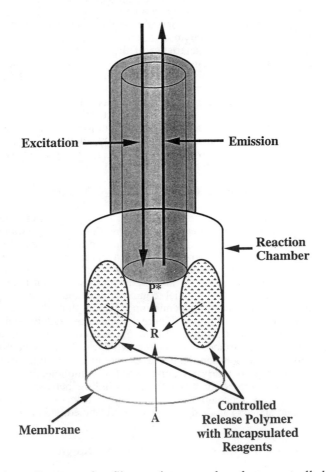

Figure 2.   Concept of a fiber optic sensor based on controlled release polymers. The reagents are regenerated continuously through release from the polymer. Analyte (A) enters the reaction chamber and reacts with the released reagent (R) to yield fluorescent product (P*). The fiber both excites the product and collects the fluorescence which is correlated to the analyte concentration. Accumulated product diffuses through the membrane into solution.

polymeric delivery system and the optical transduction mechanism. In one type of sensor we envision a combination of a homogeneous competitive assay and the use of controlled release polymers. The antigen and antibody components of the immunoassay are incorporated separately into the polymer matrix (Figure 3). Upon release from the polymer, the labeled antigen and the solution analyte compete for the limited number of antibody binding sites to produce a fluorescent signal. This signal travels through the optical fiber and is measured and correlated with the concentration of the analyte of interest. The choice of transduction mechanism depends on the characteristics of the analyte. In cases where the analyte possesses no intrinsic fluorescence, a competitive immunoassay utilizing fluorescently labeled reagents must be developed. In cases where the analyte is intrinsically fluorescent (e.g. polynuclear aromatic hydrocarbons), it is possible to use an unlabeled antibody and register a signal increase due to the analyte binding to the antibody and concentrating at the fiber surface. This approach would require an amount of antibody binding sites sufficient to allow signal generation. Depending on the affinity constants of the antibody, they can be immobilized directly on the fiber surface (low affinity antibodies) or continuously released from polymer microspheres (high affinity antibodies).

Homogeneous Immunoassay Based on Energy Transfer (ET)

The most general approach to the development of homogeneous fluorescent immunoassays for optical sensors is the use of nonradiative energy transfer (12). In this case the antibody and the antigen are each labeled with different fluorescent dyes with specific spectroscopic properties. One dye, the "donor", is excited and the energy is transferred to the second dye, the "acceptor". As a result, a decrease in donor fluorescence intensity is observed that is proportional to the extent of energy transfer. An essential requirement for the donor-acceptor pair is good spectral overlap between the emission wavelength of the donor and the excitation wavelength of the acceptor (Figure 4). The most widely used pairs are fluorescein and rhodamine derivatives (Figure 5). A second critical requirement for successful ET is the distance between the two dyes. It has been established that the critical distance between the two dyes should be less than 100 Å (12). This requirement is particularly advantageous for immunoanalysis since ET occurs in dilute solutions only when there is binding between antigen and antibody. Using this mechanism we prepared a fiber-optic sensor using fluorescein labeled anti-IgG antibody and Texas Red labeled IgG (11) .

Controlled-release Polymers Used in Sensor Design

The polymers used for controlled-release can be either degradable or non-degradable. They can be prepared with different molecular weights and monomer compositions that affect the release rate of the entrapped substances. The selection of the particular type of polymer enables the system to be tailored to the requirements of the particular application. The release from non-degradable polymers proceeds only via diffusion through the polymer matrix (13). Release of large molecular weight reagents, i.e. proteins, occurs at a very slow rate. As a result, sensors will have a slow response time and inadequate sensitivity. This problem can be partially overcome by increasing the surface to volume ratio by formulating the polymers into small particles. A second approach is to use degradable polymers, such as lactide-glycolide copolymers (8,14). In this case the release is accomplished by both diffusion out of the polymer matrix (through micropores) and by release through polymer degradation. These polymers also

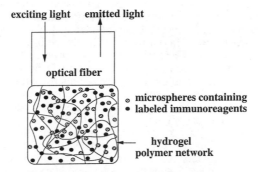

Figure 3.   Diagram of immunosensor based on controlled release polymers. The immunoreagents are entrapped in degradable microspheres and dispersed in the matrix of a hydrogel.

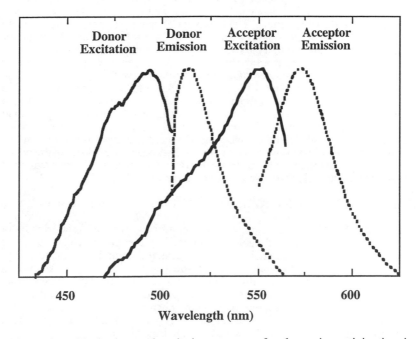

Figure 4.   Excitation and emission spectra of a dye pair participating in energy transfer. The area between donor emission and the acceptor excitation is defined as the spectral overlap and determines the efficiency of energy of energy transfer.

can be formulated into microspheres resulting in increased and more uniform release rates.

Microsphere preparation parameters have been researched widely with respect to their use as drug delivery systems. Different methods of entrapment exist for dispersion of the substance of interest in the polymer and for the formulation of microspheres. In our work we apply the use of double emulsion systems. The substances are dissolved in water (water phase 1) and emulsified in a solution of polymer (oil phase). This emulsion is further dispersed in a secondary water phase to obtain discrete polymer microspheres. The process is completed by solvent evaporation, sphere collection and drying. Microspheres of different sizes can be obtained by modifying the preparation parameters.

The use of degradable polymers might complicate the sensor design since the monomers resulting from the degradation process can change the microenvironment and thus affect the fluorescent signal of the released reagent. For example, the monomers released from the degradation of polylactide-glycolide are glycolic and lactic acids that create an acidic environment. Thus, the fluorescence of pH sensitive dyes will be altered due to the increase in local acidity. These changes should be accounted for in the sensor design and may result in a decrease of sensitivity.

Attachment of Microspheres to the Fiber Surface. The attachment of microspheres to the optical fiber can be accomplished via dispersion in a hydrogel matrix fixed to the fiber surface. The polymers of choice are polyacrylamide and poly(hydroxyethylmethacrylate). The reagents are released into the polymer hydrogel and migrate through passive diffusion. A competing unlabeled analyte diffuses into the polymer from the environment and competes with the labeled antigen for a limited number of antibody binding sites. When ET is the transduction mechanism, there will be a low background of donor fluorescence in the absence of competing antigen due to efficient energy transfer. When competing antigen is present, the donor fluorescence will increase in proportion to the amount of competing antigen. Fast and accurate response can be accomplished through careful engineering of release and mass transfer rates(15). It is important to adjust the permeability of the hydrogels by controlling the cross-linking in order to minimize the diffusion barrier. This diffusion barrier reduction is especially important in the case of antibodies and proteins.

Continuous Sensors for Antigens with Intrinsic Fluorescence

Some analytes possessing a conjugated ring system are intrinsically fluorescent. One class of such chemicals are the polynuclear aromatic hydrocarbons. With these analytes, sensor design is simplified considerably: there is no need to label the antibody or provide a competing antigen. Analysis is achieved after the fluorescent antigen binds to the antibody thereby concentrating the analyte at the fiber surface. Continuous detection can be achieved by the release of a high affinity antibody from microspheres or the immobilization of a low affinity antibody at the distal end of the fiber. The accumulation in the matrix of the hydrogel allows a detectable signal level to be attained.

Design of Continuous Immunosensors Based on Imaging Fiber Bundles

Environmental analysis of sites contaminated by several chemicals calls for continuous monitoring of multiple analytes. Monitoring can be achieved by using imaging fibers for the fabrication of sensors. Imaging fibers are made by melting and drawing several thousand individual optical fibers. The fibers are drawn coherently so that the position of an individual fiber in the bundle at one end (distal) corresponds to the identical position at the other end (proximal). By coupling imaging fibers to a charge coupled device (CCD), one has the ability to spatially discriminate the distal end of the fiber.

The distal face of the imaging fiber can be illuminated selectively for photo-polymerization of an analyte-sensitive polymer matrix (Figure 6). The distal end of the fiber is cleaned and then functionalized to permit covalent attachment of the polymer matrix to the fiber. The photoinitiation light is focused onto the proximal end of the fiber in a very discrete region. The distal end of the fiber is then placed in a monomer solution, with photoinitiator, and is irradiated for a fixed time. Polymerization occurs only at the illuminated area of the imaging fiber. After illumination and removal of the residual monomer, the initiation light is focused onto a different region of the fiber. The process is then repeated to give the next analyte-sensitive polymer matrix.

A way to achieve continuous multianalyte immunosensing is by combining controlled release techniques and imaging based sensing. Incorporating controlled release microspheres of different immunoreagents into individual analyte-sensitive polymer matrices will allow for the detection and monitoring of several analytes with a single imaging fiber (Figure 7). The immunoreagents are regenerated continuously through release from the microspheres which eliminates the need for a regeneration step and thus provides a continuous immunosensor. When using antibodies with high affinity constants, we can fabricate the individual immunosensors by uniformly suspending the preformed controlled release microspheres containing the immunoreagents in the prepolymer solution and subsequently depositing the polymer matrix containing the microspheres.

A continuous multianalyte immunosensor may also be fabricated using fast off-rate antibodies and imaging based sensing. The antibodies must be immobilized within or on the polymer matrix. Immobilization is achieved either by covalent bonding or entrapping the antibody within the interstices of the polymer matrix, which can be modified according to the size of the antibody. Thus the use of fast off-rate antibodies allows for continuous monitoring because the antibody binding site is regenerated when the complex dissociates.

Many antibodies have been produced that have a high degree of cross-reactivity. These antibodies have limited use in immunoassays because signals may be due to the analyte of interest or to a high concentration of a cross-reactive species. However these antibodies may be utilized to fabricate multianalyte immunosensors that can give both qualitative and quantitative information for families of closely related antigens. Immunosensors would be fabricated in an analogous manner to the techniques described above according to the antibody's binding properties. Immobilization of a variety of cross-reactive antibodies, in a multianalyte array could be used in conjunction with a neural network to deconvolute the observed signals into the component analytes by employing pattern recognition strategies.

Figure 5.   Structures of dye labels commonly used for energy transfer.

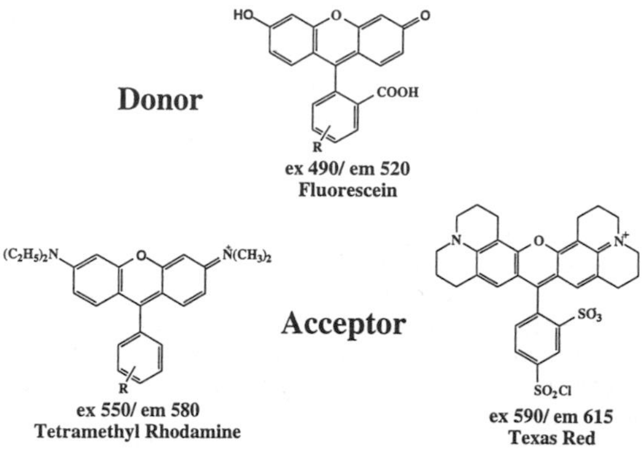

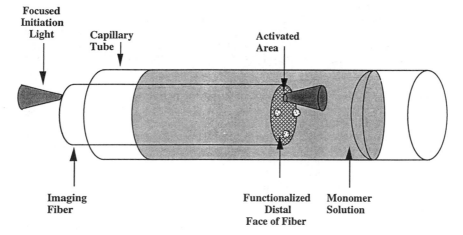

Figure 6.   Photodeposition of polymer matrices on an optical imaging fiber provides spatially resolved sensing regions.

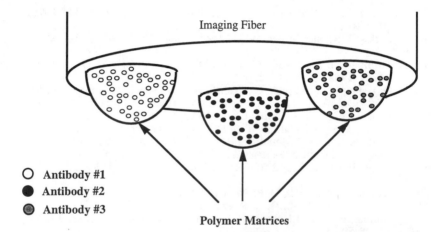

Figure 7. Principle of a multianalyte immunosensor incorporating controlled release microspheres dispersed in hydrogel matrices. Reagent is released from microspheres and reacts with solution analyte within the hydrogel. Adducts diffuse into solution while fresh reagent is released to replenish the sensing chemistry.

Acknowledgments

This work was supported in part by grants from the Environmental Protection Agency, administered by Tufts Center for Environmental Management and the USDOE, Office of Technology Development.

Literature Cited

1.   Chemistry and Technology of Evanescent Wave Biosensors. R. B. Thompson and F. S. Ligler, in *Biosensors with Fiberoptics*, D. L. Wise and L. Wingard, Humana Press, **1991**, pp. 111-138.
2.   Evanescent Wave Immunosensors for Clinical Diagnostics. Bluestein, B. Craig, M., Slovacek, R., Stundner, L., Urciuoli, C., Walczak, I., and A. Ludere, in *Biosensors with Fiberoptics*, D. L. Wise and L. Wingard, Humana Press, 1991, pp. 181-223.
3.   Bright, F. B., Betts, T. A., and Liwiler, K. S. *Anal. Chem.* **1990**, *62*, 1065-1069.
4.   Astles, J. R., and Miller, W. G. *Anal. Chem.* **1994**, *66*, 1675-1682.

5.   Tromberg, B., Sepaniak, M. J., Vo-Dinh, T., and Griffin, G. D. *Anal. Chem.* **1988**,*59*, 1226-1230.
6.   Oroszlan, P., Thommen, C., Wehrli, M., Duveneck, G., and Ehrat, M. *Analytical Methods and Instrumentation* **1993**, *1*, 43-51, .
7.   Luo, S., Walt, D. R. *Anal. Chem.* **1989**, *61*, 174.
8.   Agayn, V. I., and Walt, D. R. *In Diagnostic Biosensor Polymers;* Usmani, A. M.and Akmal, N., Eds.; ACS Symposium Series 556; American Chemical Society: Washington, DC, 1994; pp 21-33.
9.   Ekins, R., Chu, F. and Biggart, E. *Anal. Chim. Acta* **1989**, *227*, 73-96.
10.   Barnard S. and Walt, D. R. *Nature* **1991**, *353*, 338.
11.   Barnard S. and Walt, D. R. *Science* **1991**, *251*, 927.
12.   Ullman, E. F., and Khanna, P. L. *In Methods in Enzymology* **1981**, *74*, 28-59.
13.   Langer, R. *Science* **1990**, *249*, 1527-1533.
14.   Agayn, V. I. and Walt, D. R. *ImmunoMethods* **1993**, *3*, 112-121.
15.   Walt, D. R. *Chemtech* **1992**, *22*, 658-663.

RECEIVED December 1, 1994

Chapter 14

Fiber-Optic Immunosensors for Detection of Pesticides

Mohyee E. Eldefrawi[1], Amira T. Eldefrawi[1], Nabil A. Anis[2], Kim R. Rogers[3], Rosie B. Wong[4], and James J. Valdes[2]

[1]Department of Pharmacology and Experimental Therapeutics, University of Maryland School of Medicine, Baltimore, MD 21201
[2]Biotechnology Division, U.S. Army Research Development and Engineering Center, Edgewood, MD 21010
[3]Exposure Assessment Research Division, Environmental Monitoring Systems Laboratory, U.S. Environmental Protection Agency, P.O. Box 94378, Las Vegas, NV 89193
[4]American Cyanamid Agricultural Research Division, P.O. Box 400, Princeton, NJ 08543

A reusable fiber optic enzyme biosensor provided rapid detection of acetylcholinesterase (AChE) inhibitors and fast regeneration of the sensor for reuse. However, while highly sensitive in detection of oxyphosphate AChE inhibitors, it was insensitive in detection of the less active thiophosphates. It was generic in its identification and did not identify the chemical structure of the analyte. A fiber optic immunosensor, using polyclonal antiparathion antibodies (Abs) was very selective (could differentiate between parathion and paraoxon) and more sensitive, but too slow and nonreusable. A new strategy was developed, using fluorescent pesticide derivatives and polyclonal or monoclonal Abs to construct reusable biosensors with faster turn around time. An immunosensor was developed to assay for imazethapyr herbicide, that was highly sensitive and selective for imidazolinone compounds, unaffected by soil extract matrix and capable of repeated usage. Advantages of biosensors over ELISA are simplicity, speed and reduced need for sample pretreatment.

Immunoassays have been used for years to detect pesticide residues in soil, water and plants. However, the ELISA type assays used are usually time consuming and many of the antibody (Ab) types utilized are either polyclonal or are not highly selective. There is a growing need for assays that are rapid, cost effective and highly sensitive, without giving false negative or false positive results. Two major technological developments are making these objectives within reach. One is in the biosensor field, with advances in a variety of transducers and biological sensing elements, which makes it technically feasible to detect almost any analyte. The other is in the field of molecular immunology, which allows the engineering of an Ab with the exact affinity needed for each pesticide or metabolite of interest, that would provide sensitivity, selectivity and reversibility, and guarantees a stable continuous supply by transfecting the whole or fragments of the Ab gene(s) in *E. coli* (Ward, 1992).

0097–6156/95/0586–0197$12.00/0

Acetylcholinesterase-Based Biosensors

We have utilized a fiber optic evanescent fluorosensor (Rogers et al., 1991) and a potentiometric sensor (Fernando et al., 1993) to detect acetylcholinesterase (AChE) inhibitor insecticides. The fiber optic biosensor has the advantage of no direct electrical connections, no drift problem, suitability for continuous monitoring and easier applicability for field use. The fiber-optic evanescent fluorosensor instrument, designed and built by ORD, Inc. (North Salem, NH), is a portable fluorometer that is adaptable to field work. Components of this instrument, which were described in detail by Block and Hirschfeld (1986) and Glass et al. (1987), include a 10-W Welch Allyn quartz halogen lamp, a Hamamatsu S-1087 silicon detector, an Ismatec fixed speed peristaltic pump, a Pharmacia strip chart recorder, and bandpass filters and lenses as indicated in Fig. 1. The quartz fibers, onto which are immobilized the sensing elements, are 1 mm in diameter and 6 cm long with polished ends. The fiber optic evanescent fluorosensor makes use of the evansecent wave effect by exciting a fluorophore just outside the waveguide boundary (excitation wavelength = 485/20 nm; the latter number representing full width at half maximum). A portion of the resultant fluorophore emission then becomes trapped in the waveguide and is transmitted through the fiber. This is detected after transmission through 510 nm LP and 530/30 nm filters. The quartz fiber is inserted in a flow cell which allows its center 47 mm to be immersed in 46 ml, which is exchanged every 14 sec.

Initially, a fluorosensor was developed, with immobilized nicotinic acetylcholine receptor on quartz fibers. It was found to be effective in detecting fluorescein isothiocyanate (FITC)-labeled receptor-specific α–neurotoxins (e.g. α–bungarotoxin, α–cobratoxin), as well as receptor agonists (e.g. nicotine) and antagonists (e.g. d-tubocurarine) (Rogers et al., 1991). The pH-dependence of the quantum yield of fluorescence, suggested that fluorescein may also be used as a proton (H^+) sensor. Because AChE hydrolysis of acetylcholine (ACh) produces acetate and protons, we developed a potentiometric biosensor to detect AChE inhibitors in Krebs physiological solution buffered with fairly low molarity (0.1 mM) Na phosphate (Dawson and Elliot, 1959). This increases sensitivity since it is the generated protons that quench the fluorescence. The biosensor was constructed by immobilizing FITC-tagged AChE on the quartz fiber (Fig. 1) and monitoring its activity. The pH-dependent fluorescent signal, in the evanescent zone on the fiber surface, was quenched by the protons produced during ACh hydrolysis (Fig. 2). The addition of the substrate ACh to the buffer perfusion medium resulted in quenching of the steady-state fluorescence. The AChE activity was assayed by interrupting the flow of the perfusate around the quartz fibers by turning the pump off, which allowed the local pH in the vicinity of the FITC-labeled enzyme to drop (Fig. 2B), and measuring the percent decrease in the baseline fluorescence during a 2 min period. The reduction in fluorescence was dependent upon the presence of the substrate ACh (Fig. 2B). Resuming the buffer flow allowed the equilibrium to be reestablished. The assay was very stable and could be repeated numerous times on the same fiber without loss in AChE activity (Fig. 2C).

The specific activity of the soluble FITC-AChE was 680 μmol min-1 mg-1 for the hydrolysis of achetylthiocholine. Assuming that the specific activity of FITC-AChE did not change upon immobilization, the assay of the immobilized enzyme activity by the method of Ellman et al. (1961) yielded a value of 0.48 pmol catalytic sites immobilized per fiber. There was excellent substrate specificity, using various choline esters, compared to manometric data. Whereas the reversible AChE inhibitor edrophonium (0.1 mM) reduced fluorescence quenching, that recovered immediately

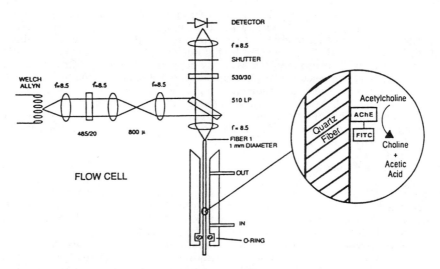

Figure 1. Schematic presentation of AChE biosensor with the optical system to measure fluorescence, showing the lenses, detector and the flow cell with the optic fiber. Inset shows the fiber with immobilized FITC-AChE and the chemical reaction that occurs. f=focal length (Reproduced with permission from Reference 9. Copyright 1991 Society of Toxicology.)

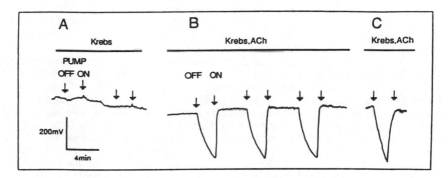

Figure 2. The change in fluorescence as a result of AChE activity. (A) Steady-state fluorescence in the absence of ACh was unaffected by interruption in the beffer flow. (B) In the presence of 1 mM ACh, fluorescence was quenched when the pump was turned off and protons accumulated. The baseline fluorescence was quickly reestablished when the pump was turned on and the excess protons were removed by the perfusing substrate solution. Enzyme activity was measured by the amplitude of signal quenching after 2 min. (C) The response was reproducible after 2 h. (Reproduced with permission from Reference 9. Copyright 1991 Society of Toxicology.)

upon its removal (Fig. 3), the carbamate insecticides bendiocarb and methomyl inhibited the biosensor response, but recovery was much slower. Pre-exposure of the fiber to the organophosphate (OP) antiChEs echothiophate and paraoxon irreversibly inhibited AChE and accordingly the quenching (Fig. 4). However, the OP-inhibited AChE biosensor, could be reactivated by the nucleophilic 2-pralidoxime (2-PAM), which reactivates the phosphorylated AChE by dephosphorylating it. This makes the biosensor reusable for detecting AChE inhibitors and distinguishing inhibition by OPs from that caused by unrelated denaturants, such as heavy metals, whose inhibition is not reversed by 2-PAM. These effects reflected the mechanisms of action of the inhibitors with AChE. The inhibition (Fig. 5) constant values, obtained by the fiber optic enzyme biosensor, were comparable to those obtained by the colorimetric method (Table I).

Antibody-Based Biosensors

Using the FITC-AChE biosensor, neither malathion, parathion nor dicrotophos could be detected even at 1 mM concentration, since they require bioactivation to malaoxon, paraoxon and dicrotophos oxon, respectively. This failure led to the development of a fiber optic immunosensor using, as the biological sensing element, rabbit Abs raised against bovine serum albumin-parathion conjugate (Anis et al., 1992). [In the immunosensor assay for parathion, a sandwich strategy was used, similar to that of the ELISA assay. The casein-parathion conjugate was attached to the glass fiber and exposed to the sample containing rabbit anti-parathion sera, in absence or presence of parathion. A second step was then needed to develop fluorescence by the addition of fluorescein-tagged goat anti-rabbit Ab]. FITC goat antirabbit IgG was used to generate the optical signal which was reduced, in a dose dependent manner, by the presence of parathion in the sample (Fig. 6). Parathion inhibited binding of the Ab to the fiber, thereby reducing subsequent fluorescence. This biosensor could detect 0.3 ppb parathion despite its poor potency in inhibiting AChE and was a 100 fold more selective for parathion than paraoxon. This is unlike the AChE optic fiber biosensor, which was highly selective for paraoxon. The AChE-based biosensor was generic in its detection capabilities and could not identify the chemical structure of the AChE inhibitor.

In order to simplify and speed up the detection process, a one step competitive Ab binding assay was developed. This strategy is based on the competition between the analyte (e.g. parathion) and a fluorescein conjugate of the analyte for binding to the immobilized anti-analyte Abs (Fig. 7). This speeds up detection significantly. We used it to detect the herbicide imazethapyr in soil extracts. The Ab against imazethapyr conjugate was immobilized directly onto the quartz fiber, and its binding of fluorescein-tagged imazethapyr (FHMI) in buffer solution resulted in increased total internal fluorescence (Fig. 8). The presence of imazethapyr in the sample competed with FHMI for the bound Ab, thereby reducing the rate of fluorescence in a time- and concentration-dependent manner from 0.1 to 100 µM imazethapyr (Fig. 9A). The rate of association was calculated from the slope of the fluorescence signal plot during the initial 20 second segment of the response. The IC_{50} of the dose-response curve (Fig. 9B) was calculated to be 2 µM.

An alternate displacement mode was used which improved sensitivity. Rather than determining the concentration of the analyte by the degree of reduction in the rate of fluorescence increase, it was determined by the reduction in fluorescence after it reached a steady state. Thus, the fluorescence resulting from binding of FHMI in buffer to the Ab coated fiber, that reached a steady state in about 2 minutes, was reduced by the addition of imazethapyr almost immediately. It was concentration -dependent and more sensitive for imazethapyr, being effective at 0.001 to 100 µM, giving an IC_{50} of 0.3 µM (Anis et al, 1993).

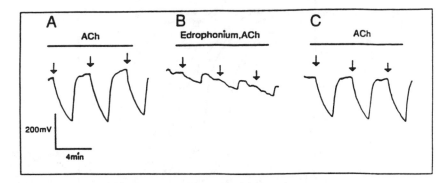

Figure 3. Reversible inhibition of the fluorescent signal generated by the biosensor in presence of 1 mM ACh (A), and after 2 min perfusion with 0.1 mM edrophonium + 1 mM ACh (B). After removal of edrophonium and reperfusion with 1 mM ACh (C) the signal was restored. Arrows indicate times when the pump was turned off. The pump was turned on again after 2 min in each case. Three measurements, 2 min apart, were recorded for each condition. (Reproduced with permission from Reference 9. Copyright 1991 Society of Toxicology.)

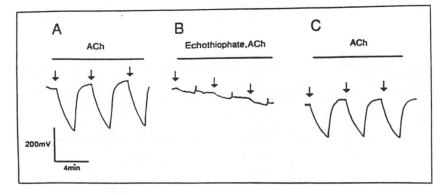

Figure 4. Inhibition of the AChE biosensor by echothiophate. (A) Control responses of the biosensor to ACh (1mM). Echothiophate (0.1 mM) was then added to the ACh-Krebs solution and after a 10-min perfusion the biosensor signal was recorded (B). Echothiophate was replaced with 1 mM 2-PAM in the ACh-Krebs solution and after a 10 min perfusion 2-PAM was removed and the biosensor response was recorded (C). (Reproduced with permission from Reference 9. Copyright 1991 Society of Toxicology.)

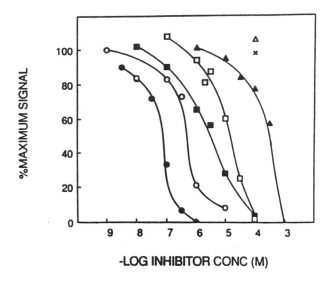

Figure 5. Concentration-dependent inhibition of the AChE biosensor by echothiophate (●); paraoxon (O); bendiocarb (■); methomyl (□); dicrotophos (▲); parathion (✕); and malathion (△). The AChE biosensor was exposed to the indicated compound for 10 min prior to the introduction of ACh and subsequent assay of activity. The symbols are means of at least three measurements. (Reproduced with permission from Reference 9. Copyright 1991 Society of Toxicology.)

Table I. Comparative Inhibition of Immobilized and Soluble AChE by Organophosphates and Carbamates, Assayed by the Fiber-Optic Biosensor and Colorimetric Assays. (Reproduced with permission from Rogers et al., 1991).

Compound	Fiber-optic biosensor assay[a] IC_{50} (M)	Colorimetric assay[b] IC_{50} (M)
Echothiophate	3.8×10^{-8}	3.5×10^{-8}
Paraoxon	3.7×10^{-7}	4.0×10^{-7}
Bendiocarb	2.2×10^{-6}	6.4×10^{-6}
Methomyl	9.0×10^{-6}	1.5×10^{-5}
Dicrotophos	3.3×10^{-4}	1.1×10^{-4}

[a] AChE biosensor was incubated in the presence of each compound for 10 min prior to introduction of ACh (1 mM) and subsequent assay of activity in the presence of inhibitior.
[b] Soluble AChE was incubated in the presence of each compound for 10 min and then assayed using the method of Ellman et al. (1961).

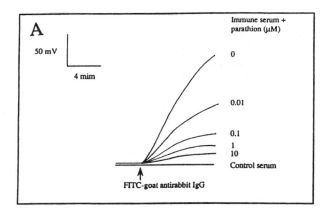

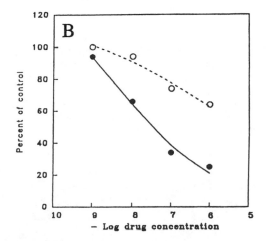

Figure 6. Use of optic fiber immunosensor for detection of parathion. (A) Inhibition of the optical signal generated by binding of FITC goat anti-rabbit IgG to fibers precoated with casein-parathion then incubated in the rabbit antiparathion IgG (1/500 diluted serum), by different concentraions of parathion. Control fiber was coated with casein-parathion but not incubated with the immune serum. The amount of parathion in the sample is reflected in reduction of fluorescence. (B) The dose effect of the presence of parathion (●) or paraoxon (O) in the medium, which competes for the fluorescent-labeled complex and prevents its binding, on the signal generated by binding of FITC-goat antirabbit IgG to the antigen and Ab coated fiber. The 100% control level is the rate recorded in absence of parathion or paraoxon. Symbols are means of triplicate measurements, made on three separate fibers, with standard errors of <5%. The flow cell was washed with 1% SDS in PBS for 2 minutes between measurements. (Reproduced with permission from Reference 1. Copyright 1992 Marcel Dekker.)

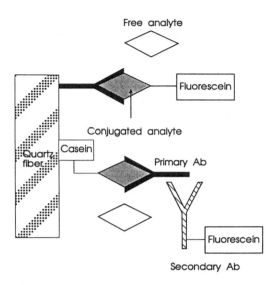

Figure 7. The two strategies used in immunosensing. Top: the Abs are immobilized directly onto the quartz fiber and the signal is generated by binding of the fluorescein conjugated analyte. The presence of analyte in the medium competes for the immobilized Abs, thereby reducing the fluorescence. Bottom: The casein-conjugate analyte is immobilized onto the quartz fiber, to which the primary Abs bind. The optical signal is generated by the fluorescein-tagged secondary Ab (e.g. goat anti-rabbit Ab).

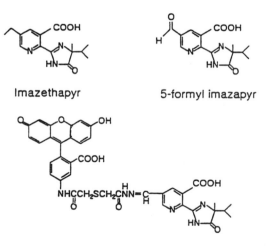

Figure 8. Structures of imazethapyr, 5-formyl imzapyr, the analog compound for fluorescein label preparation, and fluorescein hydrazino methylene imazapyr (FHMI). (Reproduced with permission from Reference 2. Copyright 1993 American Chemical Society.)

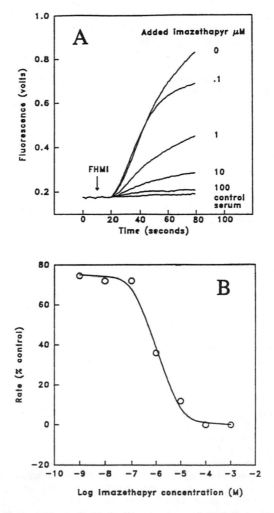

Figure 9. Concentration-dependent displacement of FHMI from the optic fiber with immobilized imazethapyr Abs by imazethapyr. (A) The effect of imazethapyr dose on fluorescence observed in seconds. Varying concentrations of imazethapyr were added to the perfusion solution (25 nM FHMI in casein/PBS). Each perfusion solution reacted with a new Ab-coated fiber. All fibers were prepared on the same day. (B) Dose-response curve of imazethapyr on FHMI binding. The rate of fluorescence change due to bound FHMI, without any addition of imazethapyr to the perfusate was taken as 100%. Addition of imazethapyr to the perfusate decreased fluorescence to a new level, which was calculated as percent value of the zero imazethapyr rate. The rate percent values on the y axis were plotted against the concentrations of imazethapyr on the x axis. (Reproduced with permission from reference 2. Copyright 1993 American Chemical Society.)

One of the significant features of this biosensor is the ability to use the same fiber for multiple measurements without significant loss in sensitivity (Fig. 10). The rate of displacement of fluorescence by 1 μM imazethapyr was highly reproducible (<5% variance in 6 measurements). In one experiment, a single fiber was regenerated repeatedly to measure imazethapyr in seven samples at concentrations from 0.001 to 100 μM. This is due to the reversible binding of both FHMI and imazethapyr to the Ab-coated fiber. The apparent small decrease in the steady state level of fluorescence with repeated used may be due to physical detachment of the adsorbed Ab on the fiber with repeated perfusion. This did not affect the rate of change in fluorescence signal used to calculate the analyte concentration. Covalent binding of Abs to the glass fiber is the strategy currently used. Matrix material in soil extracts, which may present problems in ELISA with its prolonged exposure time, seems to have minimal effects on the biosensor.

In order to determine the biosensor selectivity, the effect of three imidazolinone chemicals (imazapyr, imazaquin and imazamethabenz methyl) were compared to three different agrochemicals with some common structures (Fig. 11). The polyclonal Ab used was highly selective for the three imidazolinones, but did not differentiate much between them. The generic nature of this Ab is to be expected considering the multitude of Ab species it contains. It also emphasizes the importance of using monoclonal Abs in biosensors.

Concluding Remarks

Biosensors can be extremely useful analytical devices for detection and quantitation of pesticide residues. Among their many attributes are portability, short turn-around time, simplicity, cost-effectiveness and sensitivity. The fiber optic fluoresensors that we have used for detection of pesticides demonstrate many of these attributes. A major advantage of optic fluorosensors is their potential applicability for use with crude samples directly without pretreatment; a property that is critical for field use. Absorption and fluorescence arising from the sample matrix are not likely to interfere with the signal observed, primarily because the evanescence effect is measured only at the surface of the sensor and not in the bulk solution. Neither does color or certain chemicals interfere, such as in detection of cocaine metabolities by the immunosensor in 5-fold diluted urine samples (Eldefrawi, unpublished data). These and the speed of detection are the primary advantages of the biosensor over the ELISA technology.

The sensitivity and selectivity of a biosensor depend, in large part, on the properties of the immobilized protein rather than the physical transducer. For example, the inhibition of AChE by irreversible inhibitors shows a time-dependent cumulative effect. The AChE biosensor response mimics this physiological response. Furthermore, the inhibition of AChE varies by 1000 fold in its selectivity among various AChE inhibitor insecticides, and again the biosensor response parallels the physiological response (Table I). Consequently, although this method can be used to determine relative toxicological responses, it cannot be used to identify specific chemical structures. On the other hand, when detection of a specific compound is the required task, highly selective Abs become the preferred choice as biological sensing elements.

The one step competitive binding assay between an analyte and its fluorescent conjugate requires a detection time of only seconds. This can be accomplished by measuring the assocation rate of binding over a period of a few seconds rather than at steady state. This assay protocol, which was applied successfully to the detection of imzethapyr, can be used as a standard protocol for detection of pesticides in general. Cloning Abs using recombinant technology, adds an important dimension because it allows the screening of a large number of clones quickly to select the desired one(s) for transfection in *E. coli*, which would provide a uniform large supply of the sensing element with the required affinity for the biosensor.

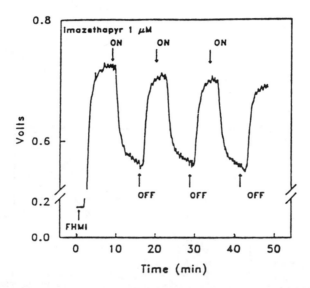

Figure 10. Reusability of the fiber optic biosensor. A 25 nM FHMI in casein/PBS solution was perfused to reach a steady state of binding (in about 5 min), at the point indicated as "ON"; 1 μM imazethapyr was introduced in the perfusate (FHMI, casein/PBS). As a considerable amount of FHMI was displaced, the perfusion solution was switched back to 25 nM FHMI casein/PBS (indicated as "OFF"). The ON and OFF process was repeated. (Reproduced with permission from Reference 2. Copyright 1993 American Chemical Society).

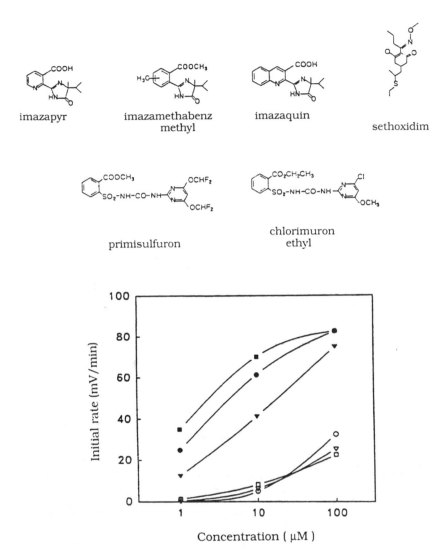

Figure 11. Concentration-dependent FHMI displacement by imidazolinone and non-imidazolinone compounds, whose chemical structures are shown. Three concentrations (1, 10, and 100 μM) of each compound were used for displacement of signal from steady-state bouund levels of fluorescence (Top). The imidazolinones displaced the signal in a concentration-dependent manner, while the non-imidazolinone, even at 100 μM concentration, did not. Each symbol represents a mean of three measurements (SD + <10%) (Bottom). O, chlorimuron ethyl; ∇, primisulfuron; □, sethoxydim; ●, imazamethabenz methyl; ▼ imazapyr; ■, imazaquin. (Reproduced with permission from reference 2. Copyright 1993 American Chemical Society.)

Acknowledgements and Notice

Much of the underpinnings of the above work was funded by U.S. Army contract No. DAAA15-89-C-0007 and ERDEC IPA to M. E. and senior associateship to N.A. Some of the information in this writeup was funded in part by the U.S. Environmental Protection Agency under cooperative agreement No. CR820460-0, with the University of Maryland at Baltimore. It has not been subject to Agency's peer review; therefore, it does not reflect the views of the Agency.

Literature Cited

1. Anis, N.A.; Wright, J.; Rogers, K.R.; Thompson, R.G.; Valdes, J.J.; Eldefrawi, M.E. *Analyt. lett* .1992, *25*, 627-635.
2. Anis, N.A.; Eldefrawi, M.E.; Wong, R.B., *J. Agr. Food Chem* . 1993, *41*, 843-848.
3. Block, M.J.; Hirschfeld, T.B. 1986, U.S. Pat. 4,582,809.
4. Dawson, R.M.C. and Elliott, W.H. In *"Data for Biochemical Research"* (Dawson R.M.C., D.C.; Elliott, W.H. Jones, K.M., Eds) Oxford Univ. Press, Oxford, 1959, 192-209.
5. Ellman, G.L.; Courtney, K.D.; Andres, V. Jr.,; Featherstone, R.M. *Biochem. Pharmacol.* 1961, *7*, 88-95.
6. Fernando, J.C.; Rogers, K.R.; Anis, N.A.; Valdes, J.J.; Thompson, R.G.; Eldefrawi, A.T.; Eldefrawi, M.E. *J. Agr. Food Chem*. 1993, *41*, 511-516.
7. Glass, T.R.; Lackie, S.; Hirschfeld, T.; *Appl. Opt.* 1987, *26*, 2181-2187.
8. Rogers, K.R.; Eldefrawi, M.E.; Menking, D.E.; Thompson, R.G.; Valdes, J.J. *Biosensors & Bioelectronics* 1991, *6*, 507-516.
9. Rogers, K.R.; Cao, C.J.; Valdes, J.J.; Eldefrawi, A.T.; Eldefrawi, M.E. *Fund. Appl. Tox.* 1991, *16*, 810-820.
10. Ward, E.S. *The FASEB J.* 1992, *6*, 2422-2427.

RECEIVED November 1, 1994

Chapter 15

Liposome-Amplified Immunoanalysis for Pesticides

Stuart G. Reeves, Sui Ti A. Siebert, and Richard A. Durst

Analytical Chemistry Laboratories, Department of Food Science and Technology, Cornell University, Geneva, NY 14456–0462

Liposome-amplified competitive immunoassay systems for the herbicide alachlor, in both laboratory-flow injection and field immunomigration formats, are described. The preparation and characteristics of analyte-tagged liposomes, along with details of both types of assay are given. The laboratory assay is designed for automation and measurement of a large number of samples in the laboratory, whereas the field assay is designed for rapid field screening of large numbers of samples by non-technical personnel. The advantages of liposome immunoassay compared to more traditional formats are discussed, as are the problems posed by liposome curvature.

The use of immunoassays for pesticide monitoring is now well established, and a variety of laboratory and field assays have been developed (1,2). These immunoassays are mainly in the form of microtiter plate or tube ELISAs (enzyme-linked immunosorbent assays), where measurements are made of the color produced from a chromogenic substrate by the action of an enzyme conjugated to either an antibody or an analyte molecule. Such assays have been developed and are commercially available for the herbicide alachlor (3, 4), which is sometimes found as a contaminant of well water.

Previous studies (5-7) have demonstrated the advantages of liposome-encapsulated dye rather than enzymatically produced color to enhance the signal obtained in the competitive binding reaction of an immunoassay. In this communication the use of liposomes in a field assay format using immunomigration techniques and in an automated flow-injection immunoassay system is demonstrated, using alachlor as a model analyte. Liposomes provide instantaneous, rather than time-dependent, enhancement and offer considerable potential for both automated and field assays, for generic rather than specific assay reagents, and for multi-analyte assays.

The use of liposomes in immunoassay

Liposomes are lipid bilayer vesicles that are formed spontaneously when lipids are dispersed in water. During this formation they encapsulate a portion of the aqueous solution in which they were dispersed, and if this solution contains a marker molecule such as a dye, this will be present in the aqueous core of the liposome. Furthermore, if the analyte of interest is conjugated to a lipid, this can be incorporated into the liposome surface. A diagram of such liposomes is shown in Figure 1.

The principle of a competitive liposome immunoassay using such liposomes is shown in Figure 2. The tagged liposomes and the sample containing the analyte are passed over a solid surface to which antibody to the analyte of interest has been immobilized. Competition occurs between the free analyte molecules and the analyte molecules conjugated to the liposomes, and the number of liposomes that bind to the antibodies is inversely proportional to the amount of free analyte present. Unbound liposomes, which are directly proportional to the sample analyte, move out of the antibody region, and can be measured by an appropriate downstream detector. Alternatively, the bound liposomes can be measured *in situ*, or a detergent can be added in a flowing stream to release the marker which is then measured downstream.

The use of liposomes instead of the more usual enzyme-produced marker has several advantages. The lipid composition can be varied to give the liposomes different physical characteristics and almost any water-soluble marker can be encapsulated, giving rise to a broad range of possibilities for detection. The size of the liposome and the surface concentration of analyte tag can be varied and controlled accurately, giving more control over the experimental parameters of the assay. Enhancement is instantaneous, removing the requirement for a timed enzymatic incubation step, and the whole process lends itself to automation.

In this study, the widely used herbicide alachlor has been used as a model analyte. In order to raise antibodies, the analyte molecule was conjugated to a protein (BSA) (8), and to prepare alachlor-tagged liposomes it was conjugated to a lipid (DPPE - dipalmitoyl phosphatidyl ethanolamine) (9). In both cases available amino groups in the BSA and DPPE were thiolated and subsequently conjugated to the alachlor by nucleophilic displacement at the chloroacetamide group. This is shown in Figure 3, using conjugation to DPPE as the example.

Liposomes were then prepared by the reverse-phase evaporation method using a mixture of DPPC (dipalmitoyl phosphatidyl choline), cholesterol, DPPG (dipalmitoyl phosphatidyl glycerol), and alachlor-DPPE conjugate in a molar ratio of 5:5:0.5:0.01 (8-11). To improve their size distribution and homogeneity, the liposomes were passed through polycarbonate filters of 1.0 and 0.4 μm. They were then purified by gel filtration and dialysis.

Characteristics

Stable liposome preparations have been made containing a variety of detectable compounds, but the work reported in this paper was carried out with 100 mM sulforhodamine B as the encapsulant. In the case of the flow-injection assay the fluorescence was measured, whereas in the strip assay the high molar absorbance was utilized, and the color was estimated by eye or by densitometry.

These liposomes were stable for over 18 months when stored in the dark at 4 °C. They have also been stored for over 9 months at 25 °C and 6 months at 35 °C, again in the dark. Light was deleterious to membrane integrity, and caused dye leakage (or lysis) at a rate of 15-25% per month at room temperature. This is probably caused by energy dissipation effects from light absorbed by the dye inside the liposomes. This is supported by the fact that rapid lysis of the liposomes can be observed under the

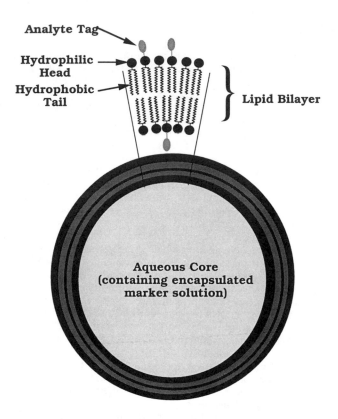

Figure 1.   The Structure of an Analyte-tagged Liposome

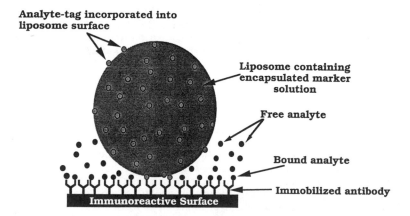

Figure 2.   The Competitive-binding Reaction between Liposomes and Analyte
                 Molecules for Antibody Sites on an Immunoreactor Solid Surface

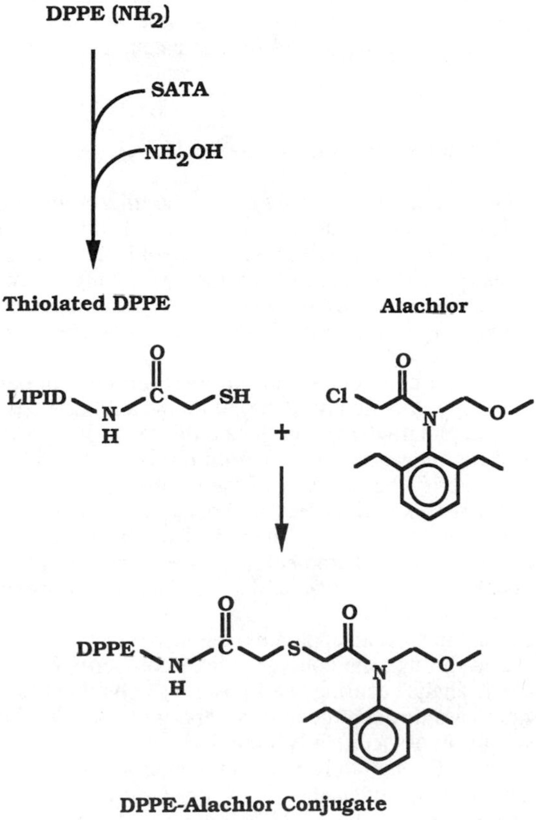

Figure 3.   Conjugation Reaction of Alachlor to DPPE

microscope when the illumination is focused on the sample, and also by the observation that light-induced leakage does not occur in preparations where potassium ferrocyanide is the encapsulant (A. Edwards, unpublished data).

The diameters of the liposomes were measured by laser scattering in a LA-900 Particle Size Distribution Analyzer (Horiba Inc., Irvine, CA), using the manufacturers method, except that the usual sonication step was omitted to avoid lysis (rupture) of the liposomes (9). This gave an average diameter of 0.7 μm. A second determination, using a different instrument (a Coulter LS 130), gave a diameter of 0.4 μm. Using these figures, the concentration of DPPE-alachlor tag in the lipid mixture, and with published data for the cross-sectional area of lipids in membranes (12), certain parameters could be calculated (9) for both possible diameters. Experiments are underway to resolve this discrepancy, and figures calculated from both diameters are given in Table 1.

Flow-Injection Liposome ImmunoAssay (FILIA)

The majority of conventional immunoassays use the microtiter plate format. These plates are cumbersome to use in an automated assay, and furthermore, each individual well needs to be coated with antibody, leading to unavoidable well-to-well and plate-to-plate variations. These problems can be overcome by utilizing a flow-injection method with a reusable, regenerable immunoreactor column. The arrangement of the system in current use in our laboratory is shown in Figure 4 A. This system was modified from a previously published design (6).

The immunoreactor column contains an inert support, (in our case glass beads silanized with aminopropyl trimethoxy silane), with the antibody conjugated to it (anti-alachlor conjugated via glutaraldehyde in this study) (13). In a flow-injection system all samples pass through the same, reusable column, and calibration standards can be included in the runs to verify the accuracy of the results. It has been found possible to use such a column for weeks without any deterioration in performance. During this period there was no apparent leakage of antibody from the column, but after a certain period of time there would be a sharp decline in the response of the column. At this point the column packing material was replaced. Such a continuous flow system is very easy to automate.

In a single assay in such an apparatus, the sample and liposomes in the carrier stream are first passed through the column, where competitive binding occurs, with higher levels of sample analyte causing less liposomes to bind to the column. Unbound liposome and sample pass through the column and go to waste. The column is then washed to remove any non-specifically bound liposomes and other contaminants. Next, detergent is passed through to lyse the liposomes and release the marker, which is detected downstream and quantitated. Finally, carrier is once more passed through to regenerate the antibody by washing off bound analyte, presumably by a simple exchange mechanism driven by mass action effects, and then the column is ready for the next run. Because the regeneration was mild, there was no measurable loss of activity during this step. A typical trace from a single run is shown in Figure 5. This was based on alachlor-tagged liposomes containing sulforhodamine B, and utilizing fluorescence detection. The small initial peak (A) is from the quenched fluorescence of intact liposomes that do not bind to the column, with the second, larger peak (B) produced by the released dye being the one that is used for quantitation. With such a system it is possible to quantitate alachlor down to the 10 ppb level, as shown in Figure 6, with a total assay time of 5 minutes. This level of detection is not as low as that obtained by a microtiter plate ELISA (0.1 ppb), and also higher than the MCL (Minimum Contaminant Level) for alachlor as defined by the EPA.

A. Current Design

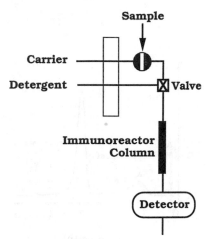

B. Future Design

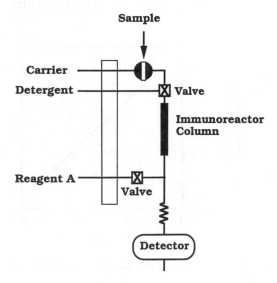

Figure 4.   Schematic Diagram of the Flow-Injection System
A. The current FILIA design
B. A potential future design, using post-column addition and a
mixing/holding coil.

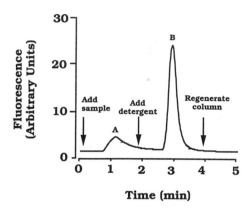

Figure 5.   Trace of a Typical FILIA Run

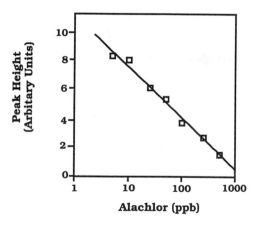

Figure 6.   Alachlor Dose-Response Curve, FILIA System

Various factors can be manipulated to alter the characteristics of the assay. Studies are currently underway to examine the effects of flow rate, column-bound antibody concentration, liposome size and surface concentration of the analyte tag on the assay. The object is to produce an assay with the required sensitivity which takes the minimum possible time, so that a large number of assays can be carried out each day. Recent experiments (G. Rule, unpublished data) suggest that the required limit can be reached under the correct conditions. Furthermore, other possible detection methods are being studied, e.g. chemiluminescence, and this would require addition of other reagents after the competition step. The design of the FILIA system allows such additions, as is common in flow-injection systems, and an example of how this could be arranged is shown in Figure 4B.

The exact amount of amplification due to the liposomes is currently unknown. If one liposome bound to one antibody, then it would be in the order of one million, the number of molecules entrapped per liposome (Table 1). However, multiple binding and shadowing (see later) reduce this considerably. At this point there is no fluorescent analog of alachlor available, so this control cannot be run. However, such an analog is currently being synthesized, and this will allow these measurements to be made.

Field immunomigration strip assay

An extra-laboratory assay for alachlor has been developed, using immunomigration techniques (9). A liposome and alachlor mixture is allowed to migrate up a strip of absorbent material on which anti-alachlor and egg white avidin zones have been immobilized. The competition occurs in the antibody zone, and the amount of liposome bound (quantified as the amount of color from encapsulated sulforhodamine B) is inversely proportional to the amount of analyte in the solution. A second (capture) zone containing egg white avidin to bind the liposomes, collects all of the liposomes that do not bind to the antibody zone, and thus the amount of color in this zone is directly proportional the amount of analyte in the sample. Quantitation in either zone is possible, with the use of the second zone being more intuitive and preferred. This assay has the potential for rapid field screening of analytes of interest, and is simple enough to be used by non-technical personnel.

In this assay, a protein-binding membrane with a plastic backing to provide rigidity was required, and porous nitrocellulose membranes (> 3μm pore size) supported in this manner were found to be the most suitable. An automatic thin-layer plate sample applicator (Camag Linomat IV) was used to dispense the antibody and egg white avidin solutions onto spatially separate and well-defined zones of the membrane for immobilization. The membrane sheet was 8.0 cm high and 15.5 cm wide for later subdivision into strips 5 mm in width. The protein-coated membrane was blocked with 2% PVPP (polyvinylpyrrolidone) and 0.002% Tween-20 in TBS (Tris buffered saline) to prevent non-specific binding. The prepared sheets were vacuum dried and stored at 4°C in the presence of silica gel desiccant until ready for use. They were cut into 5 mm wide strips using a paper cutter when needed. The final strips had a 5 mm long antibody zone 15 mm above the bottom of the strip and a similar egg white avidin zone 15 mm above the antibody zone.

The prototype assay consists of a reagent containing alachlor-tagged liposomes and sample, and a test strip as described above. The assay is performed by dispensing 2 drops of the sample or control solution and 1 drop of a three-times concentrated buffer into a 10 x 75 mm glass test tube, mixing the contents, and adding 1 drop of a liposome solution. The test tube is shaken mildly to mix the contents and the test strip is inserted into the tube; the strip is left in the tube until the solution front reaches the end of the strip (about 9 min); the strip is removed and air dried. The color intensity of

Table 1. Liposome Characteristics

All parameters were calculated using an analyte-tag concentration of 0.1%, and one of the two measured values for diameter. The sulphorhodamine B concentration in the liposomes was assumed to be the same as that used in the encapsulating procedure, and the lamellarity was assumed to be 1.

Mean Diameter	$0.7\ \mu m$	$0.4\ \mu m$
Volume of Each Liposome	$1.8 \times 10^{-10}\ \mu L$	$3.4 \times 10^{-11}\ \mu L$
Liposome Concentration	1.2×10^8 per μL	6.0×10^8 per μL
SRB (Concentration)	100 mM	100 mM
SRB (molecules per liposome)	9.6×10^6	1.8×10^6
Alachlor (molecules per liposome)	3.5×10^3	1.1×10^3

the antibody zone and the egg white avidin capture zone are estimated either visually or by scanning densitometry.

Figure 7 shows diagrammatically the results obtained with a series of alachlor standards (A=0, increasing B<C<D), with the decrease in color of the antibody zone with increasing concentrations of added alachlor, and the concomitant increase in the color of the capture zone.

Dose-response data obtained by scanning densitometry of strips run in the presence of various concentrations of alachlor are shown in Figure 8. The response in both the antibody and avidin zones varied logarithmically with alachlor concentration, and both were estimated to be able to detect 5-10 ppb alachlor. When these strips were assessed visually, a similar determination could be made, but at low levels of added alachlor it was somewhat easier to detect increases of red color over a white control (capture zone) than decreases in color intensity (antibody zone). There are still difficulties with the reproducibility of the strips. These seem to be due to heterogeneity in the supplied sheets of plastic-backed nitrocellulose, and this problem is currently being addressed. As in the case of the FILIA, the detection limit is not as low as is required, but this can be improved down to the 1 ppb level by manipulation of various parameters (T. Siebert, unpublished data).

Figure 9 is a diagram of how the prototype strip could be developed into a user-friendly field device, using visual estimation of the color by comparison with a color card, or by measurement in a calibrated reflectometer (14). It both cases it is anticipated that the capture zone would be used for measurement rather than the antibody zone. A further possibility is for the production of multianalyte strips. A series of antibody zones, each to a different analyte, could be applied to the strip, and a suitable mixture of liposomes applied.

The effect of liposome size and analyte-tag density

In any liposome immunoassay system, the analyte density (concentration) on the surface of the liposome will have an obvious effect on the binding and competitive reactions. What is less immediately obvious is that liposome size can also have an effect.

A scale drawing of the surface view of a liposome with the concentration of analyte tag that has been routinely used, along with scale drawings of antibodies, is shown in Figure 10 A. Assuming random distribution of the tag molecules on the

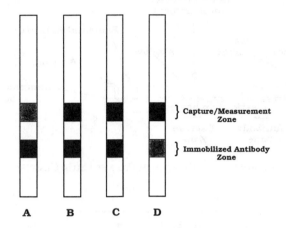

Figure 7.   Diagram of a Series of Strip Assays

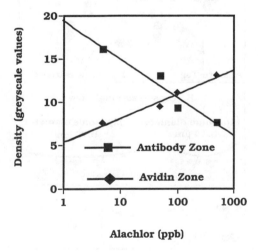

Figure 8.   Alachlor Dose-Response Curve, Strip Format

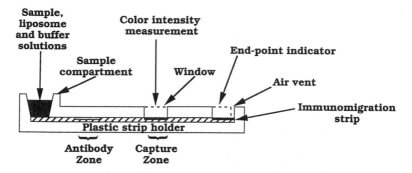

Figure 9.   Diagram of Proposed Strip Cassette

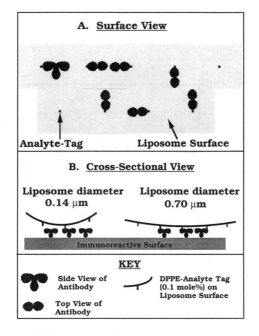

Figure 10.   Diagram of Liposome Surface and Analyte/Antibody Interaction (Drawn to Scale)

liposome surface, then the spacing is such that decreasing the spacing between analyte molecules (i.e. increasing the surface concentration) will presumably increase the number of binding interactions between the antibody and analyte tag on each individual liposome until the liposomes form a monolayer and no more can be packed on the surface. The number of alachlor-tag to antibody interactions per liposome will logically affect the number of free molecules required to "out compete" a liposome, and thus should affect the sensitivity of the assay. However, this diagram does not take into account the curvature and deformability of the liposome, or the mobility of DPPE-alachlor in the bilayer.

If we consider the immunoreactive surface, in either the FILIA or strip assay, to which the antibody is bound as a flat surface, then scale drawings of cross-sections can be made, and these are shown in Figure 10 B. If one assumes that there is no deformation of the liposomes (certainly not the case), then as the diameter decreases, the chances of multiple interactions of a single liposome with several antibodies decreases.

These spatial effects, along with the fact that the system, like many flow-injection systems, is not at equilibrium, must be taken into consideration when attempting to modify and develop both the immunomigration and FILIA assays.

Acknowledgments

This work was supported in part by the Cornell Biotechnology Program, the National Institute of Environmental Health Sciences, National Institutes of Health, DHHS, under the Superfund Research and Education grant no. 1 P42 ES05950-01, and by the CSRS/USDA Water Quality Special Grants Program #93-34214-8846. Any opinions, findings, conclusions or recommendations expressed in this publication are those of the authors and do not necessarily reflect the views of the Cornell Biotechnology Program, the U.S. Dept. of Health and Human Services or the U.S. Dept. of Agriculture.

References

1   Kaufman, B. M.; Clower, M. Immunoassay of pesticides. *J. Assoc. Off. Anal. Chem.* **1991,** *74,* 239-247.
2   Immunochemical Methods for Environmental Analysis. (1990) Van Emon, J. M. and Mumma, R. O., eds, ACS, Washington, DC.
3   Feng, P. C., Wratten, S. J., Horton, S. R., Sharp, C. R. and Logusch, E. W. (1990) Development of an Enzyme-Linked Immunosorbent Assay for Alachlor and Its Application to the Analysis of Environmental Water Samples. *J. Agric. Food Chem.* *38*, 159-163.
4   Lawruk, T. S., Hottenstein, C. S., Herzog, D. P and Rubio, F. M. (1992) Quantification of Alachlor in well water by a novel magnetic-particle based ELISA. *JBull. Environ. Contm. Toxicol 48,* 643-650.
5   Locascio-Brown, L., Plant, A. L., Horvath, V. and Durst, R. A. (1990) Liposome Flow Injection Immunoassay: Implications for Sensitivity, Dynamic Range, and Antibody Regeneration. *Anal. Chem. 62*, 2587-2593.
6   Durst, R. A., Locascio-Brown, L. and Plant, A. L. (1990) Automated Liposome-Based Flow Injection Immunoassay System. In *"Flow Injection Analysis Based on Enzymes or Antibodies"*, R.D. Schmid, ed., GBF Monograph Series, **14**, VCH Publishers, Weinheim, FRG, 181-190.
7   Yap, W. T., Locascio-Brown, L., Plant, A. L., Choquette, S. J., Horvath, V., and Durst, R. A. (1991) Liposome Flow Injection Immunoassay: Model Calculations of Competitive Immunoreactions Involving Univalent and Multivalent Ligands, *Anal. Chem. 63*, 2007-2011.

8   Reeves S. G., Roberts, M. A., Siebert, S. T. A. and Durst, R. A. (1993)
     Investigation of a Novel Microtiter Plate Support Material and Scanner
     Quantitation of Proteins, Phospholipids and Immunoassays. *Anal Letters, 26*,
     1461-1476.
9   Siebert, S. T. A., Reeves, S. G. and Durst, R. A. (1993) Liposome
     Immunomigration Field Assay Device for Alachlor Determination. *Anal. Chim.
     Acta, 282*, 297-305.
10   Szoka, S., Olsen, F., Heath, T., Vail, W., Mayhew, E. and Papahadjopoulos,
     D. (1980) Preparation of Unilamellar Liposomes of Intermediate Size (0.1-0.2
     μm) by a Combination of Reverse Phase Evaporation and Extrusion Through
     Polycarbonate Membranes. *Biochim. Biophys. Acta, 601*, 559-571.
11   O'Connell, J. P., Campbell, R. L., Fleming, B. M., Mercolino, T. J., Johnson,
     M. D. and McLaurin, D. A. (1985) A Highly Sensitive Immunoassay System
     Involving Antibody Coated Tubes and Liposome Entrapped Dye. *Clin. Chem.
     31*, 1424-1426.
12   Israelachvili, J. N. and Mitchell, D. J. (1975) A Model for the Packing of Lipids
     in Bilayer Membranes. *Biochim. Biophys. Acta, 389*, 13-19.
13   Reeves, S. G., Rule, G. S., Roberts, M. A., Edwards, A. J., Siebert, S. T. A.
     and Durst, R. A. (in press) Flow-Injection Liposome ImmunoAnalysis (FILIA)
     for Alachlor. *Talanta.*
14   Durst, R. A., Siebert, S. T. A. and Reeves, S. G. (1993) Immunosensor for
     Extra-Lab Measurements based on Liposome Amplification and Capillary
     Migration. *Biosen. Bioelectron. 8*, xiii-xv.

RECEIVED November 17, 1994

Chapter 16

Polarization Fluoroimmunoassay for Rapid, Specific Detection of Pesticides

Sergei A. Eremin

Division of Chemical Enzymology, Department of Chemistry,
M. V. Lomonosov State University, Moscow 119899, Russia

Polarization fluoroimmunoassay (PFIA) is a simple, inexpensive, easily automated screening method for pesticide residues in large numbers of environmental samples. PFIA measures the increased polarization of fluorescence when a fluorophore-labeled hapten (tracer) is bound by specific antibody, and the decreased signal when free analyte competes with the tracer for binding. No separation of free and bound analyte is required. PFIAs for 2,4-dichlorophenoxyacetic acid (2,4-D), 2,4,5-trichlorophenoxyacetic acid (2.4,5-T), simazine, and atrazine were automated on an Abbott TDx Analyzer. Ten water samples of 0.05 mL can be analyzed in 7 minutes, with detection limits of 100 ng/mL for 2,4-D and 5 ng/mL for simazine, and coefficients of variation < 5%.

The monitoring of pesticide residues in ground water, surface water, soil, and other environmental samples has gained increasing importance worldwide. Established procedures for detecting pesticides include high-pressure liquid chromatography (HPLC) and gas chromatography (GC). These require extraction of the samples to concentrate the residues and remove interfering matrix materials. Over the last ten years immunochemical methods such as enzyme-linked immunosorbent assay (ELISA) have been increasingly used for detection of pesticides (1). ELISA and related methods have several advantages and facilitate analysis of large numbers of samples. ELISAs are much less expensive to run and their detection limits can be as good or better than those of instrumental methods. However, ELISAs are difficult to automate and standardize. They generally require several washing steps, a step in which the free and bound analyte is separated, and the approach to equilibrium binding may be relatively slow (30 to 120 min). From our point of view, simplifying the assay and minimizing the analysis time per sample are the primary goals in developing screening methods for large numbers of samples. Polarization fluoroimmunoassay (PFIA) is a "homogeneous" immuno-chemical method, i.e., it does not require washing or separation of the free and bound analyte. The principle and some of the critical factors in design and automation of PFIA for pesticides are summarized below. Details of the theory and application of PFIA may be found in recent reviews (2,3).

PFIA originated from experiments in the early 1960s by Dandliker and his colleagues, in which antigen-antibody reactions were monitored by changes in fluorescence polarization (4). Subsequently, PFIA became widely used in clinical chemistry because of its simplicity and precision. There are now clinically accepted PFIAs for monitoring the administration and effects of about 100 therapeutic drugs (5). The first application of PFIA to pesticide detection was reported by Colbert and Coxon, who developed a PFIA for paraquat in serum samples and adapted it to run on

0097–6156/95/0586–0223$12.00/0

the Abbott TDx Analyzer, an instrument that was specially designed to automate PFIA (6). The TDx Analyzer can be set up to perform sample pre-treatment, add fluorescein-labeled hapten and antiserum, measure the signals, and calculate and report the results. A major advantage of PFIA on the Abbott TDx Analyzer is that it is generally not necessary to run standards every time an assay is repeated over periods of one week to a few months. This substantially reduces costs and workload (3). The reproducibility is due primarily to the stability of the small molecular weight fluorophore-labeled hapten tracers in solution, and the way that the response is measured. Fluorescence polarization units are a ratio of intensities of different polarized components of the fluorescence, so they are relatively independent of time and nonspecific fluorescence caused by the sample matrix. Subsequently we published preliminary results of a PFIA for 2,4-D and other pesticides (7,8).

Principles of PFIA. PFIA is a competition method based on detection of the difference of fluorescence polarization between a small fluorescent-labeled antigen and its immuno-complex with specific antibody (2). PFIA depends upon the difference in the signal given by a relatively small fluorescein-labeled hapten when it is in the free form as compared with the much higher polarization value when it has been bound by its specific antibody. Eliminating the need to separate the free and bound tracer is a considerable advantage, as it simplifies the assay, often improves its precision, and makes it much easier to fully automate. The polarization of fluorescence (P) is determined by exciting the mixture of antibody, sample, and tracer with vertically polarized light and measuring the intensity of both the vertically (I_v) and horizontally (I_h) polarized components of the emitted fluorescence. The P value is defined as the ratio of difference and sum of these components:

$$P = (I_v - I_h) / (I_v + I_h)$$

It is convenient to use "milliunits" of p (mP values) such that mP = 1,000(P).

Several factors influence the p values. The most significant of these is the size of the fluorescein-labeled tracer. Because it is present in limiting amounts in the PFIA, much more of the tracer is bound than is free. Other important variables are the length and type of bridge between the fluorescein and the analyte moiety of the tracer, and the temperature and viscosity of the reaction mixture. As with any analytical method, PFIA has disadvantages. These include poorer detection limits than obtained with the best ELISAs, and the cost and limited availability of instrumentation to detect polarized fluorescence. In addition, PFIAs do not work for high molecular weight (> 1000 dalton) analytes. PFIAs are susceptible to interference from substances in some sample matrices such as plant extracts, or serum. However, ground water and surface water samples are generally free of interfering compounds.

PFIAs are particularly suitable for routine pesticide contamination tests where the most sensitive limit of detection is not needed, or where it is possible to extract and concentrate the analyte prior to assay. In Russia 2,4-dichlorophenoxyacetic acid (2,4-D) and simazine are two of the most widely used herbicides. The regulatory action levels in surface water are 100 ng/mL for 2,4-D and 2.4 ng/mL for simazine. For preliminary screening of surface water it is only necessary to semi-quantitatively detect these amounts, but the assays must be highly reliable. To monitor these two herbicides in water samples throughout the agricultural areas of Russia we would need at least one million assays. Even semi-quantitative screening at this level of sensitivity would be of great value in detecting dioxin, which is a trace contaminant of 2,4,5-trichlorophenoxyacetic acid (2,4,5-T). In Russia, many water samples positive for dioxin are also contaminated with 2,4,5-T.

Here we briefly report our most recent results adapting pesticide PFIA for 2,4-D, 2,4,5-T, simazine, atrazine and related compounds so that they could be run on the

Abbott TDx Analyzer. Other polarization fluorimeters such as the Perkin-Elmer LS-50, Merck VITALAB, and the Roche COBAS FARA II can be adapted to run PFIA. In the process of developing our PFIAs, we studied how the structure of the labeled antigen affects sensitivity. The sensitivity was greatest using the shortest chemical "bridge" between the antigen and the fluorescent label. Labeled antigens that were structurally homologous or heterologous to the primary target analytes were investigated. Our results indicated that competitive-binding PFIAs are more sensitive when structurally heterologous tracers are used.

Materials and Methods

2,4-D. Antiserum was raised in a rabbit immunized with 2,4-D conjugated via the carboxyl group to bovine serum albumin (BSA) (*9*). The mouse monoclonal antibody designated E2/G2 to 2,4-D was derived using the same immunogen at the Veterinary Research Institute (Brno, Czech Republic), and provided as unpurified ascites fluid (*10*). Fluorescent tracers were prepared by synthesizing an N-hydroxysuccinimide ester of 2,4-D and conjugating it to fluoresceinthiocarbamyl derivatives of 1,2-diaminoethane (n=2), 1,4-diaminobutane (n=4), and 1,6-diaminohexane (n=6) (Figure 1) (*9*). These derivatives were synthesized by a direct reaction of $NH_2(CH_2)_nNH_2 \cdot 2HCl$ with fluorescein isothiocyanate, according to established procedures (*11*). The tracers were purified by thin-layer chromatography and their concentrations were estimated spectrophotometrically using the published molar extinction coefficient for fluorescein (*11*).

2,4,5-T. Rabbit polyclonal antiserum was raised using 2,4,5-T conjugated via its carboxyl group to BSA or keyhole limpet hemocyanin (KLH) (*12*). The tracer was prepared from 2,4,5-T and a fluoresceinthiocarbamyl derivative of 1,2-diaminoethane (n=2) by the same method used for the 2,4-D tracer.

Simazine. Reaction of 2,4-dichloro-6-(ethylamino)-1,3,5-triazine with 6-aminohexanoic acid yielded simazine derivatives which were conjugated to KLH using the carbodiimide method (*13*). The resulting immunogen was used to raise antiserum in rabbits (Figure 4). The tracer was prepared from the same simazine derivatives and fluoresceinthiocarbamyl derivatives of 1,2-diaminoethane (n=2) (Figure 4) (*14*).

Atrazine. A derivative of atrazine with thiopropionic acid substituted in place of the chlorine atom (*13*) was conjugated to KLH. This conjugate was used to raise antiserum in sheep (Figure 5). A tracer with the homologous structure was synthesized from the same triazine hapten and fluoresceinthiocarbamyl derivatives of 1,2-diaminoethane (n=2). Tracers with heterologous structure were prepared from 2,4-dichloro-6-(isopropyl)-1,3,5-triazine by condensation with fluoresceinthiocarbamyl derivatives of $NH_2(CH_2)_nNH_2$ (n=2,4,6) (Figure 5) (*15*).

PFIA Analyzer. A TDx Analyzer (Abbott Laboratories, USA) was used to measure fluorescence polarization in milli-units (mP). To perform the measurements, up to ten TDx glass cuvettes were loaded into the special "Photo Check" carousel. Measurement and calculations were performed automatically and printed by the instrument. The total time for measurement of 10 samples was about 7 min.

PFIA Procedures. Sodium borate buffer (0.05M, pH 8.6) was used as the diluent in all experiments. To determine antibody concentrations usable for competition PFIA, 0.5 mL of various dilutions of antiserum was mixed with 0.5 mL of tracer at 10

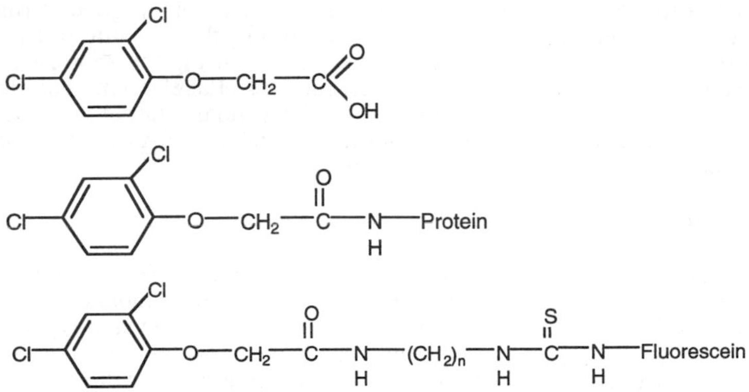

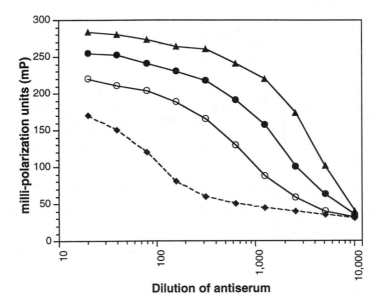

Figure 1. Structures of 2,4-D (top), the hapten conjugate used for antibody production (center), and fluorescent tracers of differing bridge length (n=2,4, or 6; bottom).

Figure 2. Rabbit anti-2,4-D serum dilution curves obtained using tracers of differing bridge length. Dilutions of the antisera were incubated with each tracer and fluorescence polarization was measured. (O) 2 carbon; (●) 4 carbon; (▲) 6 carbon. The dotted line (--◆--) shows nonspecific binding with normal rabbit serum (similar for all tracers).

nmol/L in TDx glass cuvettes at room temperature, and fluorescence polarization was measured immediately. The competition PFIA was performed in TDx cuvettes by sequentially adding 50 μL of standard or sample, 0.5 mL of tracer solution (10 nmol/L), and 0.5 mL of antibody at a dilution that gave about 70% of maximum binding of tracer (determined from the antibody dilution curve). After the measurement of fluorescence polarization in mP as described above, the standard curves were plotted as mP vs. logarithm of the concentration of analyte.

Results

The PFIA can be used to rapidly measure the relative binding of different antibodies to a particular tracer, and to compare tracers of different structure using a particular antibody or serum. Figure 2 shows the antibody dilution curves for polyclonal anti-2,4-D serum with tracers that have different bridge lengths between the 2,4-D and fluorescein, as shown in Figure 1. This antiserum had a higher titer for the tracer with the longest bridge, but displacement of 2,4-D was significantly greater with the tracer that had the shortest bridge (Figure 3). The most sensitive assay, in terms of the minimal detectable concentration at the 95% confidence level, was 0.1 μg/mL (5 ng of 2.4-D in the 50 μL sample) using tracer with the shortest bridge. The detection limit using monoclonal antibodies (MAbs) with the same tracer was also 0.1 μg/mL, but the specificity with the MAbs was greater (Table I).

A very similar PFIA was developed for 2,4,5-T. Polyclonal antisera to 2,4,5-T were raised in three rabbits using KLH conjugates as immunogen, and in three other rabbits using BSA conjugates of the same hapten. The rabbits immunized with the KLH conjugates developed higher titer sera against 2,4,5-T than those immunized with the BSA conjugate. The specificity of the 2,4,5-T PFIA, using the best tracer and polyclonal anti-2,4,5-T-KLH serum, was comparable to that of the monoclonal 2,4-D PFIA (Table I). The detection limit was 0.1 μg/mL, equivalent to 5 ng of 2,4,5-T in the sample of 50 μL. This is comparable to the detection limit for GC of 2,4,5-T, making PFIA potentially suitable for screening water samples.

Table II shows 50% inhibition of tracer binding and percent cross-reactivity for PFIA of simazine using polyclonal antiserum and tracer with a structure homologous to the immunogen (Figure 4). The sensitivity for simazine was greater than for 2,4-D (5 ng/mL in a 50 μL sample or 250 pg per test), making the simazine PFIA one of the best we have tested. However, this serum cross-reacted about equally with simazine and atrazine (Table II).

To develop a more specific PFIA for atrazine we designed an immunogen with a thio-group analog of s-triazines. The polyclonal antiserum to atrazine bound very well (titer 1/2000) with an atrazine tracer structurally homologous to the immunogen (Figure 5). However, use of this tracer resulted in a poor competitive PFIA (results not shown). Accordingly we synthesized and tested heterologous tracers (Figure 5) for better competition with atrazine using this polyclonal antiserum. A sensitive competition PFIA for atrazine was developed using the heterologous tracers (Table III). As in the assay for 2,4-D, the most sensitive competition PFIA for atrazine - a detection limit of 10 ng/mL in a 50 μL sample - was obtained using the tracer with the shortest bridge (Figure 6). Herbicides structurally related to atrazine were tested and the cross-reactivities are given in Table III. This PFIA was about 100-fold more sensitive to atrazine than to simazine. This means that our polyclonal antibodies recognize the isopropyl group in s-triazines. The antisera actually detected propazine, ametryne, and prometryne better than atrazine in this assay. Ametryne and prometryne, which are structurally similar to the immunogen, could be detected at 5 ng/mL with this assay. However, propazine, ametryne, and prometryne are not commonly used in Russia.

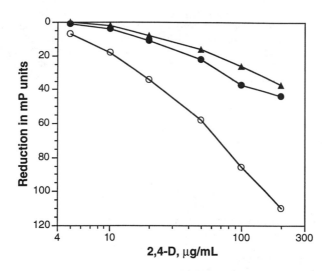

Figure 3. Standard curves for PFIA of 2,4-D using tracers with differing bridge length: (O) 2 carbon; (●) 4 carbon; (▲) 6 carbon.

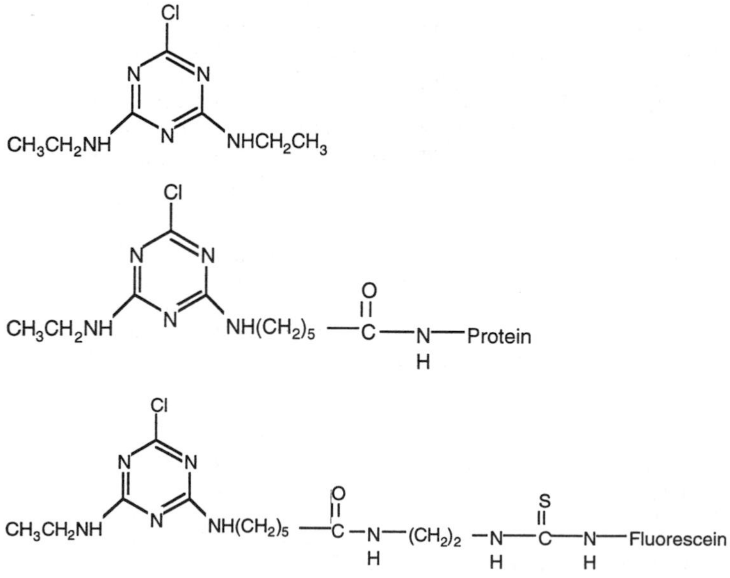

Figure 4. Structures of simazine (top), the immunizing conjugate (center), and the tracer for PFIA (bottom).

Table I. Relative cross-reactivity of some compounds structurally related to 2,4-D in PFIAs with polyclonal or monoclonal antibodies, and in PFIA of 2,4,5-T with polyclonal antibodies.

No.	R2	R3	R4	R5	R6	Z	Cross-reactivity (%) [a]		
							2,4-D (poly)	2,4-D (mono)	2,4,5-T (poly)
1	Cl	H	Cl	H	H	OCH_2COOH	100	100	5.6
2	Cl	H	Cl	H	H	$OCH(CH_3)COOH$	2.5	- [b]	5.4
3	Cl	H	Cl	H	H	$O(CH_2)_3COOH$	41	3.6	15
4	CH_3	H	Cl	H	H	OCH_2COOH	7.2	2.1	9.3
5	CH_3	H	Cl	H	H	$OCH(CH_3)COOH$	1.0	-	5.0
6	CH_3	H	Cl	H	H	$O(CH_2)_3COOH$	12	-	21
7	CH_3	H	Cl	H	Cl	OCH_2COOH	2.6	-	4.0
8	CH_3	H	H	H	H	OCH_2COOH	1.0	-	4.0
9	CH_3	H	H	H	Cl	OCH_2COOH	1.0	-	3.6
10	Cl	H	H	H	H	OCH_2COOH	7.7	2.8	2.4
11	Cl	Cl	H	H	H	OCH_2COOH	34	-	-
12	H	Cl	Cl	H	H	OCH_2COOH	8	-	-
13	Cl	H	Cl	Cl	H	OCH_2COOH	59	5.0	100
14	Cl	H	Cl	H	H	CH_2COOH	0.1	-	-

[a] Percent cross-reactivity (%CR) is defined as the ratio of mP units at 10 ppm (10 μg/mL) of 2,4-D to mP units for the indicated analyte.

[b] - Not tested

Compounds tested:

1. 2,4-Dichlorophenoxyacetic acid
2. 2-(2,4-Dichlorophenoxy)propionic acid
3. 4-(2,4-Dichlorophenoxy)butyric acid
4. 4-Chloro-*o*-tolyloxyacetic acid
5. 2-(4-Chloro-*o*-tolyloxy)propionic acid
6. 4-(4-Chloro-*o*-tolyloxy)butyric acid
7. 4,6-Dichloro-*o*-tolyloxyacetic acid

8. *o*-Tolyloxyacetic acid
9. 6-Chloro-*o*-tolyloxyacetic acid
10. *o*-Chlorophenoxyacetic acid
11. 2,3-Dichlorophenoxyacetic acid
12. 3,4-Dichlorophenoxyacetic acid
13. 2,4,5-Trichlorophenoxyacetic acid
14. 2,4-Dichlorophenylacetic acid

Table II. Sensitivity and percent cross-reactivity for PFIA of simazine.

R1	R2	R3	s-triazine	I_{50}, ng/mL [a]	%CR [b]
Et	Et	Cl	Simazine	50	100
Et	iPr	Cl	Atrazine	50	100
iPr	iPr	Cl	Propazine	160	31
Et	tBu	Cl	Terbutylazine	200	25
Et	iPr	SMe	Ametryne	>1000	<1
iPr	iPr	SMe	Prometryne	>1000	<1
Et	tBu	SMe	Terbutryne	>1000	<1

[a] I_{50} is the concentration of analyte that inhibits the maximal response by 50%..
[b] Percent cross-reactivity (%CR) is defined as the ratio of I_{50} value for simazine to that of the indicated analyte.

Table III. Sensitivity and percent cross-reactivity for PFIA of atrazine.

R1	R2	R3	s-triazine	I_{50}, ng/mL [a]	%CR [b]
Et	Et	Cl	Simazine	>1000	<1
Et	iPr	Cl	Atrazine	450	100
iPr	iPr	Cl	Propazine	300	150
Et	tBu	Cl	Terbutylazine	>1000	<1
Et	iPr	SMe	Ametryne	100	450
iPr	iPr	SMe	Prometryne	80	560
Et	tBu	SMe	Terbutryne	1800	25

[a] I_{50} is the concentration of analyte that inhibits the maximal response by 50%..
[b] Percent cross-reactivity (%CR) is defined as the ratio of I_{50} value for atrazine to that of the indicated analyte.

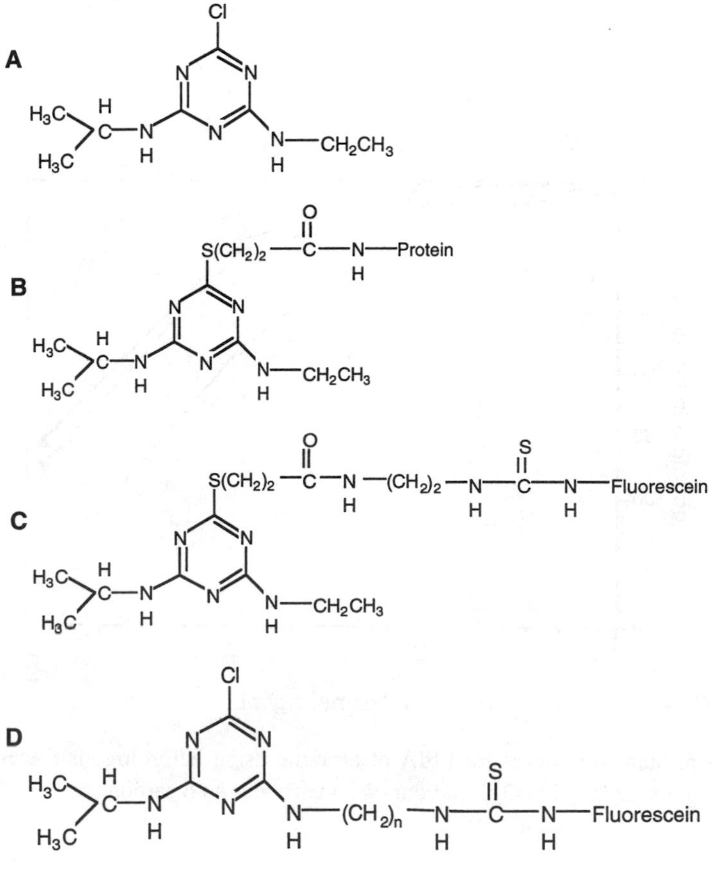

Figure 5. Structures of atrazine (A), the immunizing conjugate (B), the structurally homologous tracer (C), and the heterologous tracers of differing bridge length (n=2, 4, or 6) (D).

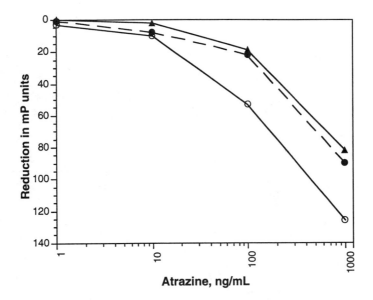

Figure 6. Standard curves for PFIA of atrazine using heterologous tracers with differing bridge length: (O) 2 carbon; (●) 4 carbon; (▲) 6 carbon.

All of the PFIAs we developed has very good accuracy, precision, and reproducibility. For example, precision of the simazine assay was assessed by measuring ten replicates each of three water samples spiked to 10, 100, and 1000 ng/mL of simazine in one assay, giving within-assay coefficients of variation (CV) of 5.1, 1.5 and 0.7%, respectively. Measurement of the same samples on 3 different days gave between-assay CVs of 6.0, 2.1 and 1.1%, respectively. In general, within-assay CVs for PFIA were not more than 5%, and between-assay CVs were less than 7% . In the PFIA for 2,4-D, the analytical recovery was assessed by adding 2,4-D to 1.0, 8.0, and 120 µg/mL in tap water. Ten replicates of the spiked samples were analyzed. The mean measured concentrations were 0.98, 8.45, and 129.4 µg/mL, and the within-assay CVs were 1.66, 0.81, and 0.41%, respectively. Recoveries were: 98, 106, and 108%, respectively. In the PFIA for 2,4,5-T, the linearity of the estimated analyte concentration with dilution was assessed using a water sample which was spiked with 2,4,5-T to 10 µg/mL, and then diluted two-, four-, and eight-fold. For ten replicates of each dilution, the mean measured concentrations were 5.07, 2.39, and 1.24 µg/mL, with within-assay CVs of 3%, 4%, and 3%, respectively. The recoveries was 101.4%, 95.6%, and 99.2%, respectively (mean recovery was 98.7%). The analytical recovery of pesticide added to samples or serial diluted usually ranged from 80% to 120%. Water samples spiked to 2 µg/mL with 2,4,5-T were also spiked with amounts of the structurally related herbicides 2,4-D and MCPA (4-chloro-2-methylphenoxyacetic acid) in eight-fold to ten-fold excess of the 2,4,5-T. The 2,4-D and MCPA had no effect (at the 95% confidence level) on the concentration of 2,4,5-T estimated by PFIA.

The stability of the fluorescent tracers may be the most important factor for the reproducibility of PFIA. We have kept a fluorescein-2,4,5-T tracer in methanolic solution at 4 °C for more than three years with no significant change in its properties. Working dilutions of 10 nmol/mL in borate buffer can be used for at least a week with no effect on the standard curves. The tracers for 2,4-D and triazines are similarly stable.

We have only recently begun to analyze herbicides in actual environmental water samples by PFIA . In our initial attempt we analyzed 14 water samples from rivers in the Moscow area for 2,4-D and compared the PFIA results with those obtained by GC. By PFIA, two of these samples had concentrations of 0.1 and 0.2 µg/mL of 2,4-D, near the detection limit. These samples were also positive by GC analysis.

Discussion

PFIA has many of the most desirable characteristics for a method of routine screening of large numbers of environmental samples for pesticide residues. It uses small amounts of reagent, and it is fast, simple, highly reproducible, and easy to automate. Synthesis of tracers is relatively easy, and the tracers are chemically stable. A major drawback of the present technology is that the detection limits of some PFIAs are not as low as those obtained in ELISA and similar immunoassays that use amplification in the detection stage. However, the sensitivity with some assays is sufficient for some environmental and commodity screening applications, especially when it is possible to extract and concentrate analyte from samples. The sensitivity may be further improved as new polarization fluorimeter or/and some fluorescent labels with greater extinction coefficients become available.

To achieve optimal sensitivity for PFIA, as for ELISA (*16*) and any immunoassay method, one must select the best combination of antibodies and tracer. Our results with the assays for 2,4-D, 2,4,5-T, and simazine are consistent with the evidence cited by Ekins (*17*) that the most sensitive assay is obtained when the antibody has similar affinity for the tracer and the free analyte. However, we could not get competitive binding of analyte in our PFIA for atrazine unless we used a tracer that was structurally different from the immunizing hapten. We have not determined the affinity of our atrazine antiserum for the tracer, but we speculate that it may be less than the affinity for atrazine. Use of the heterologous hapten may also account for the greater selectivity of the atrazine assay.

The sensitivity of all four of the PFIAs we tested was greatest using fluorescent tracers with the shortest chemical bridge between the antigen and the fluorescent label. This is consistent with similar observations made with a PFIA for methamphetamine (18). We do not yet know the basis of this effect. The antigen-binding pockets of antibodies are rich in aromatic amino acids, so there may be a possibility of fluorescence energy transfer when the fluorescent label is suitably close.

We are presently testing the potential usefulness of the automated PFIAs as the initial screening method for regulatory monitoring of the pesticides described in this chapter.

Acknowledgments

The excellent technical work of Jean V. Samsonova and Isabella M. Polak is greatly appreciated.

Literature Cited

1. Sherry, J.P. In *Critical Reviews in Analytical Chemistry.*; CRC Press, Boca Raton, FL, 1992, Vol. 23, 217-300.
2. Gutierrez, M.C.; Gomez-Hens, A.; Perez-Bendito, D. *Talanta* **1989,** *36,* 1187-1201.
3. Williams, A.T.R.; Smith, D.S. In Methods of Immunological Analysis; Albert, W.H.W.; Staines, N.A. Eds.; VCH, Publishers: Weinheim, Germany, 1991, Vol.1. pp. 466-475.
4. Dandliker, W.B.; Schapiro, H.C.; Meduski J.W.; Alonso, R.; Feigen, G.A.; Hamrick, J.R. *Immunochemistry* **1964,** *1,* 165-191.
5. Eremin, S.A. *Mendeleev Chemistry Journal*, Allerton Press, Inc., ISSN 0025-925x, **1989,** 67-75.
6. Colbert, D.L.; Coxon, R.E. Clin. Chem.. **1988,** *34,* 1948.
7. Eremin, S.A.; Landon, J.; Smith, D.S.; Jackman, R. In *Food Safety and Quality Assurance: Applications of Immunoassay Sstems;* Morgan, M.R.A.; Smith, C.J.; Williams, P.A. Eds.; Elsevier Publishers, London 1992, pp. 119-126.
8. Eremin, S.A.; Moreva, I. Y.; Dzantiev, B.B.; Egorov, A.M.; Franek, M. *Voprosy meditsinskoi khimii* (in Russian). **1991,** *37,* 93-95.
9. Eremin, S.A.; Lunskaya, I.M.; Egorov, A.M. *Bioorganicheskaya khimiya* (in Russian). **1993,** *19,* 836-843.
10. Lunskaya, I.M.; Eremin, S.A.; Egorov, A.M.; Kolar, V.; Franek, M. *Agrokhimiya* (in Russian) **1993,** 2, 113-118.
11. Pourfarzaneh, M.; White, G.W.; Landon, J.; Smith, D.S. *Clin. Chem..* **1980,** *26,* 730-733.
12. Eremin, S.A.; Melnichenko, O.A.; Tumanov, A.A.; Sorokina, N.V.; Molokova E.V.; Egorov, A.M. *Voprosy meditsinskoi khimii* (in Russian), **1994,** *40,* 57-60.
13. Goodrow M.H.; Harrison R.O.; Hammock B.D. *J. Agric. Food Chem..* **1990,** *38,* 990-996.
14. Samsonova, J.V.; Egorov, A.M.; Eremin, S.A. *Agrokhimiya* (in Russian), **1994,** *1,* 97-102.
15. Samsonova, J.V.; Eremin, S.A.; Egorov, A.M. *Voprosy meditsinskoi khimii* (in Russian), **1994,** *40,* 53-56.
16. De Boever, J.G.; Kohen, F.; Bosmans E. *Anal. Chim. Acta* **1993,** *275,* 81-87.
17. Ekins, R.P. In *Alternative Immunoassays,;* Collins, W.P., Ed.; John Wiley & Sons, Ltd., Chichester, U.K., 1985, pp. 219-237.
18. Colbert, D.L.; Eremin, S.A.; Landon J. *J. Immunol. Meth..* **1991,** *140,* 227-233.

RECEIVED January 9, 1995

Chapter 17

Immunoaffinity Chromatography Applications in Pesticide Metabolism and Residue Analysis

Rosie B. Wong[1], Joseph L. Pont[1], David H. Johnson[1], Jack Zulalian[1], Tina Chin[2], and Alexander E. Karu[2]

[1]American Cyanamid Agricultural Research Division, P.O. Box 400, Princeton, NJ 08543
[2]Department of Plant Pathology, University of California, Berkeley, CA 94720

The imidazolinone compounds are a new class of herbicides which are environmentally safe and effective at very low application rates. Because of the shared imidazolinone ring structure amongst several herbicides, generic antibodies were prepared which recognized all the compounds in this class (1). These antibodies are useful not only for developing immunoassays for the detection of low levels of residues in soil or in plant extracts, they are also useful for preparing antibody affinity columns used for sample clean-up in metabolism and residue studies. Imidazolinone-compound containing extracts from wheat plant, corn grain, corn fodder, goat urine, and goat kidney tissue were tested on the antibody columns. It was found that the analytes were bound to the antibody column after some simple sample preparation procedures. The bound analytes were easily eluted with a solution of 30% methanol in water. It was further demonstrated that metabolites of imidazolinone compounds which retain the imidazolinone ring structure can be purified through a simple antibody affinity chromatography procedure and be identified thereafter by mass spectrometry. A comparison of mono-and poly-clonal antibody columns indicates that the monoclonal antibody with its uniform affinity and more restricted binding epitope is better for antibody affinity chromatographies. A lower amount of matrix is bound to the monoclonal antibody column, thus producing a cleaner sample than a polyclonal antibody column.

The specificity of antibodies has been utilized extensively in immunoassays where analytes can be quantified in relatively crude matrices. This property can be further harnessed in affinity chromatography. Preparation of monoclonal antibodies allows an unlimited supply of mono-specific antibodies for large scale production of affinity

0097–6156/95/0586–0235$12.00/0
© 1995 American Chemical Society

columns. The first application of monoclonal antibody affinity chromatography was reported by Secher and Burke in 1980 (2) on the purification of human leukocyte interferon. Numerous articles on immunoaffinity methods for hormones, vaccines, and growth factors (2-12), as well as reviews (13-17) have since been published. Most of these reports dealt with large proteins. Applications of this technology for small molecules, on the other hand, has not enjoy as much popularity. A survey of literature reported by Farjam (18) showed less than thirty reports involved molecules under 1,000 molecular weight. There were affinity methods for steroids such as testosterone, nortestosterone, cortisol (19-22) and carcinogenic compounds such as afflatoxins and ochratoxin (23-26). Among the techniques reported, immobilized antibodies have been used for on-line sample pre-treatments in liquid and gas chromatographies (10, 27), liquid chromatography/ mass spectrometry (21, 29) and manual affinity chromatography for residue sample clean-up. (30).

In situations where an antibody recognizes a certain structural feature in an analyte which is maintained during the metabolic process, this antibody can conceivably be used for metabolite purification and identification. Successful demonstrations of this approach was reported by Groopman and Donahue (24), for various afflatoxins and their protein and DNA adducts, and Goto et al., (31) in the isolation of methyl group containing pharmaceuticals. We demonstrate that the antibodies against the class of imidazolinone herbicides (Figure 1) are useful in affinity chromatography not only for the entire class of imidazolinone compounds but are also suitable for metabolite binding and purification. The metabolites purified from such antibody affinity columns are sufficiently clean that no further purification was required before compound identification by mass spectrometry. We also show that the level of analyte in the sample as well as the nature of the matrix governed the extent of sample pretreatment required before antibody affinity chromatography. To further investigate the matrix effect, we compared monoclonal and polyclonal antibody affinity columns in their ability to purify a wheat plant extract which contained several imidazolinone ring containing metabolites.

This is the first demonstration of an antibody affinity chromatography application for an agrochemical and its metabolites. The application of this technology in residue analysis and environmental monitoring will be discussed.

Methods.

Antibody Column Preparation. Polyclonal antibody was produced in sheep as described before (1), details of the monoclonal antibody production will be presented elsewhere (Chin, T, Karu, A. Pont, J. and Wong R. B.).

The monoclonal antibody was purified by binding the mouse IgG to a HiTrap 5 mL protein G column (Pharmacia, Piscataway, NJ.) according to the manufacturer's recommendation. Chemical cross-linking of the antibody to the protein G with dimethylpimelimidate was carried out according to the method of Schneider et al (32). Briefly, ascites solution was diluted 10 fold with 20mM sodium phosphate buffer pH 7.0 (PB) and applied onto a HiTrap G column via a peristaltic pump at flow rate of 0.3 mL/min. After the entire sample had been applied, the column was exhaustively washed with PB to remove non-specifically bound protein. The bound antibody was eluted with 0.1 M glycine buffer pH 2.5. The glycine eluant was neutralized with 1 M Tris-HCl buffer pH 9.0 and dialyzed against PB. The activity of the purified antibody was confirmed by enzyme immunoassay (1) and the protein concentration was determined by absorption at 280 nm.

To prepare an antibody column, a solution of purified antibody equivalent to 3.5 mg of protein was applied onto a HiTrap G 1 mL column. The column was equilibrated with 20mM ethanol amine pH 8.2 followed with 15 mL of 30 mM dimethyl pimelimidate pH 8.2 to cross-link the antibody. The column was washed

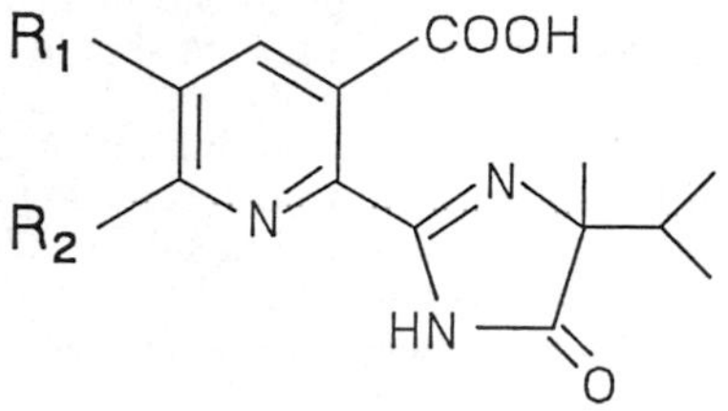

$R_1 = H$	$R_2 = H$	imazapyr
$R_1 = C_2H_5$	$R_2 = H$	imazethapyr
$R_1 = R$	$R_2 = H$	Compound A

Figure 1. Chemical Structures of Imidazolinone herbicides.

with PB and the imazethapyr binding capacity of the column was determined by using radiolabeled imazethapyr following the procedure described in the affinity chromatography section. The binding capacity of the column was calculated from the amount of radioactivity present in the 30% methanol fraction based on a known specificity activity.

Since the sheep IgG does not bind tightly to either Protein A or Protein G columns, an Avid AL column (Unisyn Technologies.) was used to purify sheep IgG. The binding and elution buffers and purification method were those provided by the manufacturer. Briefly, the serum was diluted with equal volume of binding buffer and loaded onto the column. After washing the column with binding buffer until no protein remained in the solution, the antibody was eluted with the elution buffer. Buffer exchange was performed on the purified antibody by passing through a desalting column equilibrated with 0.1M sodium phosphate buffer pH 7.0. Antibody was chemically linked to the hydrzide gel support through the carbohydrate moiety of the antibody. Periodate oxidation of the carbohydrate moiety on the purified antibody was achieved following the CarboLink kit (Pierce) recommendations. Briefly, 5 mg of meta-periodate was added to 2 ml of 1 mg/mL purified antibody for 30 minutes at room temperature with gentle mixing. (Meta periodate is an oxidizing agent thus should be kept away from combustible materials. One should exercise caution and prevent contact with skin.) The oxidized antibody was desalted using a desalting column and coupling buffer provided in the CarboLink kit. A 2 mL CarboLink gel was packed in a column and the antibody from the desalting column was added directly into the gel column. After gently mixing with side to side motion for 6 hours at room temperature, excess reactive sites on the CarboLink gel were blocked with the wash buffer provided in the kit and the column was equilibrated with PB for affinity chromatography.

Affinity Chromatography. All affinity chromatographies were carried out with the monoclonal antibody columns unless otherwise indicated. PB was the buffer used throughout. Samples were applied by using a peristaltic pump with the monoclonal columns or by gravity with the polyclonal columns. The flow rate was between 0.3 to 0.5 mL per minute. After sample application, the column was washed with PB to remove unbound material, the bound analyte was eluted with 30% methanol in water (V/V). The columns stored at 4^0 C in 30 % methanol retained their binding capacity over six months. Over twenty cycles have been carried out with the monoclonal columns without reduction in binding.

Sample Preparation. Because the nature of the matrices as well as the residue levels in the samples are different, the sample preparation methods before affinity chromatography also varied. Individual sample treatments are described as indicated below. All extraction procedures have been previously optimized using ^{14}C labeled incurred samples from metabolism studies.

Urine and Kidney Sample Preparation. [14]C ring labeled Compound A was fed to goats for seven consecutive days. Urine was collected within 22 hours after the last dosing before sacrifice. Kidney was collected for metabolite identification. Due to the high level of residue in the urine, the sample was applied directly to the affinity column after diluting with PB. Frozen ground kidney samples were extracted with 80% methanol: 20% water (V/V). After removing the methanol, the aqueous solution was partitioned with methylene chloride. The aqueous phase was further purified by passing through a 10,000 dalton centrifugal filter. The filtrate was diluted with PB and used for affinity chromatography. (Methylene chloride is listed both by the National Toxicology Program and the International Agency for Research on Cancer as a highly toxic chemical, appropriate safety precautions should be followed when handling and disposing this material.)

Corn Sample Preparation. A corn metabolism study using ^{14}C labeled imazapyr applied post emergence at four leaf growth stage was carried out in a field plot. Harvested fodder and grain were found to contain radioactivity. A 62 day fodder sample was extracted with aqueous methanol, concentrated to remove the organic solvent and adjusted to pH 1.7 for C-18 solid phase extraction. At the acidic pH of 1.7, the compound was uncharged and was bound to the C-18 cartridge. After eluting the radioactivity from the cartridge with 2% and 30% methanol, the combined methanol effluent was dried and redissolved in PB for affinity chromatography. Corn grain was extracted with hexane to remove oil followed by acid methanol extraction. The extract was concentrated and acidified to pH 1.7 before subjecting to C-18 solid phase extraction. A 30 % methanol solution removed all the bound radioactivity. This methanol effluent was dried and redissolved in PB for affinity chromatography.

Wheat Plant Extract Preparation. A wheat metabolism study was conducted using ring-labeled ^{14}C Compound A applied post-emergence to wheat at the 1-3 tiller stage in a field plot. Pulverized frozen green plant sample was extracted with 80% methanol: 20% water (v/v). After evaporation of the methanol, it was partitioned with methylene chloride. The aqueous phase, which contained all the radioactivity, was further processed through a 10,000 dalton centrifugal filter or a combination of 10,000 and 1,000 dalton filters. The resulting samples were diluted in PB and applied to the affinity column.

High Performance Liquid Chromatography (HPLC). HPLC for the wheat plant extract, kidney extract and urine samples were performed using a 1 X 25 cm C-8 column with an isocratic mobile phase of 35% acetonitrile: 0.48% phosphoric acid in water for 20 minutes at 2 mL/min. Absorbance was measured with a flow-through detector at 240 nm and fractions collected at 30 second intervals. Radioactivity was monitored by scintillation counting.
For the corn extracts, a 0.5 X 25 cm C-18 column with a gradient mobile phase was used. The program of mobile phase consisted of 5 minutes of 100% water pH 2.1, a gradient of acetonitrile from 0 to 50 % in 50 minutes followed by an isocratic 100% acetonitrile for 5 minutes. The flow rate was 1 mL/min and the absorbance was measured at 254 nm. Fractions were collected at 1 minute intervals and radioactivity monitored by scintillation counting.

Results.

Urine, Kidney, Corn Grain and Corn Fodder Samples. Comparison of UV and radioactivity profiles of crude and affinity chromatography purified samples from HPLC dramatically demonstrated the power of antibody in sample purification. Figures 2A, 2B 3A and 3B show the results of goat urine, goat kidney, corn grain, and corn fodder samples. The arrows represent the position where reference standards of the analytes eluted. The radioactivity peaks of all the affinity chromatography purified samples coincided with that of the reference standards by HPLC, providing tentative identifications of the metabolism products. These purified samples were later subjected to GC/MS and were confirmed as imazapyr from the corn material and Compound A from the goat kidney and urine samples.

Wheat Metabolites. A crude wheat metabolism sample extract was subjected to HPLC chromatography. As shown in Figure 4, the radioactivity profile contained the parent Compound A and possible metabolites Compounds B, C, and D. Since all the proposed metabolites contain the imidazolinone ring structure, it was postulated that the antibody column may bind the parent as well as the proposed metabolites. When the crude plant extract was subjected to affinity chromatography, none of the radioactivity, including parent Compound A was bound. This indicated a matrix effect in the sample. The centrifugal molecular sizing filtration purification schemes were developed to remove the matrix effect on the antibody columns.

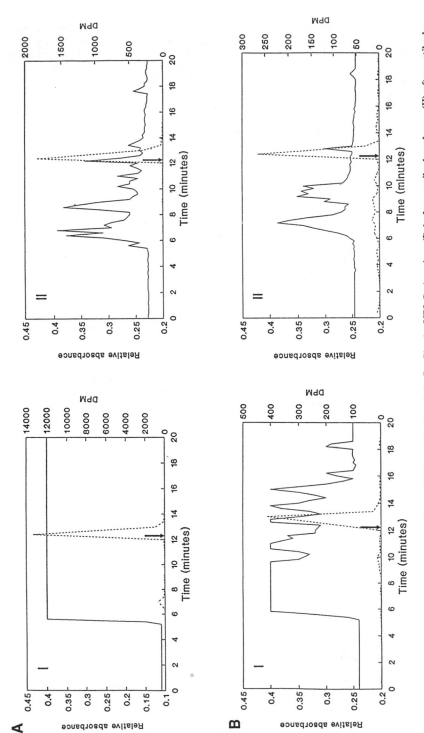

Figure 2. Goat Urine and Kidney, Comparison of UV and Radioactivity Profiles by HPLC. A: urine, (I) before antibody column, (II) after antibody column; B: kidney, (I) before antibody column, (II) after antibody column. Arrows mark the position for standard Compound A. Solid lines: absorbance at 240 nm; dashed lines: radioactivity, disintegrations per minute (DPM).

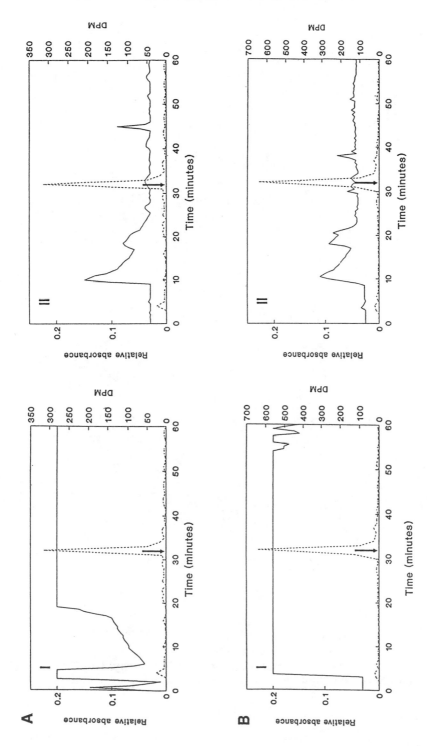

Figure 3. Corn Grain and Fodder, Comparison of UV and Radioactivity Profiles by HPLC. A: grain, (I) before antibody column, (II) after antibody column. B: fodder, (I) before antibody column, (II) after antibody column. Arrows mark the position for standard imazapyr. Solid lines: absorbance at 254 nm; dashed lines: radioactivity (DPM).

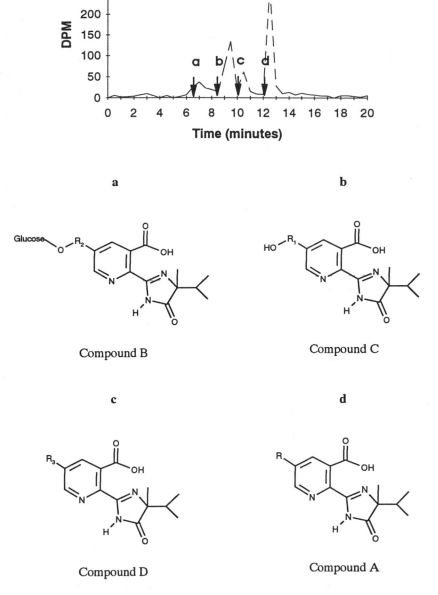

Figure 4.Crude Wheat Extract Radioactivity Profile by HPLC and Proposed Metabolites Structures. The Rfs of the metabolites are indicated by arrows.

Centrifugal Molecular Size Exclusion Filtration of Wheat Samples. Since molecular size exclusion filtration is simple and can accommodate a large number of samples in one centrifugation step, and the metabolites are less than 1,000 dalton in molecular weight, filtration can be a simple and useful method of purifying sample extracts. A comparison study was conducted where wheat extract samples were processed through 10,000 dalton and 1,000 dalton filtrations followed by monoclonal or polyclonal antibody affinity chromatography. All the affinity chromatography effluents were subjected to HPLC using the conditions described above. The UV and radioactivity profiles of all HPLC runs were obtained. Figure 5A compares the HPLC profiles of 10,000 dalton filtration product from the monoclonal and polyclonal antibody affinity chromatographies. The total amount of radioactivity recovered from the peaks represented 69.5% and 49.7% of the input samples from the mono- and poly-clonal columns respectively. When the unbound radioactive material from the polyclonal column was reapplied onto the polyclonal column, a radioactive peak corresponding to Compound C was found upon HPLC (data not shown). This accounts for the low recovery of 49.7% from the first polyclonal column affinity chromatography indicating that the 10,000 dalton filter did not remove all the interfering matrix which prevented some imidazolinone containing material from binding to the polyclonal antibody column.

In Figure 5B, the 10,000 dalton followed by 1,000 dalton filtration products are affinity purified and the HPLC profiles are compared. The peaks represented 74% and 60% of the input from the monoclonal and polyclonal antibody columns respectively. These values are comparable to the 55% recovery value obtained when the original crude extract was chromatographed on HPLC. When the unbound material from the polyclonal column was reapplied, no more radioactivity could be bound, indicating a more extensive purification of the matrix was achieved with 1,000 dalton filtration.

Since samples from both filtration schemes applied to the monoclonal antibody column recovered about 70% of the applied radioactivity and no additional radioactivity could be bound when the unbound material was reapplied (data not shown), suggests that the monoclonal antibody column is more specific for the compounds and the purified sample has lower matrix interference. From the UV profiles it is also evident that, with samples processed through a 1,000 dalton filter the monoclonal antibody column produced a cleaner sample than that obtained from the polyclonal column.

Recovery of Compound B from Antibody Columns. Since the amount of Compound B in the original wheat extract was low, it was difficult to quantify the recoveries of this compound from the affinity columns. To test it further, we introduced a known amount of non-radioactive Compound B to a 1,000 dalton filtered wheat extract and subjected it to affinity chromatography on both mono- and poly-clonal antibody columns. The methanol effluents were again subjected to HPLC. As shown in Figure 6A and B, the UV absorbing peak corresponding to that of reference compound Compound B was present in samples purified through both antibody columns and the recoveries as calculated by integrated peak areas were 98% and 160% for mono- and poly- clonal columns respectively. The greater than 100% recovery from the polyclonal column may be due to integration of co-eluting matrix.

Conclusion.

We have shown that a compound class-specific antibody such as the anti-imidazolinone antibody can be useful in affinity chromatography. The samples used for this demonstration varied from animal tissue, body fluid, corn plant, grain and wheat plant extracts. Using radio-labeled samples from metabolism studies simplified the evaluation process in that the compounds can be tracked by radioactivity. Traditionally, metabolism samples contain low levels of residues and

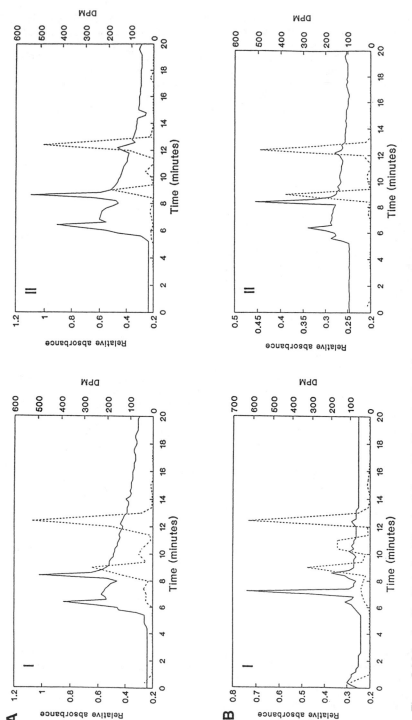

Figure 5. Molecular Size Exclusion Filtration followed by Monoclonal and Polyclonal Antibody Columns Purification, Samples Profiled by HPLC. A: 10,000 dalton filtration, (I) monoclonal column, (II) polyclonal column; B: 10,000 and 1,000 dalton filtration (I) monoclonal column, (II) polyclonal column. Solid lines: absorbance at 240 nm; dashed lines: radioactivity (DPM).

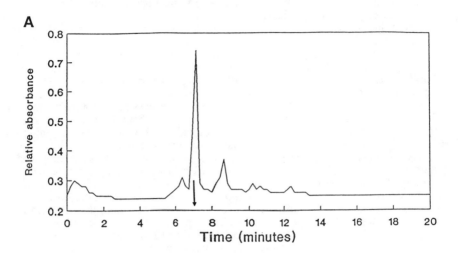

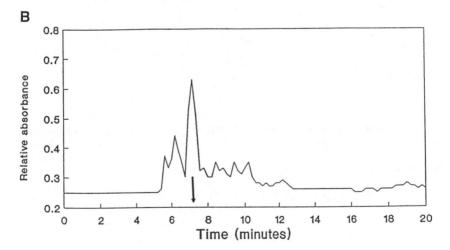

Figure 6. UV profiles of Wheat Sample Fortified with Compound B by HPLC. A: monoclonal antibody column processed, B: polyclonal antibody column processed. Arrows mark the position for standard Compound B.

numerous TLC, solid phase extraction, HPLC and solvent partition steps are required to prepare the samples for identification by mass spectrometry. Although some matrices did interfere with antibody binding, a simple filtration method was developed to overcome this difficulty. The samples purified by one passage of antibody affinity column was sufficiently clean for mass spectrometric identification. This means a significant time, labor as well as solvent savings.

The monoclonal and polyclonal antibody column comparisons showed that the monoclonal antibody may be superior in sample clean-up. One may reason that polyclonal antibodies contain antibodies with different affinities as well as variations in binding specificity. These properties may contribute to greater non-specific binding of matrix material thus lowering the column capacity for the analytes. This was supported by our data using the 10,000 dalton filtrate samples. The polyclonal column bound 49.7% of the total radioactivity through the first passing as compared to 69% using monoclonal column. Upon re-application of the unbound material, additional Compound C was recovered from the polyclonal column while an insignificant amount of material was recovered from the monoclonal column.

We demonstrated that centrifugal molecular exclusion filtration removed a majority of the complex matrices which interfere with the analyte binding to the antibody. Since this filtration method can process a large number of samples simultaneously, it can be used in sample processing for immunoassays. If an immunoassay cannot distinguish the parent from metabolites, then the filtration plus affinity chromatography steps may be useful for preparing residue samples for other analytical methods. The demonstration that the 1,000 dalton filteration and monoclonal column affinity chromatography has near base-line resolution and 98% recovery of the metabolite Compound B by HPLC indicates that this procedure may be applicable for residue analysis of parent and metabolites in a single procedure. This procedure is simpler than the traditional partition and extraction methods. Since antibody chromatography requires aqueous buffer and methanol and the columns can be used repeatedly, it is more economical to perform and safer for the operator. With some modifications to the column support, this affinity chromatography method may be automated to allow greater throughput.

Acknowledgment.

We would like to thank Drs. G. Asato, T.Whatley, F. Taylor, and A. Lee for support; M. Malik, S. Paci, C. Davis-Thomas, and F. J. Washington for technical assistance.

References

1. Wong, R. B.; Ahmed, Z. H. *J. Ag Food Chem.* **1992,** *40,* 811-816.
2. Secher, D.; Burke, D. C. *Nature* **1980,** *285,* 446.
3. Jack, G. W.; Gilbert, H. J. *Biochem. Soc. Trans,* **1984,** *12,* 246.
4. Jack, G. W.; Blazek, R. *J. Chem. Technol. Biotechol.* **1987,** *39,* 1.
5. Jack, G. W.; Blazek, R.; James, K.; Boyd, J. E.; Micklem, L. R. *J. Chem. Technol. Biotechol.* **1987,** *39,* 45.
6. Olson, K. C.; Fenno, J.; Lin, N.; Harkins, R. N.; Snider, C.; Kohr, W. H.; Ross, M. J.; Fodge, D.; Prender, G.; Stebbing, N. *Nature* **1981,** *293.* 408.
7. Yanagawa S.; Hirade, K.; Ohnota, H.; Sasaki, R.; Ahiba, H.; Ueda, M; Gotor, M. *J. Biol. Chem.* **1984,** *259.* 2707.
8. Jenny, R.; Church, W.; Odegaard, B.; Litwiller, R.; Mann, K. *Prep. Biochem.* **1986,** *16,* 227.
9. Pixley, R. A.; Procino, L. G.; Silver, L.; Colman, R. W. Clin. Res. **1985,** *33,* 550A.

10. Rybacek, L.; D'Andrea, M.; Tarnowski, J. *J. Chromatogr.* **1987,** *397,* 355.
11. Limentsni, S. A.; Furie, B. C.; Poiesz, B. J.; Montagna, R.; Wells, K.; Furie, B. *Blood* **1987,** *70,* 1312.
12. Berry, M. J.; Davies, J.; *J. Chromatogr.* **1992,** *597,* 235.
13. Pfeiffer, N. E.; Wylie, D. E.; Schuster, S. M. *J. Immuno. Methods.* **1987,** *97,* 1.
14. Chase, H. A. Chem. Eng. Sci. **1984,** *39,* 1099.
15. Bazin, H.; Malache, J.-M. *J. Immunol. Methods.* **1986,** *88,* 19.
16. Boschetti, E.; Egly, J. M.; Monsigny, M.*Trends Anal, Chem.* **1986,** *5,* 4.
17. Hall, R.; Hunt, P. D.; Ridley, R. G. *Methods in Molecular Biology,* **1993,** 21, 389.
18. Farjam, A. *The Use of Immobilized Antibodies for Selective On-Line Sample Pretreatment in Liquid and Gas Chromatography,* Ph.D. thesis, Vrije Universitat Amsterdam, Offsetdruck Paletti, Aachen, 1991.
19. Nilssson, B. J. Chromatogr. **1983,** *276,* 413.
20. Tagliaro, R.; Sorizzi, R.; Lafosca, S.; Stefani, L.; Ferrari, S.; Plescia, M; Marigo, M. *Spectroso. Int. J.* **984,** *3,* 311.
21. Van Ginkel, L. A.; Stephany, R. W.; Van Rossum, H. J.; Van Blitterswijk, H.; Zoontjes, P. W; Hooijschuur, P. W.; Zuydendorp, J. *J. Chromatogr.* **1989,** *489,* 95.
22. Webb, R.; Baxter G.; McBride, D.; Nordblom, G. D.; Shaw, M. P. K. *J. Steroid Biochem.* **1985,** *23,* 1043.
23. Groopman, J. D.; Trudel,L. J.; Donahue, P. R.; Marshak-Rothstein, A.; Wogan, G. N. *Proc. Natl. Acad. Sci. U.S.*. **1984,** *81,* 7728.
24. Groopman, J. D.; Donahue, K. *J. AOAC.* **1988,** *71,* 861.
25. Sharman, M.; Patey, A. L.; Gilbert, J. *J. Chromatogr.* **1989,** *474,* 457.
26. Nakajima, M.; Terada, H.; Hisada, K.; Tsubouhi, H.; Yamamoto, K.; Uda, T.; Itoh, Y.; Kawamura, O.;Ueno, Y. *Food and Agri. Immunol.* **1990,** 2, 189.
27. Janis, L. J.; Regnier, F. E. *J. Chromatogr.* **1988,** *444,* 1.
28. Haasnoot, W.; Ploum, M. E.; Paulussen, J. A. Schilt, R.; Huf, F. A. *J. Chromatogr.* **1990,** *519,* 323.
29. Rule, G. S.; Henion, J. D. *J. Chromatography.* **1992,** *582,* 103-112.
30. Kakouri E. I.; Christodoulidow, M.; Constantinidou, E. Food Safety and Quality Assurance Conference, El Escorial, Spain, 1993. Poster
31. Goto, J.; Myairi, S.; Awata, N.; Shimada, H. Japanese Patent JP05310603 A2 **1993.**
32. Schneider, C, *J. Biol. Chem.* **1982,** *257,* 10766-10769

RECEIVED October 25, 1994

Chapter 18

Development of Assay for Analysis of Hg^{2+} Based on Sulfur-Containing Ligands

Ferenc Szurdoki, Horacio Kido, and Bruce D. Hammock

Departments of Entomology and Environmental Toxicology,
University of California, Davis, CA 95616

We have developed a simple analytical method for the detection of mercuric ions at low ppb levels. We combined inexpensive ELISA methodology with the selective, high affinity recognition of Hg^{2+} ions by dithiocarbamate chelators. Thus, we employed sulfur-containing complexing agents instead of antibodies as analytical tools. One of our assay formats comprises a sandwich chelate formed by a chelator doped on the surface of the ELISA plate, the mercuric ion, and a chelator coupled to a reporter enzyme. Another format is based on the competition of the analyte ions with a mercury containing reagent (e.g., enzyme tracer) for the binding to a chelate linked reagent (e.g., immobilized chelator). Our preliminary results demonstrate that the high sensitivity and selectivity of our assay systems hold promise for the monitoring of mercury in the environment.

Toxicological Significance of Mercury. Mercury compounds were long used in medicine as bactericidal and diuretic agents; they have also been widely applied as fungicides (*1-3*). However, the toxicity of mercury derivatives has gained world-wide attention because of several serious incidents of environmental pollution since the 1950s. In Sweden a mass extermination of birds took place due to the excessive use of methylmercury dicyandiamide for seed-dressing during the 1960s (*3*). Methylmercury poisoning of humans occurred

0097–6156/95/0586–0248$12.00/0
© 1995 American Chemical Society

in Japan at Minamata bay in 1953-60 (*1, 4, 5*). The most serious symptoms of the "Minamata disease" were paralysis and irreversible or fatal neurological disorders (*3*). In this case, the origin of the mercury was industrial waste discharged into the bay.   Mercury was then concentrated in seafood consumed by humans.   These events resulted in a reduction of the use of mercury compounds as agricultural pesticides in many countries since the sixties.

The most important target organs of inorganic and organic mercury are the central nervous system and the kidneys.   Mercuric ion and organomercury derivatives interfere with enzyme reactions and membrane permeability by binding to sulfhydryl groups.   Mercury compounds may also interact with phosphoryl groups of cell membranes and amino and carboxyl groups of enzymes as well (*3*). Alkylmercury compounds are considered to be the most harmful mercury derivatives; they can readily absorb through contact with skin, digestion, and respiration due to their solubility properties and volatility. The metabolism of these compounds is slow; they accumulate in the liver, brain, and red blood cells.   Even moderate poisoning with alkylmercury derivatives may cause typical neurotoxic symptoms such as tremors, ataxia, difficulty of hearing and vision (*1*). Prenatal exposure to low doses eventually results in mental retardation in children (*1, 2, 5*). Adverse immunological consequences associated with exposure to mercurials have also been well recognized (*5, 6*).

One of the most important sources of the mercury in the environment is the natural degassing of the earth's crust through the soil and water bodies, including also volcanic and geothermal production (*1*). However, recent assessments demonstrated that the global anthropogenic emission of this toxic metal is similar to or even higher than the natural one (*4, 7*). Mercury pollution resulting from its use in agriculture, paper and chlor-alkali industries has diminished in some industrialized countries since the sixties; however, the total anthropogenic emission of mercury has increased during this period. In Europe, major sources of anthropogenic environmental pollution by mercury, mostly emitted to the atmosphere, are fossil fuel combustion (69%), chlorine production (18%), waste incineration (7%), and non ferrous metal industry (6%) (*7*). The most important man-made forms of mercury released to the atmosphere are the volatile Hg^0 and Hg^{2+} (mostly as $HgCl_2$, "sublimate"); the contribution of the so called particulate mercury (e.g., HgO in ash) is only minor (*7*).

Metallic mercury in the atmosphere represents a major pathway of the global cycle of mercury (*1*). Mercury is transported to aquatic ecosystems by atmospheric deposition and surface runoff (*1*). Metallic mercury may be oxidized to mercuric ion in the aquatic environment (*1*). However, the opposite mass transfer process also occurs at the same time. In aquatic sediments, bacterial activity converts HgS, the main mercury ore, *via* soluble Hg^{2+}-salts to Hg^0 that eventually appears in the atmosphere (*8*). Methyl-mercury (CH_3Hg^+), produced by bacterial methylation of Hg^{2+} in aquatic sediments, is the most toxic, persistent, and most commonly occurring organic mercury species in natural waters (8, 9). It is about 100 times more toxic than Hg^{2+} (inorganic mercury) (8). Methylmercury accumulates in fish and it is amplified through the food chain (5). Thus, it presents very serious human and environmental health hazards. The potent toxicity of methylmercury highlights the importance of trace mercury analysis for agriculture and food industry. The concentration of methylmercury in natural waters is one of the most important water quality parameters (10). The abundance and the toxicological significance of Hg^0 and Hg^+ in aquatic environments is only marginal.

The average mercury level is about 0.5 ppm in soil (11). The usual total mercury concentration (organic plus inorganic mercury) ranges about 2 to 10 ppt in lake water (12) and about 0.01 to 1 ppb in river water (11). In a number of countries the maximum permitted mercury level in environmental water samples is 1 ppb (11, 13). However, waste water occasionally has significantly higher mercury levels than this limit value.

Literature Methods for the Analysis of Mercury and Some Other Toxic Metals. Numerous approaches for transition/heavy metal analysis with sensitivity in the ppm/high ppb range can be found in the literature (14). Most of these methods are based on the detection of suitable metal derivatives by molecular absorption spectrophotometry (e.g., UV-VIS, IR) or by low-sensitivity electro-analytical techniques. Many of these methods employ chelates (i.e., a cyclic compounds involving the metal atom) derived from the metal ions of the analyzed sample. Metal ions in complex matrices have also been measured in the form of chelates by gas- and liquid chromatography (15).

Simple mercury analyses based on the detection of chelates with optical techniques (mostly UV-VIS photometry) often display interferences by foreign ions (16), and they usually lack the necessary high sensitivity (13, 17). Although these methods are frequently combined with chelate extraction (preconcentration), detection limits below 0.1 ppm have rarely been reported (13).

Relatively few simple, reliable, and selective techniques are available for the detection of transition/heavy metals at ppb/ppt levels. Most of them are instrumental analyses (e.g., flameless atom absorption spectroscopy, special electroanalytical techniques) (18-23). These methods have some drawbacks that are typically encountered when using instrumental analyses, e.g., the sample throughput is limited, they require expensive apparatus and highly qualified analyst, they are not suitable for on-site analysis in the field (24).

For two decades, in environmental and clinical laboratories, the most popular instrumental method for the determination of mercury has been cold vapor atomic absorption spectrometry (CVAAS) (18, 23, 25-28). However, this technique (18) often lacks high precision at top sensitivity and needs relatively large sample volumes to achieve high sensitivity (24). Chemical speciation of mercury (e.g., CH_3Hg^+ versus Hg^{2+}) is of utmost importance in the medical and environmental sciences (see above) (5). The CVAAS technique detects Hg^{2+} (inorganic mercury). However, the quantification of organic mercury (usually mostly methylmercury) is also possible indirectly by decomposition of the organic species into Hg^{2+} and then by the determination the total mercury content (organic mercury = total - inorganic mercury) (*18, 25-28*). Other common ways of mercury speciation combine chromatographical separation of mercury derivatives with various detection systems (*5, 9, 26, 29*). Some recent mercury trace analyses, e.g., atomic fluorescence spectroscopy (*11, 30, 31*), ICP-MS (*32, 33*), and neutron activation analysis (*34*), are based on highly sophisticated and costly instrumentation.

Interesting alternative methods for detection of certain lanthanide elements and heavy metals are immunoassay procedures (*24, 35, 36*). These novel techniques involve antibodies that were raised against metal chelates. Reardan et al. (*35*) reported the selective recognition of an EDTA-type indium chelate by monoclonal antibodies generated by the same metal complex attached to a carrier protein. The antibodies showed the highest affinity for the indium-chelate. However, only 10-1000 times less affinities for chelates of some other heavy metal ions were displayed. The lack of high selectivity might create difficulties in the analysis of real samples containing other metal ions in moderate to high concentrations. Wylie et al. (*24, 36*) immunized mice with a $HgCl_2$-glutathione-KLH-conjugate. It was demonstrated that some of the resulting monoclonal antibodies had high affinity for the mercuric ions either chelated to glutathione-BSA or alone. The latter observation suggests that the antibodies may have recognized the metal ion as an independent epitope with high selectivity. (In the immunizing glutathione-complex, the mercuric ion was supposed to be only partly shadowed by the chelator.) One of the resulting ELISAs based on these antibodies showed linear relationship between the optical density and $\log[Hg^{2+}]$ in the range of 0.5-10 ppb. Generally low cross-reactivities with other metal ions were reported. Immunoassay technology makes possible the analysis of multiple samples at the same time by a simple procedure, it does not require costly instrumentation, and it is also adaptable as a cheap field-portable assay. However, these methods require highly specific monoclonal antibodies which may be expensive to obtain.

Results and Discussion

Assay Based on Hg^{2+} Sandwich Chelates. Our first new approach is based on the formation of so called sandwich chelates of the ion to be detected. It means that the analyte forms an aggregate simultaneously with two complexing agents. Such an association is termed as sandwich complex with the target ion in the middle of it. In this assay, one chelator is immobilized on a solid carrier, while the other is linked to a reporter system, e.g., enzyme (first assay format, Figure 1). This arrangement forms a highly selective, sensitive, and convenient system for quantitative detection of the target analyte. It combines the specific interaction of the target ion with the sandwich chelating agents with the great signal amplification offered by

enzymes (ELISA technology).   Again, this procedure needs no specific
antibodies for recognition.

Many sulfur containing chelators display strong affinities for certain
transition/heavy metal ions and also for some of the organometallic derivatives of these metals (e.g., cadmium(II), lead(II), mercury(II), methylmercury)
(*37-41*).   Dithiocarbamates form complexes of very high thermodynamic
stability with several heavy metal ions, especially with mercury(II) (*40, 42,
43*).   Bond and Scholz (*40*) developed a new voltammetric method for the
measurement of thermodynamic data and determined the conditional stability
constants (β_2 values) of a number of mercury and lead dithiocarbamate
complexes in water for the equilibria $M^{2+} + 2dtc^- = M(dtc)_2$. The stability
constant is defined as: $\beta_2 = [M(dtc)_2]/([M^{2+}] \times [dtc^-]^2)$, where dtc^- is a
substituted dithiocarbamate anion, $RR'N-CS_2^-$. (This important information is
rather difficult to obtain by other techniques because of the very poor
solubility of the chelates.)   The β_2 value of each studied mercury bis-
dithiocarbamate was found to be extremely high, much higher than that of the
corresponding lead compound.   For instance, the $\log\beta_2$ value for the mercury
bis-diethyldithiocarbamate ($R = R' = C_2H_5$) is 38.2, while that for the
corresponding lead derivative is 17.7. In addition to that, the substitution of
Hg^{2+} for other metal ions in the form of dithiocarbamate complex was
reported to be extremely fast and quantitative for a number of metals (e.g.,
Cd, Co, Cu, Fe, Mn, Ni, Pb, Zn) (*44*). These and other literature data (*43*)
suggested that an assay system which is highly selective and sensitive for
Hg^{2+} ions might be developed using dithiocarbamate chelators. It appears
that only some noble metals form chelates of this type with similar or even
higher thermodynamic stability (*45*); however, it is somewhat difficult to
compare literature data obtained by different methods under different
experimental conditions. Dithiocarbamates obtained from secondary amines
are known to be fairly chemically stable under nonacidic conditions in the
absence of oxidizing agents.   A recent study demonstrated that primary
amines would form unstable dithiocarbamates prone to decomposition to
reactive isothiocyanates which in turn might result in extensive cross-linking
of the conjugated proteins and enzymes (*46*). We thus decided to employ
dithiocarbamates derived from secondary amines in the assay for chelation.
The chemical reactions used to prepare the enzyme linked and the plate
coating chelators were similar. Secondary amino groups were generated on
the surface of the macromolecules (e.g., proteins, enzymes), then carbon
disulfide was added to the basic solution of the protein or enzyme to obtain
the corresponding dithiocarbamates (Figures 2, 3).

For the preparation of the coating chelator (Figure 2), conjugate with
secondary amino groups (1) were formed by reductive alkylation (*47*) of
primary amino groups (i.e., ϵ-amino groups of the lysine residues) on the
surface of the protein by isobutyraldehyde and sodium tetrahydroborate.
Alternatively, a *N*-protected secondary amine bearing another group for the
conjugation to biopolymers can also be employed in the synthesis of the
chelators. This approach was used to prepare the enzyme tracer L-prolyl

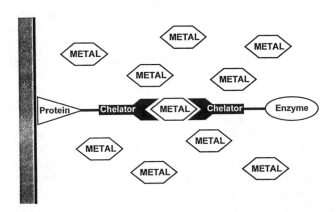

Figure 1. Schematic presentation of the sandwich chelate assay principle.

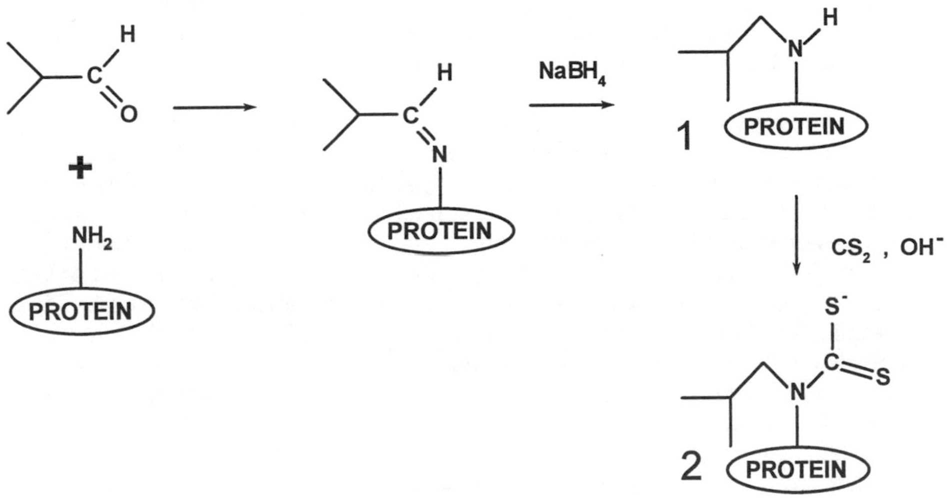

Figure 2. Synthesis of the 2-CONA conjugate.

derivative (Figure 3). *N*-Trifluoroacetyl-L-proline (3) (*48*) was coupled by using the mixture of water soluble carbodiimide and *N*-hydroxysulfosuccinimide. The efficiency of this novel reagent combination was demonstrated in conjugation reactions of various haptenic carboxylic acids to proteins (*49-52*). The application of this coupling method is of particular advantage in case of highly lipophilic acids (*52*). Our method of the removal of the *N*-trifluoroacetyl (TFA) group of conjugate 4 is based on the report by Weygand and Frauendorfer (*53*) on the special ease of the reductive cleavage of the *N*-terminal protecting group in *N*-TFA-prolyl peptides. The attachment of the blocked proline and then the splitting of the TFA group by sodium tetrahydroborate under the reaction conditions of choice was qualitatively confirmed by F-NMR-spectroscopy in a model experiment.

The chelators linked to certain proteins, e.g., conalbumin (CONA), can be simply doped on the surface of ELISA plates making use of the common plate coating immunoassay methodology. (The possibility of using other polymer carriers for coating or direct covalent binding of chelators to plates is being investigated.) In case of sequential incubation, first the plates were coated with the chelating protein conjugate (2), then the analyzed sample containing mercuric ions was added to the wells and incubated, and the unbound components were removed by washing the plates. After treatment with the chelator (6) linked to the reporter enzyme, alkaline phosphatase (AP), and washing the plates, the enzyme substrate was added, and finally, the optical density (OD) was measured (Figures 4, 5). Alternatively, in case of simultaneous incubation, the chelator coated plates were treated with the mixture of the sample and the enzyme linked chelator (6); the rest of the procedure was identical.

The optical density pattern of the two site (sandwich) immunoassays exhibits the so called "hook effect" (*54*). The basic principle of our sandwich chelate based assay is similar to these immunoassays; thus, one may expect a bell shaped standard curve (OD versus lg[concentration]) also in our case. The explanation of this effect is as follows: the higher concentration of the analyte ion, the increased amount of a sandwich complex composed of two second (mobile) sandwich chelators and the metal ion is formed. This complex is then washed off, resulting in decreasing optical density at increasing ion concentrations. The descending part of the curve is usually seen only at very high analyte concentrations. In most analyses, the assay's increasing, low-concentration, linear range is utilized. Thus, the unknown samples must be tested at various dilutions or other preventive measures (*55*) have to be taken to insure that the optical density actually falls within this interval.

Our experiments showed that Hg^{2+} ions can be conveniently detected at ppb concentrations by the assay system, based on sandwich chelate formation, characterized above. For instance, the standard curve (Figure 5) of our sequential incubation 2-CONA/6-AP assay has an IC_{50} value of 13 ppb (65 nM) of mercuric ion concentration. The relative high variance and blank readings are due to the lack of detergent in the assay buffer. (Longer color

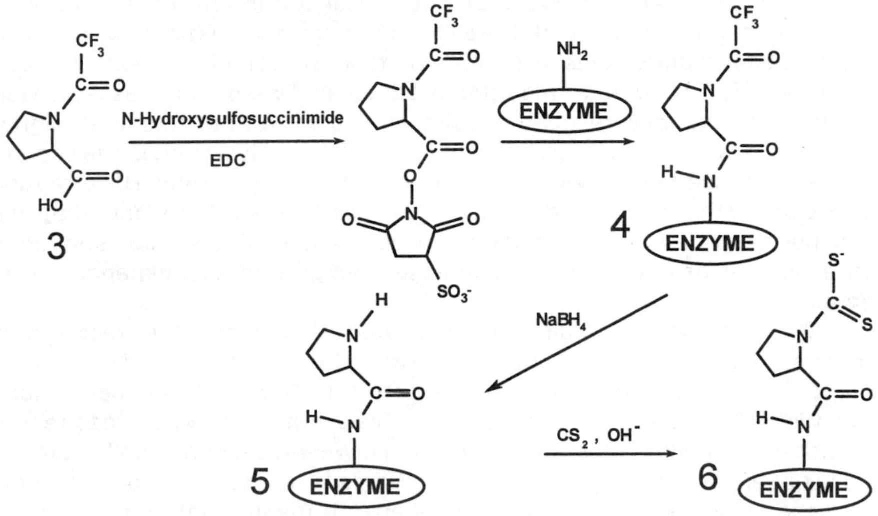

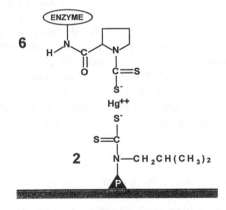

Figure 3. Synthesis of the 6-AP conjugate.

Figure 4. Structures of the reagents used in the 2-CONA/6-AP assay.

development periods, higher regent concentrations resulted in higher absorbance values; however, the background became excessive.) As it is usual in the ELISA methodology, hydrophobic (aspecific) interactions between the tracer and the solid phase increased the background signal also in our assays. We suspected that Tween-20 and similar surfactants commonly used to reduce background might interfere with the detection of mercury, because polyoxyethylene derivatives were shown to form complexes with various metal ions (56, 57). In fact, the attempted use of Tween in the assay based on sandwich chelates resulted in diminishing *both* the zero and higher concentration signals in our preliminary experiments. However, in the second assay format (see below) we could manage to employ Tween 20 to reduce background without serious difficulty of this kind. Thus, it appears that this effect needs to be studied separately in each case. Further investigations with a number of detergents to decrease background and variance are in progress.

Our preliminary results demonstrated that most the metal ions investigated (e.g., Al^{3+}, Ca^{2+}, Cd^{2+}, CH_3Hg^+, Co^{2+}, Cr^{3+}, Mg^{2+}, Mn^{2+}, Ni^{2+}, Pb^{2+}, Zn^{2+}) did not display signals at least up to 3,000 nM concentrations which would interfere with the mercury analysis, but low level of cross-reactivity with Ag^+ and Pd^{2+} ions was detected. However, silver and palladium are usually not expected as abundant contaminants in most environmental media where the presence of mercury is a concern. It means that in a number of potential applications (e.g, some water analyses), this selectivity seems to be satisfactory. If the interference from these or other metals becomes problematic, a cleanup procedure (e.g., selective chelate extraction, ion-exchange chromatography) *prior to* the mercury analysis may be necessary. Alternatively, the use of masking agents for interfering ions with this assay system or application of even more selective chelators as reagents is also possible. It is worth noting that this assay system did not exhibit any cross reactivity with methylmercury. This important characteristic of the assay allows the quantification of methylmercury (organic mercury) by assessing the difference in response between a sample aliquot in which the organic mercury species have been decomposed into Hg^{2+} (total Hg) and an untreated aliquot of the same sample (organic Hg = total Hg - inorganic Hg). Other mercury species may be ignored, as detailed above, because most of the mercury in aqueous environmental samples is either in the form of methylmercury (CH_3Hg^+) or mercuric (Hg^{2+}) ions. Should their analysis become important, there are numerous procedures to convert other species of mercury to Hg^{2+}.

Assay Based on Competition with Mercury-Linked Reagents. A different assay based on the competition of mercuric ions of the sample with the binding of a mercury-linked reagent to a chelator-containing reagent was also developed (Figure 6). Thus, in this second assay format, mercury is attached to a reporter enzyme (conjugate 7) while the sulfur containing chelator is linked to the plate coating protein (Figure 6). (The opposite arrangement, i.e., the metal atom is immobilized on the surface of the plate while the chelator

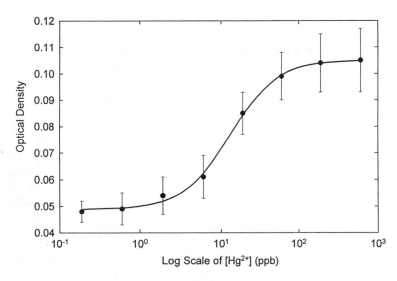

Figure 5. *Standard curve of the 2-CONA/6-AP assay performed with sequential incubation. Zero concentration absorbance: 0.029±0.003. The decreasing part of the curve at higher concentrations is not presented here.*

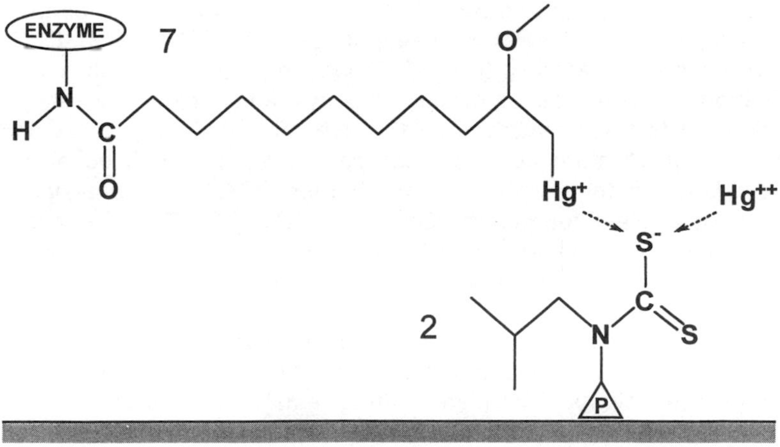

Figure 6. *Structures of the reagents used in the 2-CONA/7-AP assay.*

is bound to the reporter enzyme, is also conceivable. Work to explore this possibility is also in progress.)

10-Undecenoic acid (8) was subjected to methoxymercuration to obtain organomercury compound, $^+Hg\text{-}CH_2\text{-}CHX\text{-}(CH_2)_8\text{-}CO_2H$ (9, X: OCH_3) (Figure 7). Homologues and similar compounds having different X-substituents can be prepared similarly (58, 59). The mercury containing acid (9) was then linked to AP by means of the combination of water soluble carbodiimide and N-hydroxysulfosuccinimide (52). The obtained enzyme conjugate (7) proved to be a very useful tracer in our second assay format. The structure of this organomercurial (7) was confirmed indirectly by the fact that after treatment with sodium tetrahydroborate and dialysis there was no color development at all while the obtained conjugate was applied in the assay. Sodium tetrahydroborate is known to effect the facile reductive cleavage of mercury-carbon bond in this type of compounds with concomitant formation of Hg^0 (58, 59). During this transformation and also throughout all the syntheses of our enzyme tracers, the enzymatic activity of AP was conserved.

Our preliminary experiments demonstrated that mercuric ions can be selectively detected at low ppb concentrations by the system 2-CONA/7-AP. The standard curve shown on Figure 8 has an IC_{50} value of 8 ppb (40 nM), a linear range of 2 ppb to 20 ppb, and a detection limit of about 1 ppb (5 nM) of mercuric ion concentration. (The detection limit is the concentration at the absorbance value of the standard curve which is less than the zero concentration signal by three times the standard deviation of the blank readings.) We found generally low interferences with foreign ions. No or very little cross-reactivity (CR) was detected with Al^{3+}, Ca^{2+}, Cd^{2+}, Co^{2+}, Cr^{3+}, Cr^{6+}, Fe^{2+}, Fe^{3+}, Mg^{2+}, Mn^{2+} Ni^{2+}, Pb^{2+}. These metal ions might interfere with the mercury analysis only if they present in higher than 3,000 nM concentrations. The following metals gave moderate responses: Au^{3+} (CR: <1%), Cu^{2+} (CR: 10%). High cross-reactivity (CR: 82%) was encountered only with Ag^+. Investigations on the use of masking agents for foreign ions have been undertaken. We hope that through the use of these compounds possible interferences from more complex matrices (e.g., soil) will be effectively reduced. Unexpected responses, i.e., higher ODs with increasing metal concentrations, were found with Zn^{2+} and CH_3Hg^+. This upward trend, however, was not observed when somewhat different assay protocol was applied. Further studies have been performed to solve these problems.

Specific Aspects of the Development of the Chelate Based Assays. In our studies, CONA was employed for the preparation of the coating chelator. Use of this native protein itself for plate coating resulted in only very weak color development in both formats. Similarly, there was only a background intensity signal with hardly any slope when the intermediate secondary amine (1) derived from CONA was used for coating in both formats. Thus, it appears that the role of "aspecific" binding of Hg^{2+} or the alkylmercury moiety of tracer 7 by other chelating moieties (e.g., histidine, tryptophan) of CONA is only marginal under the assay conditions. CONA (conalbumin,

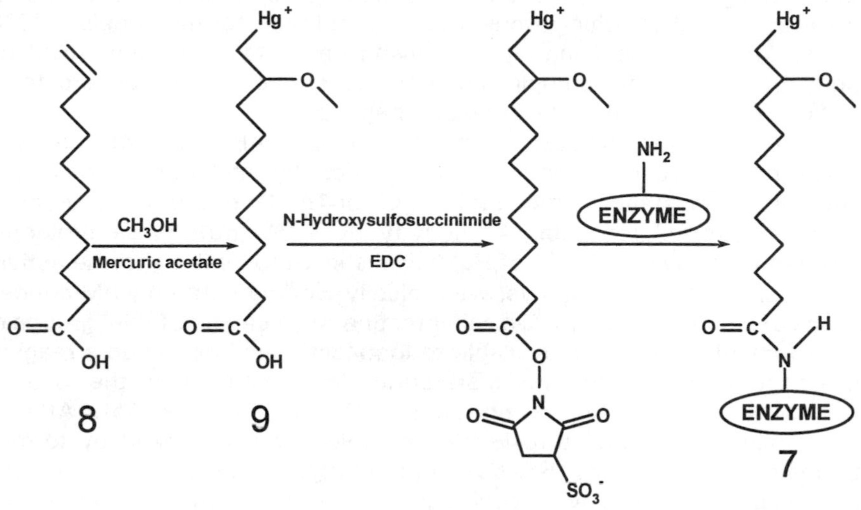

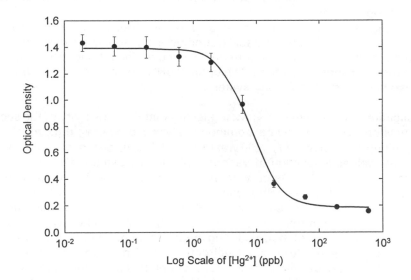

Figure 7. Synthesis of the 7-AP conjugate.

Figure 8. Standard curve of the 2-CONA/7-AP assay. Zero concentration absorbance: 1.440 ± 0.023.

ovotransferrin) is a well characterized protein with a molecular weight of about 78,000 Da. All the 30 cysteine residues of CONA form *internal disulfide bridges* (*60*, *61*); there is no free thiol group (unpaired cysteine) on the surface of CONA which would have high affinity for mercurials. CONA has two known iron-binding sites involving no cysteine residues (*62*); the affinity of these metal complexing moieties to mercurials appears to be significantly lower than that of dithiocarbamates.

We found that, in accord with literature data, AP, the reporter enzyme applied in our investigations was not significantly inhibited by *trace level* mercurials. For instance, either 2 mM $HgCl_2$ or 1mM *p*-hydroxymercurybenzoate was reported to inhibit AP activity by 17% (*63*); while prolonged treatment of AP with 0.1 mM of Hg^{2+} was shown to inhibit enzyme activity only by 2% (*64*). In our assays, we typically worked with only nM concentrations of mercuric ions. Native calf intestine AP is devoid of thiol groups on the surface of the enzyme available to iodoacetic acid or Ellman's reagent; cysteine residues with free thiols are buried (*65*). Removal of the essential Zn^{2+} from the active site does not unmask the hidden thiols (*65*). AP of *E. coli* contains only cysteine residues with thiol groups blocked by forming interchain disulfide bonds (*65*, *66*). Apparently, neither cysteine nor other amino acid residue displaying very high affinity for mercurials is involved in the known active sites of several APs (*67*, *68*). When native bovine intestinal mucosa AP was used instead of identical amounts of our AP-derivative tracers (6, 7) under assay conditions of either the first or the second format, there was no signal observed.

Slightly acidic acetate buffers were used as assay medium in our investigations. Under strongly acidic conditions the dithiocarbamate chelators would decompose, while under more basic conditions it might be difficult to keep mercuric ions in solution. The optimal pH-value and ionic strength must be established for each assay system.

Applications. Literature overview suggested that the sensitivity of the second assay is promising for several common analytical problems, e.g., monitoring of drinking (*13*), river (*11*, *13*), waste (*11*, *13*, *69*), and mine runoff water (*70*), as well as inspection of various hazardous wastes with high mercury content and soil (*11*). Mercury bound to organic matter has to be liberated to analyze the total mercury content of some of these matrices. Mild procedures (see, e.g., references *27*, *69*) using no concentrated mineral acids are preferred for the treatment of water samples to avoid extensive dilution with large volumes of buffer during neutralization. Some environmental applications require ultra-trace level detection of mercury. In such cases, preconcentration of the samples (e.g., chelate extraction, ion-exchange chromatography) would be necessary. In our preliminary studies, tap water samples spiked with 5 different spike levels (1-100 nM) of Hg^{2+} were analyzed by our second assay system in blind fashion; good correlation was found between the determined and spiked concentrations (Figure 9). Further work with several matrices is in progress; these environmental-analytical

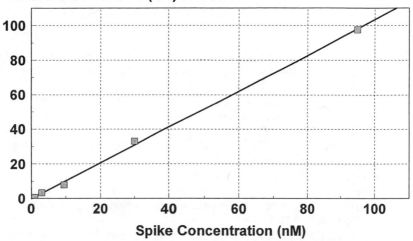

Figure 9. Analysis of spiked tap water samples. y: concentration found (nM). x: spike concentration (nM). y = -0.1025 + 1.0346x (n = 5, r²: 0.999).

studies will be presented elsewhere. The possibility of the application of our assay principles for modified and new chelators, for analysis of further targets, and for other reporter systems and formats (e.g., sensors) is currently being investigated.

Acknowledgements

Early aspects of this research were supported by the U. C. Systemwide Toxic Substances Program. This work was supported in part by the California Department of Food and Agriculture, the NIEHS Superfund Grant 2 P42 ES04699, the U. S. EPA Cooperative Agreement CR-814709-01-0, and the U. S. EPA Center for Ecological Health Research at U. C. Davis (R819658). H. K. received a fellowship from NIEHS Center for Environmental Health Sciences ES05707. B. D. H. is a Burroughs-Wellcome Scholar in Toxicology.

Literature Cited

1. *Casarett and Doull's Toxicology. The Basic Science of Poisons. 3rd ed.;* Klaassen C. S.; Amdur, M. O.; Doull, J., Eds.; Macmillan Publishing Co.: New York, NY, 1986.
2. Hodgson, E.; Levi, P. E. *A Textbook of Modern Toxicology;* Elsevier Science Publishing Co.: New York, NY, 1987.
3. Matolcsy, G.; Nádasy, M.; Andriska, V. *Pesticide Chemistry;* Studies in Environmental Science, Vol. 32; Elsevier Science Publishing Co.: Amsterdam, Netherlands, 1988; pp 283-313.
4. Mukherjee, A. B. *Water Air Soil Pollut.* **1991,** *56,* 35-49.
5. *Advances in Mercury Toxicology;* Suzuki, T.; Imura, N.; Clarkson, T. W., Eds.; Rochester Series on Environmental Toxicology; Plenum Press: New York, NY, 1991.
6. Zelikoff, J. T.; Smialowicz, R.; Bigazzi, P. E.; Goyer, R. A.; Lawrence, D. A.; Maibach, H. I.; Gardner, D. *Fundam. Appl. Toxicol.* **1994,** *22,* 1-7.
7. Pacyna, J. M.; Münch, J. *Water Air Soil Pollut.* **1991,** *56,* 51-61.
8. Brock, T. D.; Madigan, M. T. *Biology of Microorganisms;* Prentice Hall: Englewood Cliffs, NJ, 1991; pp 651-652.
9. Palmisano, F.; Zambonin, P. G.; Cardellicchio, N. *Fresenius J. Anal. Chem.* **1993,** *346,* 648-652.
10. Bloom, N. S.; Effler, S. W. *Water Air Soil Pollut.* **1990,** *53,* 251-265.
11. Janjic, J.; Kiurski, J. *Wat. Res.* **1994,** *28,* 233-235.
12. Meili, M.; Iverfeldt, Å.; Håkanson, L. *Water Air Soil Pollut.* **1991,** *56,* 439-453.
13. Mariscal, M. D.; Galban, J.; Urarte, M. L.; Aznarez, J. *Fresenius J. Anal. Chem.* **1992,** *342,* 157-162.
14. Blanco, M.; Coello, J.; Iturriaga, H.; Maspoch, S; Bertan, E. *Mikrochim. Acta* **1992,** *108,* 53-59.
15. Lajunen, L. H. J.; Eijarvi, E.; Kenakkala, T. *Analyst* **1984,** *109,* 699-700.
16. Sharma, R. L.; Singh, H. B. *Talanta* **1989,** *36,* 457-461.

17. Kamburova, M. *Talanta* **1993**, *40*, 719-723.
18. Magos, L.; Clarkson, T. W. *J. Assoc. Off. Anal. Chem.* **1972**, *55*, 966-971.
19. Daniels, R. S; Wigfield, D. C. *Sci. Total Environ.* **1989**, *89*, 319-339.
20. Shiowatana, J.; Matousek, J. *Talanta* **1991**, *38*, 375-383.
21. Wang, J.; Tian, B. *Anal. Chem.* **1993**, *65*, 1529-1532.
22. Turyan, I.; Mandler, D. *Nature* **1993**, *362*, 703-704.
23. Liang, L.; Bloom, N. S. *J. Anal. Atomic Spectr.* **1993**, *8*, 591-594.
24. Wylie, D. E.; Carlson, L. D.; Carlson, R.; Wagner, F. W.; Schuster, S. M. *Anal. Biochem.* **1991**, *194*, 381-387.
25. Magos, L. *Analyst* **1971**, *96*, 847-853.
26. Lind, B.; Body, R.; Friberg, L. *Fresenius J. Anal. Chem.* **1993**, *345*, 314-317.
27. Guo, T.; Baasner, J. *Anal. Chim. Acta* **1993**, *278*, 189-196.
28. Vahter, M.; Mottet, N. K.; Friberg, L.; Lind, B., Shen, D. D.; Burbacher, T. *Toxicol. Appl. Pharmacol.* **1994**, *124*, 221-229.
29. Bulska, E.; Emteborg, H.; Baxter, D. C.; Frech, W.; Ellingsten, D.; Thomassen, Y. *Analyst* **1992**, *117*, 657-663.
30. Tanaka, H.; Morita, H.; Shimomura, S.; Okamotu, K. *Anal. Sci.* **1993**, *9*, 859-861.
31. Winfield, S. A.; Boyd, N. D.; Vimy, M. J.; Lorscheider, F. L. *Clin. Chem.* **1994**, *40*, 206-210.
32. Powell, M. J.; Quan, E. S. K.; Boomer, D. W. *Anal. Chem.* **1992**, *64*, 2253-2257.
33. Shum, S. C. K.; Pang, H.; Houk, R. S. *Anal. Chem.* **1992**, *64*, 2444-2450.
34. Kucera, J.; Soukal, L.; Horáková, J. *Fresenius J. Anal. Chem.* **1993**, *345*, 188-192.
35. Reardan, D. T.; Meares, C. F.; Goodwin, D. A.; McTigue, M.; David, G. S.; Stone, M. R.; Leung, J. P.; Bartholomew, R. M.; Frincke, J. M. *Nature* **1985**, *316*, 265-268.
36. Wylie, D. E.; Lu, D.; Carlson, L. D.; Carlson, R.; Babacan, K. F.; Schuster, S. M.; Wagner, F. W. *Proc. Natl. Acad. Sci. USA* **1992**, *89*, 4104-4108.
37. Irving, H. M. N. H. *Dithizone;* The Chemical Society, Burlington House: London, Great Britain, 1977.
38. Casas, J.S.; Jones, M. M. *J. Inorg. Nucl. Chem.* **1980**, *42*, 99-102.
39. Maeda, M.; Okada, K.; Wakabayashi, K.; Honda, K.; Ito, K.; Kinjo, Y. *J. Inorg. Biochem.* **1991**, *42*, 37-45.
40. Bond, A. M.; Scholz, F. *J. Phys. Chem.* **1991**, *95*, 7460-7465.
41. Robert, J. M.; Rabenstein, D. L. *Anal. Chem.* **1991**, *63*, 2674-2679.
42. Kemula, W.; Hulanicki, A.; Nawrot, W. *Rocz. Chem.* **1964**, *38*, 1065-1072.
43. Russeva, E.; Budevsky, O. *Talanta* **1973**, *20*, 1329-1332.
44. Lo, J. M.; Yu, J. C.; Hutchison, F. I.; Wal, C. M. *Anal. Chem.* **1982**, *54*, 2536-2539.

45. Sachsenberg, S.; Klenke, T.; Krumbein, W. E.; Zeeck, E. *Fresenius J. Anal. Chem.* **1992**, *342*, 163-166.
46. Valentine, W. M.; Amarnath, V.; Graham, D. G.; Anthony, D. C. *Chem. Res. Toxicol.* **1992**, *5*, 254-262.
47. Means, G. E.; Feeney, R. E. *Biochemistry* **1968**, *7*, 2192-2201.
48. Steglich, W.; Hinze, S. *Synthesis* **1976**, 399-401.
49. Staros, J. V.; Wright, R. W.; Swingle, D. M. *Anal. Biochem.* **1986**, *156*, 220-222.
50. Anjaneyulu, P. S. R.; Staros, J. V. *Int. J. Pept. Protein Res.* **1987**, *30*, 117-124.
51. Rowe, L. D.; Beier, R. C.; Elissalde, M. H.; Stanker, L. H.; Stipanovic, R. D. *Synth. Commun.* **1993**, *23*, 2191-2197.
52. Bekheit, H. K. M.; Lucas, A. D.; Szurdoki, F.; Gee, S. J.; Hammock, B. D. *J. Agric. Food Chem.* **1993**, *41*, 2220-2227.
53. Weygand, F.; Frauendorfer, E. *Chem. Ber.* **1970**, *103*, 2437-2449.
54. Nomura, M.; Imai, M.; Usuda, S.; Nakamura, T.; Miyakawa, Y.; Mayumi, M. *J. Immunol. Methods* **1983**, *56*, 13.
55. Cole, T. G.; Johnson, D.; Eveland, B. J.; Nahm, M. H. *Clin. Chem.* **1993**, *39*, 695-696.
56. Thoman, C. J.; Habeeb, T. D.; Huhn, M.; Korpusik, M.; Slish, D. F. *J. Org. Chem.* **1989**, *54*, 4476-4478.
57. Al-Khayat, B. H.; Al-Kafaji, J. K. *British Polym. J.* **1989**, *21*, 369-373.
58. Brown, H. C.; Georghegan, P. *J. Amer. Chem. Soc.* **1967**, *89*, 1522-1524.
59. Brown, H. C.; Rei, M.-H. *J. Amer. Chem. Soc.* **1969**, *91*, 5646-5647.
60. Williams, J.; Elleman, T. C.; Kingston, I. B.; Wilkins, A. G.; Kuhn, K. A. *Eur. J. Biochem.* **1982**, *122*, 279-303.
61. Hirose, M.; Akuta, T.; Takahashi, N. *J. Biol. Chem.* **1989**, *264*, 16867-16872.
62. Sola, M. *Eur. J. Biochem.* **1990**, *194*, 349-353.
63. McComb, R. B.; Bowers, G. N.; Posen, S. *Alkaline Phosphatase;* Plenum Press: New York, NY, 1979.
64. Plocke, D. J.; Vallee, B. L. *Biochemistry* **1962**, *1*, 1039-1043.
65. Fosset, M.; Chappelet-Tordo, D.; Lazdunski, M. *Biochemistry* **1974**, *13*, 1783-1795.
66. Bradshaw, R. A.; Cancedda, F.; Ericsson, L. H.; Neumann, P. A.; Piccoli, S. P.; Schlesinger, M. J.; Shriefer, K.; Walsh, K. A. *Proc. Natl. Acad. Sci.* **1981**, *78*, 3473-3477.
67. Kim, E. E.; Wyckoff, H. W. *J. Mol. Biol.* **1991**, *218*, 449-464.
68. Watanabe, T.; Wada, N.; Chou, J. Y. *Biochemistry* **1992**, *31*, 3051-3058.
69. Munaf, E.; Takeuchi, T.; Ishii, D.; Haraguchi, H. *Anal. Sci.* **1991**, *7*, 605-609.
70. Malm, O.; Pfeiffer, W.C.; Souza, C. M. M.; Reuther, R. *Ambio* **1990**, *19*, 11-15.

RECEIVED November 1, 1994

Data Interpretation, Quality Assurance, and Regulatory Applications

Chapter 19

Interpretation of Immunoassay Data

James F. Brady

Ciba Crop Protection, P.O. Box 18300, Greensboro, NC 27419–8300

Few discussions in the field of immunoassay analysis have centered on the interpretation of immunoassay data. Personnel at Ciba Plant Protection have analyzed several thousand real-world samples by immunoassay and have reached some conclusions from this experience. This paper will discuss the various types of standard curves in use, the means by which the limits of quantitation are determined and the idea of analyte-equivalency to more clearly elucidate the problem of how to interpret immunochemical data. The goal of this paper is to answer the question, "What do the numbers generated by an immunoassay mean?"

Of all the aspects of immunoassays, few discussions have focused on interpretation of data. This is surprising given the potential impact of immunochemical methods on pesticide residue analysis. These techniques have traveled from the laboratory bench in academic surroundings to application in industrial and regulatory settings. Several companies now manufacture immunoassay kits for a variety of agrochemicals in addition to other compounds of regulatory interest. The cost-effective aspects of immunoassays are now being realized on a daily basis in Madison, Wisconsin, for example, where analysts in the State Laboratory of Hygiene screen drinking water for triazine residues for less than twenty dollars per sample. To date, personnel at Ciba Plant Protection have run several thousand real-world soil and water samples. Based on our experience, we have concluded it has become not a trivial concern to examine the interpretation of immunoassay data. By examining the various types of standard curves in use, the means by which the limit of quantitation are determined and the concept of analyte-equivalency, some insight can be gained into this issue. The goal of this paper is to answer the question, "What do the numbers generated by an immunoassay mean?"

For the sake of discussion, a typical enzyme immunoassay is illustrated in Figure 1. Selection of this example is not meant to exclude other assay variants such as the magnetic particle bead assay; since the basic principles remain the same, the following discussion should readily apply to most immunoassays used to detect pesticide residues. Sample or standard solutions are added to a well of a polystyrene microtiter plate or culture tube coated with antibodies specific to a particular test substance.   An enzyme conjugate consisting of a form of the test substance covalently bound to the enzyme is also added. The two forms of the analyte compete for the limited number of antibody binding sites over a predetermined incubation time. The reagents are then removed, the vessel washed and a colorless substrate added. The enzyme converts the substrate to a colored form and the absorbance of this colored signal is measured at the conclusion of the assay. In this case, the reporter enzyme and substrate are horseradish peroxidase and tetramethyl benzidine, respectively. The absorbance is monitored at 450 nm.

Standard curves. In the developmental phase of an immunoassay, analysts adjust the amounts of antibody and enzyme tracer used to optimize the assay to its intended use. This usually involves making the assay as sensitive as the reagents allow and utilizes some type of checkerboard experiment. These concentrations are then applied to examining inhibition over a wide range of concentrations of the test analyte. These results take the form of a sigmodial response (Fig. 2)(1,2). This response can be described as a linear dose-response region bounded by two "tails" over which varying doses yield similar responses. At the low dose end of the curve, assay response is insensitive to change as the doses applied are too small to effect a difference in the signal generated. Above 0.1 ppb, a change in the response variable correlates with increasing doses of inhibitor until approximately 10 ppb of analyte are added. At this point, the assay is saturated as the amount of antibody in the test is overwhelmed by excess inhibitor. Consequently, all doses of about 10 ppb or greater generate similar responses, regardless of the amount of inhibitor added. Since the response variable plotted on the sigmodial tails does not correlate on a one-to-one basis with a corresponding dose, the tail regions cannot be included in a standard curve expression. Ordinarily, investigators either proceed from this point and work within the constraints of the linear dose-response region as defined above (Fig. 3) or attempt to further fine-tune the reagent concentrations to achieve greater sensitivity.

A typical standard curve produced by working within the linear dose-response region uses a logarthimic abscissa and linear ordinate.   While these scales are convenient to work with, a scan of the literature shows a variety of mathematical transformations have been applied to immunoassay data (Table I).

Several workers have utilized the simplicity of the first expression.   Strictly speaking, this is not a transform as the raw data are reported and analyzed. Using the raw absorbances, an analyst generates a regression function comparing the response of the standards against the log of their concentrations. This is a simple procedure that can be easily carried out on a hand-held calculator and the results plotted on log-linear paper. This form of data treatment does not require a computer

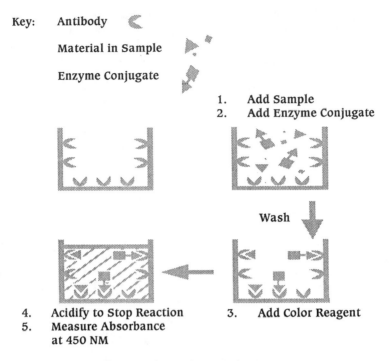

Figure 1. Principles of enzyme immunoassay.

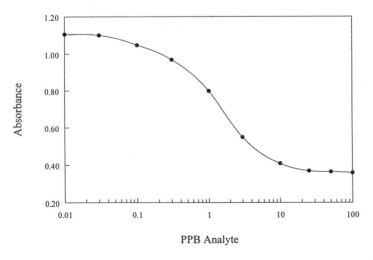

PPB Analyte

Figure 2. Typical sigmodial dose-response curve generated over a range of doses with fixed concentrations of antibody and enzyme tracer.

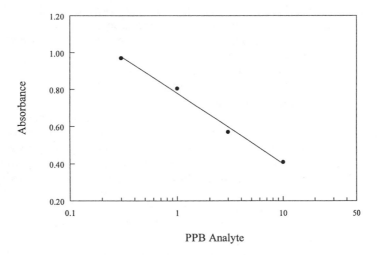

Figure 3. Standard curve derived from the linear region of the sigmodial dose-response curve.

Table I. Mathematical transformations applied to immunoassay data.

	Transform	References
1.	$y = m\log(x) + b$	3-8
2.	$\% y = m\log(x) + b$	9-11
3.	$B/B_o = m\log(x) + b$	12-14
4.	$\% B/B_o = m\log(x) + b$	15-26
5.	$\% \text{Inhibition} = m\log(x) + b$	27-30
6.	$\log(B/B_o) = m\log(x) + b$	31
7.	$\text{logit}(y) = m\log(x) + b$	32, 33
8.	$\text{logit}(B/B_o) = m\ln(x) + b$	34, 35
9.	$\text{logit}(\% B/B_o) = m\log(x) + b$	36-39
10.	$y = (a-d)\big/\left(1 + \left(\dfrac{x}{c}\right)^b\right) + d$	40-42

but should an automated system be used, software validation would be a straightforward procedure.

The second and third expressions normalize the absorbance readings relative to the response of the blank, or zero dose (Fig. 4). Although only a small operation was performed on the data, the raw data are no longer being used. An additional step is now required to verify the validity of the analytical results. A further manipulation is shown in the fourth transform which expresses the ordinate value of the third expression as a percentage. The wide use of processing data in this fashion may be due to the routine application of this transform to the results of cross-reactivity experiments to determine the parameter I_{50}, the concentration of test substance that yields half the response of the zero dose. These transformations superficially appear to have little effect on the data but can substantially affect the manner in which it is interpreted. If the range of absorbances from the responses of the standards is very small, for example 0.2 to 0.3 absorbance units, normalizing sample and standard responses can make the assay appear to respond over a broader range (43). The response of an assay that indeed spans a wide range of absorbances can, in turn, be compressed via normalization of the responses. The spread of measurement of each sample will also be compressed, thereby reducing assay variability and improving assay precision. This phenomena was observed by Thurman et al. (44).

The fifth transform subtracts the results obtained by the fourth expression from one hundred percent. The net effect is to reverse the order of the ordinate scales which produces a standard curve with a positive slope instead of the negative slope calculated by other means (Fig. 5). This approach affords no improvement to the interpretation of the data but does further increase the complexity of the calculations and, hence, the validation process.

Taking the logarthim of the normalized absorbance is not frequently done (31). This procedure introduces a greater level of complexity to the calculations and the validation process.

Logit transforms are among the most complex transformations applied to immunoassay data. Early immunochemistry methods used this transform but it is currently applied primarily to magnetic particle bead assays (31-33). The conversion to logit units is accomplished by the equation

$$\text{logit}(y) = \ln(y / 1 - y)$$

for values of y (absorbance readings) that lie between 0 and 1. To accommodate responses equal to or greater than 1, normalized responses (B/B_o) are transformed, with the values of y expressed as percentages (1). The logit transform attempts to add the tail regions of the original sigmodial response in a linear fashion to the dose-response region by weighting the responses in the tails such that the small changes in slope that occur there are given greater weight than the slope along the middle of the curve where a true dose-response exists (1) (Figs. 6, 7). Preferentially weighting the results of extreme responses, both large and small doses, is a questionable practice since assay precision is reduced at the extreme ends (Fig. 8). This is a consequence

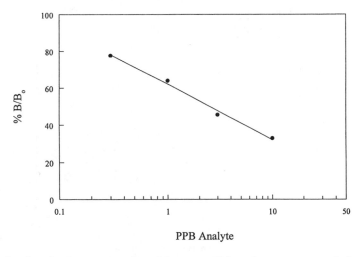

Figure 4. Standard curve produced by normalizing the responses relative to the response of the zero dose standard.

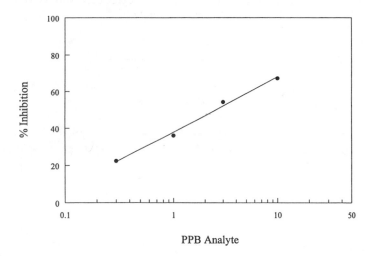

Figure 5. Standard curve produced by converting normalized responses to percentages and subtracting those percentages from 100 percent.

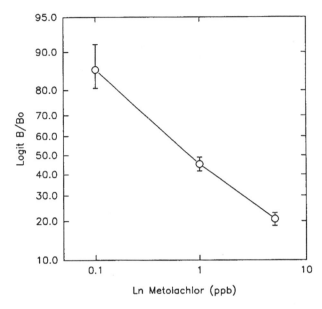

Figure 6. Standard curve using a logit transform. Note the non-linear scale on the response axis compresses the responses in the middle of the curve and expands the responses at the low and high ends of the curve. (Reproduced with permission from ref. 35. Copyright 1993 American Chemical Society.)

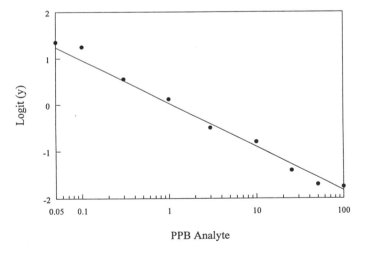

Figure 7. Standard curve using a logit transform with the scale on the vertical axis in "logit units" (1).

distributional differences between the tails and linear regions of the dose-response curve. Unless the standard curve is restricted to the central region of the dose-response curve, where distributional properties are maintained, the legitimacy of using the logit transform is in question when the root of an assay's apparent sensitivity lies in mathematics and not in physical reality or direct observation.

End-users of immunochemical pesticide residue methods using a logit transform may face two difficulties. First, if the tests are to be conducted under good laboratory practices guidelines (GLP), performing the calculations to verify instrument or computer output will be a time consuming task. More importantly, results generated from the low dose end of the transformed curve may be suspect. This may require the methodology undergo revision to ensure the data generated are valid. Although this transform has been in common use in clinical chemistry applications, immunoassays applied to pesticide residue analysis come under the umbrella of regulatory compliance chemistry, not clinical chemistry. Immunoassay developers should recognize the constraints GLP requirements place on residue chemists and design the assay to meet their needs.

The final transform listed is the four-parameter log fit (1). This method is unsuited to immunological determination of pesticide residues because it describes the entire dose-response curve including non-linear regions (Fig. 9). It also presents some daunting validation problems. Variables a, b, c and d are determined by iterative processes solved by computer software (45). Replicating those processes by manual means would indeed be challenging.

After examining the various types of standard curves in use, the question remains as to why different method developers favor particular ways of constructing dose-response curves. A presumptive reply is that a particular transform is used to linearize a data set while concurrently obtaining the most sensitive detection limit. Unfortunately, the end-user pays a greater price in conducting more arduous validation procedures as the mathematics applied become increasingly complex. The developer must always bear in mind that software applied to analyzing data must be validated by each end-user under GLP, not merely by a manufacturer or a vendor. Moreover, computer spreadsheets used to perform the validation calculations must themselves be validated prior to use.

A recommendation is simply not to transform the absorbance values at all. Keep the mathematics to a minimum and make it easy for end-users to verify their results. Tijssen enumerated positive and negative characteristics for reporting of immunoassay data and listed "easy comprehension" of data handling as his primary positive aspect. To the contrary, he decried transformation of dose-response curves as a means of compressing experimental error with apparent improvement of the data (2). Using the straightforward equation

$$y = m\log(x) + b$$

allows the analyst to perform the required calculations on a hand-held calculator if applicable computer software is not available. This permits end-users, especially

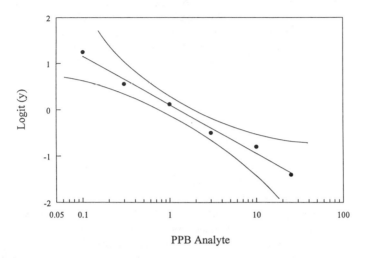

PPB Analyte

Figure 8. Plot of the confidence limits surrounding a standard curve calculated using a logit transform. Note the limits increase at the ends of the curve, indicative of reduced precision at these values. Adapted from ref. 42.

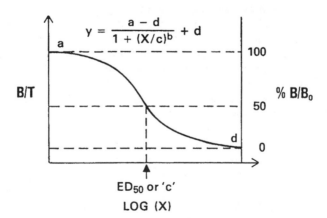

$$y = \frac{a - d}{1 + (X/c)^b} + d$$

Figure 9. Standard curve produced by application of the four-parameter log fit. Note that regions other than the linear dose-response region are included in the standard curve expression. (Reproduced with permission from ref. 1. Copyright 1981 Masson Publishing.)

those without a great of experience with immunochemical mathematics, to easily reproduce their calculations.

This approach can be documented on a basic data reporting sheet displaying sample identification codes, absorbance values and the amount of analyte found (Fig. 10). The responses of the standards used to generate the regression curve can be listed along with sample results. Other information required by GLP such as test substances analyzed for, project description or number, analyst name, date of analysis and notebook references can also be included. With all the information pertinent to the generation of results on one page, an analyst, quality assurance auditor or regulator can easily check the calculations to verify reported results. Thus, what may first appear as a simplistic approach to the treatment of the data is actually a convenient and powerful way of assuring the interpretability and quality of the results.

Limit of quantitation.   Just as pesticide residue chemists using conventional techniques have a standardized means of determining the sensitivity of an analytical method, so too must immunochemists have criteria by which the sensitivity of immunoassays can be evaluated.   To date, the means by which immunoassay sensitivity is assessed are not standardized.   This is not surprising given immunoassays have only relatively recently been adapted to residue analysis and the problem faced by an immunochemical analyst is the opposite of what a chromatographer encounters.   Moreover, the idea of the limit of detection (LOD) is often mistaken for the limit of quantitation (LOQ).

Assay sensitivity can be regarded as involving two parameters: the limit of detection, the lowest concentration of analyte that can be determined to yield a response statistically significant from that of the blank, and the limit of quantitation, the level above which quantitative results may be obtained with a specified degree of confidence (46).

It would appear superficially that analysts using chromatographic techniques must solve the same kind of problem faced by immunochemists when determining a method's limit of detection.   In a chromatographic analysis, the LOD is usually defined in reference to signal-to-noise ratios as, for example, the amount of analyte that produces a detector response twice that of background detector noise (47).   The analyst thus tries to distinguish a signal from the smallest instrumental output, the baseline signal.   In immunoassays, the situation is reversed.   The analyst instead attempts to discriminate part of the maximal signal from the maximal signal with some degree of confidence.   Put another way, an immunochemist attempts to determine what the smallest dose is that yields a signal that can be statistically distinguished from the signal arising from the zero dose standard.

Although immunochemists have used a variety of approaches to determine the sensitivity of immunoassays, relatively few methods have been described (Table II). Feng, for example, selected 80% bound, corresponding to 0.2 ppb, as the low end of his standard curve (9).   Other authors have become increasingly less conservative. Several chose 90% B/B_0 (12, 20, 25, 32, 33) and at least one researcher opted for 95% bound (31).   Schlaeppi et al. (17) and Schwalbe et al. (16) calculated the

Immunoassay Data Sheet				
Date 3-1-94	Project #101174	Analyst DHS, JFB		
Sample Matrix Water	Protocol #15-90	Notebook ref.3599/16		
Test Substance Atrazine		Set #8		
Sample name	Sample ID	Absorbance	PPB found	Comment
BAC2001	16994	.483, .474	1.24, 1.30	diluted 1:1
BAC2002	16995	.781, .794	.16, .15	
DH334	16996	.548, .572	.46, .42	sediment in bottle
DH756	16997	.458, .455	.70, .71	diluted 1:9
DJ703	16998	.577, .587	.41, .39	
SG1070900	16999	.825, .857	.13, .12	
SG1090930	17000	.959, 1.029	<0.10, <0.10	
Standards				
0		1.099, 1.202		$y = -.458 \log(x) + .578$
0.1		.860, .869		$r = -0.993$
0.3		.687, .639		
0.5		.500, .467		
1		.365, .375		

Figure 10. Immunoassay data sheet.

Table II. Methods for determining the sensitivity of immunoassays.

	Method	References
1.	80% Bound	9, 22, 26
2.	90% Bound	12, 20, 25, 32, 33
3.	95% Bound	31
4.	Standard deviations from the blank mean	16, 17
5.	Visual inspection of curve	11,19, 28, 36

standard deviation about the mean measurement of the blank dose signal and selected two and three times that value, respectively, as the minimal detectable dose. Others simply estimated the smallest amount their assays could detect by visual inspection of the dose-response curves (11, 19, 28, 36).

Although multiple criteria have been applied to assessing assay sensitivity, three problems remain. First, all of these methods address determining the LOD from the perspective of the response variable only and do not account for the variability inherent in the measurement process. Second, these approaches are used to calculate the LOD only and do not deal with how to determine the LOQ. Finally, use of multiple techniques makes it difficult for end users to discern if an assay is actually as sensitive as the claims made for it. This issue, of course, is more important to users of commercially produced immunoassay kits than to readers of academic research papers.

The methods discussed above share the common characteristic of determining the LOD from the perspective of the response variable only and fail to address the variability in the process by which those responses were obtained. The net effect of arbitrarily selecting some level, such as 80% bound, is shown in Figure 11. The response value is merely inserted into the regression function and a corresponding dose is calculated. This approach implicitly assumes the best fit line was based on point estimates, not replicate responses of standards, the measurements of which are inherently variable. As a result, a method developer could select, almost randomly, a level suited to the assay sensitivity desired without regard to whether the precision of measurement can justify that selection. This would be particularly handy to support a manufacturer's claim that a given assay is more sensitive than that of a competitor. This practice is of particular concern given the spread of measurement of most methods increases near the LOD.

By itself, the LOD is an indicator of the sensitivity of the assay. However, the LOQ is a measure of the *utility* of the test. As defined by Keith et al., "the limit of quantitation...is...the level above which quantitative results may be obtained with a specified degree of confidence." It is, consequently, "the lower limit of the useful range of measurement" (46). The LOD can thus be viewed as a theoretical limit while the LOQ is the practical, working limit, on a per substrate basis, at or above which an analyst can obtain valid analytical results.

LOQ's can be determined for immunological techniques in the same fashion as they are for chromatographic methods, by conducting recovery studies in each sample matrix to which the assay is applied. The analyst simply performs fortification experiments at the proposed LOQ in each sample matrix. A claim that an assay can quantitate to 0.10 ppb in soil, for instance, should be supported by results of analyses of replicate aliquots fortified at that level. The mean of these experiments should fall between 70-120% recovered with a relative standard deviation of $\pm$ 20%. In this light, it is interesting that most product inserts or publications do not discuss an LOQ but instead focus on the LOD. The utility of the assay is thus mistaken for the detection limit. Fortifications that are discussed, if any, are typically run at much higher concentrations. Rubio et al., for example, claim a minimal detectable concentration of 50 parts per trillion of atrazine in water but do not provide any recovery data less than 0.5 ppb (31). The true lower limit of the useful range of measurement for this method is therefore 0.5 ppb.

One solution to a standardized approach for calculating the LOD is based on Rodbard's work (48). He developed a means of determining the LOD (he refers to it as a "minimal detectable dose") by calculating the response (absorbance) corresponding to the LOD, inserting this value into the standard curve expression and solving it for the dose. This was accomplished using his equation (1a)

$$y_{\min} = \overline{y}_1 - ts\left[\frac{1}{n_1} + \frac{1}{n_2}\right]^{\frac{1}{2}}$$

in which $y_{\min}$ is the response corresponding to the LOD, $\overline{y}_1$ is the mean response of the zero dose replicates, n_1 is the number of those replicates and n_2 represents the number of replicates run of an unknown. The t statistic is the percentile of Student's t distribution for a one-sided test at 95% probability with $n_2 - 2$ degrees of freedom (df).

This equation was modified by Ciba statisticians by substituting s, an estimate of the standard deviation of the response for a dose equal to the LOD, with the root mean square error ($RMSE$), a regression parameter generated as part of a regression analysis package. (Regression analysis software packages such as those offered by SAS (49) or Microsoft Excel (50) calculate the $RMSE$; some packages refer to the $RMSE$ as the "standard error." SAS is recognized as a validated system under GLP requirements; to date, Microsoft Excel is not.) The use of $RMSE$ permits utilization of a more precise estimate of the variance by including in its derivation all of the standard replicates involved in generating the regression curve. Moreover, the eventual value determined by this process is now dependent upon on the precision of measurement. The value of n_2 becomes $n_2 - 2$ because the actual number of standard replicates must be adjusted for the df used in obtaining the regression estimates. Thus, n_2 is now equivalent to the df for the sum of squared errors (SSE) from the regression analysis ($n_2 = 10 - 2 = 8$, assuming five standard concentrations each run in duplicate). This results in a slightly more conservative estimate of the LOD than if the unadjusted numbers of standard replicates were used. The variable n_1 value in Rodbard's equation remains unchanged as the number of zero dose replicates ($n_1 = 2$). The t statistic can be found in any statistics manual using the 0.05 level with eight degrees of freedom (df for the SSE), or 1.860.

The equation resolves to

$$y_{\min} = \overline{y}_1 - 1.860\,RMSE\left[\frac{1}{2} + \frac{1}{8}\right]^{1/2}$$

or

$$y_{\min} = \overline{y}_1 - RMSE\,(1.470)$$

which can be readily calculated. The value for $y_{\min}$ is then inserted into the standard curve expression and solved for x. In practice, the value obtained is then compared to the concentrations of the standards. If it less than the smallest standard, the LOD must default to the concentration of that standard since a detection cannot be made outside of the range of measurement. Should the LOD be calculated in excess of the smallest standard, it assumes the derived value.

Utilizing Rodbard's approach as modified by Ciba statisticians, while certainly not the final solution to this problem, does serve several purposes. First, it offers a standardized method for determining an LOD where none currently exists. If all method developers evaluated their assays by equal criteria, much of the effort presently required to assess performance claims would be unnecessary. Second, the precision of measurement of the standard curve has a bearing upon the magnitude of y_{min} and its corresponding dose. Third, the LOD would by definition default to the concentration of the smallest standard, thereby eliminating attempts to measure beyond the range defined by the standard curve. Finally, this approach ensures the smallest standard yields a response that is statistically different from the response of the zero dose. In the event the LOD is found to be greater than the smallest standard, the assay is not sensitive enough to distinguish the smallest standard from background. End-users should be assured the sensitivity of measurement at the low end of the curve is authentic.

Rodbard's technique does not apply to all the mathematical treatments previously discussed. One of several assumptions is that the responses for standards and unknowns are normally distributed. Since data described by a logit transform, especially at the extremes of the curve, cannot be described by a Gaussian distribution, this method cannot be applied to data linearized by the logit method (48).

It is perhaps suitably ironic that despite the amount of attention the determination of the LOD has received, this parameter has little practical value. Regardless of whether the LOD is found to be less than, equal to or greater than the smallest standard, an assay's range of quantitation must be limited to the LOQ or above (46). Below the LOQ, experimental evidence supporting the utility of the method does not exist. Should the result of an analysis fall between the LOD and the LOQ, it should be reported as less than the LOQ. Although this warning may appear unnecessary to experienced analysts, many users of immunoassay products lack expertise in pesticide residue analysis. Baum, for instance, presented a paper in 1992 in which he claimed to detect atrazine at 0.04 parts per billion in Michigan ground water using a standard curve with a low end at 0.10 ppb, two and one-half times less sensitive than the perceived limit of detection (51).

When an immunoassay is available for use, the developer should include an explanation of how the LOD and LOQ were determined in a manner that an end-user could replicate or apply such calculations to additional substrates. However, just as a chromatographer does not perform a statistical verification for each run that the signal of the smallest standard exceeds the background signal by a given amount, so too should the immunochemist not be required to calculate the LOD each time an assay is run. The assay developer should bear the burden of ensuring the assay is statistically sound. The end user should be able to apply immunoassay techniques with the knowledge the assay has been optimized and its performance fully evaluated.

Nevertheless, the means by which the LOD and LOQ were determined should be made available.

Analysts using immunological methods should run procedural recoveries concurrently with samples that require extraction and isolation of the test substance from the sample matrix. Acceptable recoveries demonstrate the efficiency of the extraction procedure. If end-users wish to apply a previously developed immunoassay to a new substrate, recoveries should be run until acceptable results are obtained prior to actual sample analysis. While there is no set number of recovery samples required to be completed before sample analysis can begin, the statistical weight of the results increases with the number of replicates. Analysts should take advantage of the cost savings offered by immunoassays and run a substantial number of replicates. Analytical results of ten replicates, for example, hold much more weight than that of $n = 2$ or $n = 3$.

Analyte-equivalency. The basis of the equivalency concept lies in the inability of a measurement technique to directly measure an analyte. Instead, the concentration of an indicator species is measured. The analyst compares the sample response to the responses of the standards and implies that a given amount of analyte is present. The analyst thus infers the response of an undefined sample is similar, or equivalent to, the response of a defined solution containing known amounts of analyte. This is precisely the situation encountered in immunoassays (Fig. 1). Given that the responses of undefined solutions (samples, unknowns) are compared to the responses of defined solutions (standards), immunoassay results cannot be regarded as more than "analyte-equivalents."

Treating immunoassay results in this fashion is not novel. In 1992, Hammock gave a presentation on immunoassay applied to analysis of urinary biomarkers of triazine exposure in which he referred to unconfirmed immunoassay responses as "immunoreactive-equivalents (52)." In so doing, he recognized the need for additional evidence to identify the inhibitor. Lucas et al. described unidentified constituents of urine generating an immunological response as "immunoreactive material" pending structural confirmation (53). This author also termed the immunoassay signal derived from blank solutions as "atrazine-equivalents" since the results were obtained by regressing the sample results against a curve consisting of responses of atrazine standards. Itak et al. described the results of experiments evaluating the effects of dissolved salts and pH on an immunoassay for aldicarb as the "apparent aldicarb concentration" (20). Although these responses did not arise from, respectively, atrazine or aldicarb residue, these situations are analogous to those faced by users of immunoassay kits. Without confirmatory data, they cannot validly state what the cause of the inhibition was.

This situation is in sharp contrast to that confronting an analyst using a conventional chromatographic technique. In high performance liquid chromatography (HPLC), the analyte is measured as it flows through the detector cell. In gas-liquid chromatography (GC), the analyte contacts the detector (or energy field about the detector) as it is swept along in the stream of carrier gas. In GC/- or HPLC/mass spectrometry systems, a fragment of the analyte strikes the mass selective detector. All of these methods share the characteristic of measuring the analyte directly.

Immunoassays, however, measure an indicator species in lieu of the analyte. Users of the technology must bear in mind the determinative step consists of measuring the absorbance of a colored aqueous solution. The issue, therefore, is whether an absorbance value can be used to identify the cause of reduced tracer binding relative to a control known to lack an inhibitor. A reasonable response is that an absorbance value, by itself, is simply insufficient evidence to identify the inhibitor.

Recognition of this limitation is not merely a semantic or trivial distinction. It is the crux of the matter.   Failure to acknowledge that immunoassays provide insufficient evidence to identify an inhibitor can result in misinterpretation of immunoassay results. Typically, this error is in the form of claims that, for example, "Atrazine is being measured. Of course we're measuring atrazine!" Such claims are disingenuous at best and outright misleading in the worst case.

Confirmatory, "follow-up" analyses have sometimes yielded surprising results that illustrate this problem. Personnel at the Wisconsin State Laboratory of Hygiene in Madison routinely screen drinking water samples by immunoassay for triazine residues. A sample collected from a Federal Aviation Administration radar facility in Mayville, Wisconsin was determined to contain 27.3 ppb of atrazine, nearly ten-fold greater than the federal maximum contaminant level (MCL) of 3.0 ppb (Standridge, J., Wisconsin State Laboratory of Hygiene, personal communication). Analysis of an additional sample by gas chromatography with nitrogen-phosphorous detection found 0.21 ppb of atrazine, 93 ppb of prometon and 0.67 ppb of simazine. The confirmatory analysis not only indicated the presence of other triazines but demonstrated the levels of all compounds were within compliance limits (prometon and simazine have health advisory and maximum contaminant levels of, respectively, 100 and 1.7 ppb). Regulators time and energy would have been consumed by a false positive result if regulatory action was initiated solely on the basis of the immunoassay result.

This situation also points to the difficulty of immunologically distinguishing between various triazine herbicides. This shortcoming should be recognized by users of these kits because while numerous variations on the basic triazine theme make immunological detections difficult to interpret probably more immunoassays are sold to detect triazines than any other class of pesticide.

A second instance comes from monitoring the acetanilide herbicide, alachlor. Immunoassay screens of ground water samples from across Indiana, Kentucky and Ohio are performed at the Water Quality Laboratory of Heidelberg College in Tiffin, Ohio.   Samples from over 8100 private rural wells have been analyzed by immunoassay for alachlor residue. In 1992, immunoassay results suggested nearly 1.9% of all wells tested contained alachlor in excess of its MCL, 2.0 ppb (54). Monsanto personnel unsuccessfully attempted to confirm these residues by GC/MS.

Subsequent investigation showed the cause of the immunoassay response to reside in an ethanesulfonic acid metabolite of alachlor (ES).   Although this compound had not previously been observed in water, its presence was confirmed by workers at Monsanto and the University of Maine (55). Baker concluded that most of the immunological responses with respect to alachlor were false positives (54).

An antibody had previously been evaluated for cross-reactivity to ES but was shown to be poorly reactive, estimated at 2.3% (9). The ability of the metabolite to bind so strongly to the antibody used by Baker was a fortuitous combination of circumstances: environmental degradation occurred precisely at the site of conjugation of the alachlor hapten to the carrier protein in immunogen synthesis, just where an antibody reared against this molecular configuration might ignore a structural change! In fact, the alachlor antibody was subsequently estimated to be twenty-five percent cross-reactive to ES relative to alachlor (Ferguson, B.F., Immunosystems, Inc., personal communication).

A third example arises from a study screening Wisconsin ground water for atrazine residue conducted by the Wisconsin Department of Trade, Agriculture and Consumer Protection and Ciba Plant Protection. In this project, over two thousand ground water samples were analyzed for atrazine residue by enzyme immunoassay. Part of that study involved re-sampling wells whose original sample gave an immunoassay response greater than or equivalent to 0.35 ppb. Those "follow-up" samples were assayed at Ciba by GC/MS and immunoassay. A comparison of those analytical results is shown in Figure 12 (Brady, J.F., Ciba Plant Protection, unpublished data). A line of the equation $y = x$ was overlaid onto to this plot to show the differences between results of each method increase with increasing atrazine concentration. The strong positive bias of the immunoassay data compared to the chromatographic results is evident. From this plot it is apparent that an immunoassay response of 1.0 ppb, for instance, may be an overestimation of the actual amount of residue. While such positive bias can be beneficial to a screening method as it reduces the possibility of generating false negatives, a regulator using the technique should be familiar with this trend and anticipate that some false positives will be produced. In practice, postponing initiation of regulatory action until an immunological detection of regulatory concern is verified by confirmatory analysis would be a reasonable precaution.

These examples reinforce the notion that the unknown constituents of a sample solution are indeed unknown. The confirmatory analysis of the Mayville, Wisconisn sample helped to elucidate a problem typical of situations encountered by analysts screening water samples. On the other hand, confirmatory analyses failed to verify alachlor residues in mid-west water samples because the immunoassay was reacting to compound Monsanto personnel were not analyzing for. Subsequent work confirmed the presence of an unanticipated inhibitor. Recognition that the triazine immunoassay may yield inflated concentrations of atrazine was found to be necessary to properly interpret assay results in the Wisconsin well water study. In each of these cases, a careful, conservative approach to interpreting immunoassay data was needed to take appropriate action in response to the results generated. One aspect of this approach could be to designate results as "target analyte-equivalents." Doing so acknowledges that merely comparing the sample absorbance to the absorbances of the standards does not confer to the analyst the evidence to identify the inhibitor.

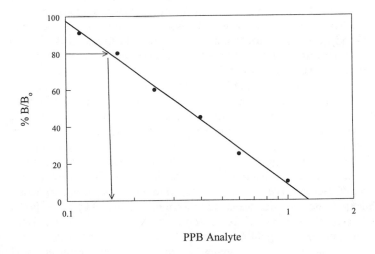

Figure 11.   Determination of the limit of detection by calculating the dose corresponding to 80% bound. This approach does not assume any variability in the measurement process.

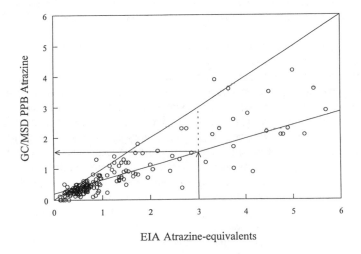

Figure 12. Comparison of immunoassay and GC/MSD analytical results for Phase II water samples from the Wisconsin well water study.   Arrows indicate an immunoassay result of 3 ppb is approximately twice that of the GC/MSD result. The magnitude of positive bias shown by the immunoassay at 3 ppb is shown by the dashed line.

Data interpretation. In light of the above discussions, the question naturally follows as to what constitutes a reasonable interpretation of immunoassay results. With an understanding of the mathematics applied to the raw absorbance data and aware that adequate performance data in the matrix at hand exists to validate the analyses, an analyst is still confronted by the question, "what do the numbers generated by an immunoassay mean?"

Negative results, as Baker put it, are unambiguous (54). An absorbance greater than that of the smallest standard is a solid indication the sample is free of test substance, or similarly reactive material, at least to that level.

Positive results, on the other hand, can arise from a variety of causes. Inhibition of enzyme tracer binding can be the result of presence of the test substance above the method's detection limit, presence of some other compound that is cross-reactive to the antibodies used or a mixture of these materials (54). Non-specific effects may also be causal factors. The increasingly disparate results between immunoassay and GC/MS in the Wisconsin well water study were not explained by mass spectrometric results since triazines known to be to cross-reactive were not found (Brady, J.F., Ciba Crop Protection, unpublished data). Wisconsin ground water is known to contain high concentrations of nitrates arising from heavy fertilizer use but the antibodies used were shown to be unaffected by up to 100 parts per million of nitrates (Brady, J.F., Ciba Crop Protection, unpublished data). Perhaps the synergy of the multitude of ionic combinations found in individual aquifers contributes to a decrease in signal. Fleeker and Cook also observed a non-specific response from the immunochemical analysis of water from mid-western sources (56). Unfortunately, their attempts to explain these effects in terms of sample conductivity or pH were unsuccessful.

What course of action to take when immunoassay positives are encountered should be determined prior to the initiation of screening samples. In an ideal world, all detects could be confirmed. Most detects, however, are very small and the expense incurred to verify the residues would not be justified. Moreover, conducting GC analyses to back up every detect would do little to realize cost savings. Confirmatory analyses, and their associated costs, should be reserved for detects of regulatory significance. What constitutes "regulatory significance," or a similar cut-off level, should be determined on a compound-by-compound basis and reflect regulatory concerns pertinent to each test substance. Detects greater than 1.0 ppb of atrazine-equivalents in water, for example, might merit confirmation given the MCL of 3.0 ppb. Since alachlor has a lower MCL, a cut-off level of 0.5 ppb of alachlor-equivalents might be more appropriate. Tailoring a water monitoring program to judiciously complement immunoassays with conventional analytical techniques can permit registrants and regulators alike to rapidly screen large numbers of samples and direct limited resources to those samples that merit further attention.

As stated at the outset, several lessons have been learned through the experience of using immunoassay techniques for the analysis of real-world samples. The foremost among them is that although immunoassay is a powerful, inexpensive screening method with enormous potential, users must recognize its limitations. In particular, analysts must be cautious not to read greater meaning into immunoassay data   than is actually there.   Given the widespread adoption of immunoassay techniques and the relative ease with which assays are performed, everyone involved with the methodology, the manufacturers, academic and industrial method developers, federal and state regulators and end-users, should be concerned with the correct generation and reasonable interpretation of immunoassay data.

Acknowledgments

The author wishes to express his gratitude to Ciba statistician Greg Pearce for his assistance with the modification of Rodbard's work and to the staff of Correspondence Center L for the final formatting of this manuscript.

Literature Cited

1.   Rodbard, D. In *Ligand Assay*, Langan, J., and Clapp, J.J. Eds., Masson Publishing: New York, **1981**: pp. 45-101.
2.   Tijssen, P. *Practice and theory of enzyme immunoassay*. Elsevier: Amsterdam, **1985**: pp. 391-416.
3.   Hunter, K.H. and Lenz, D.E. *Life Sciences*. **1982**, *30*, 355-361.
4.   Newsome, W.H. *J. Agric. Food Chem.*, **1985**, *33*, 528-530.
5.   Newsome,W.H. *J. Bull. Environ. Contam. Toxicol.* **1986**, *36*, 9-14.
6.   Forlani, F., Arnoldi, A. and Pagani, S. *J. Agric. Food Chem.*, **1992**, *40*, 328-331.
7.   Wong, R.B. and Ahmed, Z.H. *J. Agric. Food Chem.*, **1992**, *40*, 811-816.
8.   Newsome, W.H. *J. Agric. Food Chem.*, **1993**, *41*, 1426-1431.
9.   Feng, P.C.C., Wratten, S.J., Horton, S.R., Sharp,C.R. and Logusch, E.W. *J. Agric. Food Chem.*, **1990**, *38*, 159-163.
10   Brandon, D.L., Binder, R.G., Bates, A.H. and Montague, W.C. *J. Agric. Food Chem.*, **1992**, *40*, 1722-1726.
11.   Szurdoki, F., Bekheit, H.K.M., Marco, M-P., Goodrow, M. and Hammock, B. *J. Agric. Food Chem.* **1992**, *40*, 1459-1465.
12.   Lawruk, T.S., Hottenstein, C.S., Herzog, D.P., and Rubio, F.M. *Bull. Environ. Contam. Toxicol.* **1992**, *48*, 643-650.
13.   Wigfield, Y.Y. and Grant, R. *Bull. Environ. Contam. Toxicol.* **1992**, *49*, 342-347.
14.   Schneider, P. and Hammock, B.D. *J. Agric. Food Chem.* **1992**, *40*, 525-530.
15.   Rinder, D.F. and Fleeker, J.R. *Bull. Environ. Contam. Toxicol.* **1981**, *26*, 375-380.
16.   Schwalbe, M., Dorn, E. and Beyerman, K. *J. Agric. Food Chem.*, **1984**, *32*, 734-741.
17.   Schlaeppi, J-M, Fory, W. and Ramsteiner, K. *J. Agric. Food Chem.*, **1989**, *37*, 1532-1538.
18.   Harrison, R.O. and Nelson, J.O. *J. Agric. Food Chem.*, **1990**, *38*, 221-223.
19.   Giersch, T. and Hock, B. *Food Agric. Immunol.* **1990**, *2*, 85-97.
20.   Itak, J. A., Selisker, M.Y. and Herzog, D.P. *Chemosphere*, **1992**, *24*, 11-21.

21.   Schneider, P. and Hammock, B.D. *J*. *Agric*. *Food Chem*., **1992**, *40*, 525-530.
22.   Marco, M-P, Gee, S.J., Cheung, H.M., Liang, Z.Y., and Hammock, B.D. *J*. *Agric*. *Food Chem*., **1993**, *41*, 423-430.
23.   Brandon, D.L., Binder, R.G., Wilson, R.E., and Montague, W.E. *J*. *Agric*. *Food Chem*., **1993**, *41*, 996-999.
24.   Giersch, T. *J*. *Agric*. *Food Chem*., **1993**, *41*, 1006-1011.
25.   Itak, J.A., Olson, E.G., Fleeker, J.R. and Herzog, D.P. *Bull*. *Environ*. *Contam*. *Toxicol*. **1993**, *51*, 260-267.
26.   Schneider, P., Goodrow, M., Gee, S.J. and Hammock , B.D. *J*. *Agric*. *Food Chem*., **1994**, *42*, 413-422.
27.   Van Emon, J., Hammock, B. and Seiber, J. *Anal*. *Chem*. **1986**, *58*, 1866-1873.
28.   Riggle, B. and Dunbar, B. *J*. *Agric*. *Food Chem*., **1990**, *38*, 1922-1925.
29.   Hill, A.S., Beasely, H.L., McAdam, D.P. and Skerrit, J.H. *J*. *Agric*. *Food Chem*., **1992**, *40*, 1471-1474.
30.   Koppatschek, F.K., Liebl, R.A., Kriz, A.L. and Melhado, L.L. *J*. *Agric*. *Food Chem*., **1990**, *38*, 1519-1522.
31.   Rubio, F.M., Itak, J.A., Scutellaro, A.M., Seliskers, M.Y., and Herzog, D.P. *Food Agric. Immunol*. **1991**, *3*: 113-125.
32.   Newsome, W.H. and Collins, P.G. *Bull*. *Environ*. *Contam*. *Toxicol*. **1991**, *47*, 211-216.
33.   Itak, J.A., Selisker, M. Y., Jourdan, S.W., Fleeker, J. R., and Herzog, D.P.*J*. *Agric*. *Food Chem*. **1993**, *41*, 2329-2332.
34.   Lawruk, T.S., Lachman, C.E., Jourdan, S. Fleeker, J. R., Herzog, D. and Rubio, F. M. *J*. *Agric*. *Food Chem*. **1993**, *41*, 747-752.
35.   Lawruk, T.S., Lachman, C.E., Jourdan, S. Fleeker, J. R., Herzog, D. and Rubio, F. M. *J*. *Agric*. *Food Chem*. **1993**, *41*, 1426-1431.
36.   Ercegovich, C.D., Vallejo, R.P., Gettig, R.R., Woods, L., Bogus, E.R. and Mumma, R.O. *J*. *Agric*. *Food Chem*. **1981**, *29*, 559-563.
37.   Huber, S.J. and Hock, B. *Z*. *Pflanzenkrankh Pflschutz* **1985**, *92*, 147-156.
38.   Huber, S.J. *Chemosphere* **1985**, *14*, 1795-1803.
39.   Wittman, C. and Hock, B. *J*. *Agric*. *Food Chem*. **1993**, *41*, 1796-1803.
40.   Harrison, R.J., Braun, A.l., Gee, S.J., O'Brien, D.J. and Hammock, b.D. *Food Agric. Immunol*. **1989**, *1*, 37-51.
41.   Skerritt, J.H., Hill, A.S., Beasely, H.L., Edward, S.L. and McAdam, D.P. *J*. *AOAC Int*. *75*, 519-528.
42.   Lucas, A.D., Schneider, P., Harrison, R.O., Seiber, J.N., Hammock, B.D., Biggar, J.W., and Rolston, D.E. *Food Agric*. *Immunol*. **1991**, *3*, 155-167.
43.   Chard, T. *An introduction to radioimmunoassay and related techniques*. Elsevier Biomedical: Amsterdam, 1982: pp. 15-21, 207.
44.   Thurman, E.M., Meyer, M. Pomes, M., Perry, C.A., and Schwab, A.P. *Anal*. *Chem*. **1990**, *62*, 2043-2048.
45.   Rodbard, D. *Clin*. *Chem*. **1974**, *20*, 1255.
46.   Keith, L.H., Crummet, W., Deegan, J., Libby, R.L., Taylor, J.K., and Wentler, G. *Anal*. *Chem*. **1983**, *55*, 2210-2218.

47. Johnson, E.L. and Stevenson, R. *Basic Liquid Chromatography.* Varian Associates, Palo Alto, 1978. p. 272.
48. Rodbard, D. *Anal. Biochem.* **1978**, *90*, 1-12.
49. SAS Institute, Cary, NC.
50. Microsoft Excel, version 4.0, 1992, Microsoft Corporation.
51. E.J. Baum. 203rd ACS National meeting. San Francisco, CA. April, 1992, paper 173.
52. Hammock, B.D. 203rd ACS National meeting. San Francisco, CA. April, 1992, paper 185.
53. Lucas, A.D., Jones, A.D., Goodrow, M.H., Saiz, A.G., Blewett, C., Seiber, J.N. and Hammock, B.D. *Chem. Res. Toxicol.* **1993**, *3*, 107-116.
54. Baker, D.B., Bushway, R.J., Adams,S. and Macomber,C. *Environ. Sci. Technol.* **1993**, *27*, 562-564.
55. Macomber, C., Bushway, R.J., Perkins, L.B., Baker, D., Fan, T., and Ferguson, B. J. *Agric. Food Chem.* **1992**, *40*, 1450-1452.
56. Fleeker, J.R. and Cook, L.W. In *Immunoassays for Trace Analysis*; Vanderlaan, M., Stanker, L.K., Watkins, B.E. and Roberts, D.W. Eds.; ACS Symposium Series; American Chemical Society: Washington, D.C., 1991, Vol. 451: pp. 78-85.

RECEIVED October 28, 1994

Chapter 20

Guidelines to the Validation and Use of Immunochemical Methods for Generating Data in Support of Pesticide Registration

Charles A. Mihaliak[1] and Sharon A. Berberich[2]

Analytical Environmental Immunochemical Consortium,
1427 West 86 Street, #102, Indianapolis, IN 46260

Immunochemical methods are rapidly being adopted for the detection and quantitation of pesticides in the environment. The use of and necessity for development of these types of immunochemical methods is evident to pesticide registrants. Immunoassays are rapid, sensitive, easy to use, incur minimal cost, and allow simultaneous analysis of large numbers of samples compared to other, more widely used analytical methods. Although immunoassay offers many practical advantages, acceptance of these methods is dependent upon several factors, including the demonstration of quality and validity compared to more traditional methods. Currently, there are no specific guidelines for the validation and use of immunochemical environmental methods to support product registrations under Federal Insecticide, Fungicide, and Rodenticide Act (FIFRA). An ongoing dialog between the Environmental Protection Agency Office of Pesticide Products (EPA/OPP) and the users and developers of this technology is essential to ensure proper implementation of immunochemical methods for environmental applications. This document will review the current EPA guidance for proper method validation in context of FIFRA, used for acceptance of analytical methods. In addition, appropriate use of immunochemical methods for environmental applications will be discussed and a set of interim guidelines proposed for registrants who are developing and using immunochemical methods. Adoption of guidelines that incorporate concepts similar to those reviewed in this document will promote consistent validation, data reporting, and application of immunochemical environmental methods by developers and users of this technology.

Many of the opinions and suggestions presented here are those of members of the Analytical Environmental Immunochemical Consortium and not of the EPA/OPP. However, we have made an effort to present suggestions and ideas that are consistent with EPA/OPP policy.

---

[1]Current address: DowElanco, P.O. Box 68955, Indianapolis, IN 46268–1053
[2]Current address: Monsanto Company, 700 Chesterfield Parkway North–GG4A, Chesterfield, MO 63198

Immunochemical analytical methods are currently being developed by several agricultural chemical companies for use in studies performed in support of product registrations under the Federal Insecticide, Fungicide and Rodenticide Act (FIFRA) (*1*).

Immunochemical methods for pesticides were relatively unavailable when the current FIFRA guidelines were developed. They now offer several potential benefits for both registrants and regulators (Table 1) (*2,3*). The proliferation of immunoassay has necessitated the development of guidelines under which immunochemical methods can be validated and used in support of FIFRA product registrations. The objectives of this paper are to review the current guidance given to registrants by the EPA/OPP and to propose general guidelines for the validation, implementation and use of immunochemical methods. The intention is to combine the current EPA/OPP policy with the perspectives of developers of the technology into a coherent set of interim guidelines which can be used by registrants and regulators. Additionally, the guidelines must fit within the context of the regulatory framework and FIFRA guidelines.

This document should not be viewed as a finalized guideline. It is intended to generate further discussion among developers and regulators. The guidelines will require sufficient flexibility to accommodate improvements in the technology and to allow for policy modifications that might occur as the level of acceptance and use of immunochemical methods increases. The initial guidelines may have to be modified or reviewed to accommodate newer, developing immunochemical technologies (e.g., biosensors).

Interpretation of Current EPA / OPP Policy

Over the past two years, the EPA/OPP has begun to provide registrants with guidance regarding the validation and use of immunochemical methods. Most of the positions taken by the EPA/OPP have been distributed through internal memos or correspondence with registrants, immunoassay kit manufacturers, or organizations such as the Analytical Environmental Immunochemical Consortium (AEIC). Registrants who wish to use immunochemical methods to generate product registration data must demonstrate that the methods meet the all existing criteria for traditional analytical methods (*11,12*). Immunochemical methods should be supported with an acceptable written method and suitable validation data, similar to traditional methods. If the method is intended to complement an existing gas chromatography (GC) or liquid chromatography (LC) method, registrants must demonstrate that data generated by the immunochemical method is comparable to or better than data generated by an analytical method such as GC or LC (*4*). Performance-based method validation is encouraged by the EPA/OPP for all analytical methods; thus, performance data needs to be submitted for each method. For quantitative methods, confirmation of the limits of quantitation, delineation of the quantitative range, estimation of precision and accuracy, and evaluation of interferences are the minimum data requirements. The range for the result, within a stated degree of confidence, must be specified when using qualitative screens (*4*). A more detailed discussion of method validation is presented below.

The original, EPA/OPP policy for immunochemical method screening applications was developed specifically for large scale water monitoring studies and was intended to support the use of immunochemical methods to complement traditional analytical methods. It required that all positive samples and a representative set of negatives be analyzed by a traditional technique. This original policy has recently been modified (*5,6*).

The EPA/OPP now supports the use of immunochemistry complemented methods for pesticides that already have an established, traditional analytical method. The traditional method will be regarded as the primary method with immunoassay adding value to the overall procedure. An immunochemical method would be considered the primary method when a traditional method (GC, HPLC, GC/MS) is unavailable and/or immunoassay is the best available method.

Table 1. Potential Benefits from Adopting Guidelines for Validation and Use of Immunochemical Methods (A) and Acceptance of Immunochemical Methods (B).

A. BENEFITS FROM ADOPTING GUIDELINES

Regulators
- Consistency in data submissions
- Standardized criteria for evaluation of immunochemical data
- Easier and more rapid review of submitted studies.

Registrants
- Know what's expected for validation and use of immunochemical methods.
- Reduces uncertainty of "acceptability" of immunochemical data.
- Consistency in data submissions

B. BENEFITS FOR ACCEPTANCE AND USE OF IMMUNOCHEMICAL METHODS.

Registrants
- Reduced analytical costs associated with large studies.
- More timely completion of large studies (less time for analysis of each sample).
- Field screening or analysis of samples reduces sample stability problems.
- More thorough field studies (i.e., reduced costs allow for more samples)
- Reduced solvent use and hazardous waste disposal.
- Increased product stewardship opportunities.
- Provides comparable scientific quality in results

Regulators
- Allows industry to perform more thorough field studies.
- Allows industry to increase product stewardship efforts.
- Reduced monitoring and enforcement costs.

Analytical Labs
- Better delivery time per sample
- Comparable scientific quality
- Promotes higher productivity
- Better utilization of equipment; reduces need for capital outlay
- Creates synergies among analytical disciplines

When complementary immunochemical methods are used in large scale water monitoring and field studies, samples would be analyzed initially by immunoassay and a subset of samples confirmed with the traditional method (*4,6,7*). The expected degree of sample confirmation is dependent upon the type of study. When performing large scale monitoring studies, all positive results and a subset of negatives by immunoassay should be confirmed with an alternative method. A reduced rate of confirmatory analyses is expected for samples from field studies conducted in support of product registrations. When conducting field studies, a statistically representative number of both the positive and negative samples should be re-analyzed and a valid approach developed to reduce the number of confirmatory analyses as confidence in the immunochemical method increases (*5*).

Regardless of the application, type of study or matrix, registrants must demonstrate the feasibility of the technology. The EPA/OPP has suggested that industry provide case studies showing successful and appropriate applications of immunochemical methods. This will enable technical reviewers to become more comfortable with immunoassay data and raise the level of confidence in the technology. For most analytical methods used to support registration studies, general requirements exclude the use of exotic equipment (i.e., those not commercially available in the U.S.). The same requirement applies to immunochemical methods. Therefore, reagents used in immunochemical methods should be commercial products or in some way available to regulators. Widespread acceptance of immunochemical methods will depend upon the availability of required reagents and materials (*4*).

Immunochemical methods developed for analysis of soil and water in conjunction with studies required by FIFRA Subdivisions N, E, or K (*8,9,10*) should follow the guidelines detailed recently in Federal Register Notice 34613 (*11*). This document provides guidance on the expected performance data and information to be reported to the EPA/OPP regarding method validation. Immunochemical methods developed for determining residues in food and feed should follow the requirements listed in Subdivision O (*12*), PRN 88-5 (*13*), and the standard evaluation procedure for residue analytical methods (*14*). Other guidelines such as Subdivision M (*4,15*) may also apply to applications of immunochemical methods. Immunochemical methods will be accepted for identification of beta exotoxins from microbial species and taxonomic identification of microbial pest control products if supporting data has been generated (*4*). In all the above cases, the EPA guidelines require that analytical method validation be performed according to Good Laboratory Practice (*1*).

Validation of Immunochemical Methods

The most common uses of immunochemical methods will include generation of data supporting product registration and environmental monitoring. Immunochemical methods will predominantly complement traditional analytical methods such as GC, GC/MS or LC. In this case, a validated non-immunochemical, traditional analytical method for the analyte would be practical and available. Complementary immunochemical methods would typically be used to detect small molecules such as conventional herbicides, insecticides, fungicides or their metabolites or degradates. In other applications where the immunochemical method represents the best available or most appropriate analytical technique for the analyte and end-use, it would be considered as the primary method (which may be complemented by another analytical method). A non-immunochemical method might not be practical, sufficiently sensitive, or suitable for some products and applications (e.g. expression of active ingredients in genetically engineered plants or microorganisms, ultra-low volume agents, and biopesticides). In these cases, immunoassay would be the primary analytical method. Examples of complementary versus primary immunochemical methods are shown in Table 2. Prior to using either type of method, proper validation must be performed.

Table 2. Examples of Primary and Complementary Methods

Primary Method	Complementary Method
1. Practical instrument methods, e.g., GC, HPLC, GC/MS	1. Immunochemical method
2. Immunochemical method is the best available method	2. Other quantitative/semi-quantitative "confirmatory" method

---

When considering validation of an immunochemical method, it is also important to recognize the difference between a method and an immunoassay test kit. An immunoassay test kit is a packaged system containing the key or principal components (coated tubes, microplates or particles, enzyme conjugates, standards and other reagents) to be used as part of the validated analytical method. The test kit typically includes directions for use and is often a self contained analytical system. Some kits contain everything needed to perform the analysis from beginning to end. Others are one part of a multi-step process of extraction, clean-up and analysis. A single test kit may be incorporated into several analytical methods (e.g., water, soil, and crop analyses for the same analyte). Validation of the test kit is not required as part of the method validation study. Most commercially available kits are manufactured using Good Manufacturing Practices (*16*) and are therefore validated as part of the development and manufacturing process. Companies developing their own kits should perform the proper studies to ensure the stability, reproducibility, and reliability of the manufactured reagents and materials.

An immunochemical method is the complete analytical system defined by written procedure(s). The method would combine the test kit with additional processes, supporting supplies, equipment and reagents (e.g., pipettes, photometer, homogenizer, solvents), data manipulation and evaluation which allows for determination of the presence and/or the quantity of an analyte in a given matrix. The immunoassay test kit is part of the equipment used in performing the method. Many immunoassay test kits can be used to analyze water samples with little or no sample preparation. Analysis of soil, crops, food or other matrices usually requires an extraction step followed by either dilution or clean-up steps prior to the actual immunochemical analysis. The process of preparing and assaying for an analyte in a matrix constitutes the analytical method.

Method validation is the process of demonstrating that the combined procedures of sample preparation (extraction, clean-up, etc.) and analysis will yield acceptably accurate, precise and reproducible results for a known analyte in a specific matrix. Method validation includes preparing a final written method (Table 3). This document should include all of the information needed by an analyst to perform the entire analytical procedure, as well as background information and the method validation data. Some of the details of performing a method validation will differ for primary and complementary

Table 3. Information to be Included in a Written Analytical Method

A. Summary of Method
    1. Performance Claims
    2. Matrix
    3. Intended use

B. Principle of the Method

C. Analytical Procedure
    1. Materials
        a. Equipment
        b. Supplies
        c. Reagents/stability of reagents
        d. Analytical standards/calibrators
        e. Safety and Health Hazards of Materials
    2. Method
        a. Reagent preparation (including standards if applicable)
        b. Instrument settings
        c. Stepwise procedure(s)
            i.   Source/characterization of control samples
            ii.   Sample preparation
            iii. Extraction
            iv.   Fortification (if applicable)
            v.   Clean-up (if applicable)
            vi.   Instrument calibration
        d. Data interpretation (i.e., decision criteria for detection)
        e. Interferences
            i.   Cross-reactivity
            ii.   Matrix effects
            iii. Solvent and labware effects
        f. Valid non-immunochemical confirmatory method (if applicable)
        g. Time required for analysis
        h. Modifications/potential problems (e.g., critical steps)
        i. Calculations
            i.   Recovery
            ii.   Conversion/dilution factor
            iii. Statistics (standard deviation, percent coefficient of variation)
        j. Representative Raw Data
        k. Other information needed to provide thorough description of the method.

D. Results/Discussion: Summary of results of validation experiments
    1. Accuracy:
    2. Precision:
    3. Detection Limit
    4. Limit Of Quantitation (Quantitative Range)
    5. Selectivity And Specificity
    6. Correlation To Non-Immunological Method (If Applicable)
    7. Ruggedness Testing (If Performed)
    8. Limitations
E. Conclusions
F. Tables and Figures
G. References

immunochemical methods. Specifically, a non-immunological confirmatory method and correlation data demonstrating comparable performance between the immunochemical and non-immunochemical methods is essential for complementary immunochemical methods. Correlation studies are not required for primary methods. However, when a reliable, but not necessarily traditional, alternative method is available it would be appropriate to demonstrate that similar results are obtained with both methods.

When a registrant decides to use a particular method to support registration studies, whether immunochemical method or traditional, the FIFRA GLP requirements apply. Perhaps the most concise way to present the required elements for GLP validation of an immunochemical method is to discuss the requirements for reporting a method to the regulatory agency (Table 3). The performance claims, intended use, and scientific principles of the method should be included on the introductory pages of the method. This section should include information regarding the analyte(s) detected, the structure of the analyte(s) when practical, the validated limits of detection and quantitation, and the quantitative range. This section should also identify the matrix for which the method was validated and the intended uses for the method (e.g., registration studies, monitoring, enforcement, etc.). A discussion of the principles of the method should include an overview the sample extraction, clean-up, and analysis procedures as well as a brief description of how data reduction and interpretation are performed.

Description of the analytical procedure is comprised of two main sections. The first is a listing of all materials required to execute the method (including sources for purchase). Included in this section are the immunoassay test kit (or the individual reagents), the measuring device, the source and characterization of analytical standards, and all other reagents and supplies used in the study. Safety information should be provided for all hazardous materials. Special storage conditions should be noted for all perishable reagents, standards and supplies. Stability of all reagents under all storage conditions is established during development of the immunochemical method and should be confirmed for the duration of the use of each reagent lot during validation. Monitoring of reagent stability is important for ensuring the quality of data generated by immunoassay; this topic is discussed further in the accompanying paper by Rittenburg and Dautlick.

The second part of the analytical procedure is a detailed step-by-step description of sample preparation, extraction and clean-up processes as well as instrument calibration, instrument settings and data interpretation information. This section should contain details for preparation of reagents (including standards, if applicable) and fortified samples. All known interferences should be identified and any critical or unique steps in the procedure should be highlighted. Acceptable modifications or potential problems which may be encountered during execution of the method should be discussed. Any calculation which is required during routine sample analysis (e.g., percent recovery, conversion or dilution factors, statistical calculations) must be explained in detail. For non-routine calculations, examples should be presented. Examples of representative data should be included (e.g., raw data for samples/standard, calibration curve parameters, recoveries from fortified samples). All of the recovery data (i.e., expected concentration, measured concentration and % recovery for each sample) generated during the validation should be also be included in the report.

The results and discussion sections include the validation data and a detailed discussion of those results. Accuracy is demonstrated through measuring the recovery of analyte in fortified samples across the quantitative range: data points for the recovery at all target levels should be presented in the method report along with a standard deviation and other statistical analyses. Other components of accuracy that impact interpretation of data (e.g., extraction efficiency for in-grown residues, aged soil residues, and measurement of finite levels of the active ingredients in genetically engineered products) should also be reported with some measure of precision. Precision should be reported as the standard deviation or coefficient of variation for the recoveries of the accuracy samples at each fortification level (and for extraction efficiencies, when applicable). Quantitative methods should demonstrate mean recoveries distributed between 70% and 120% for targeted fortification levels and a relative standard deviation of less than 20% for measured recoveries at each fortification level (12). These requirements for precision and accuracy may not apply to primary immunochemical methods used for quantitation of non-toxic analytes in genetically engineered plants or micro-organisms since, currently, there are no FIFRA requirements for these methods and most of the active ingredients in these products are exempted from the requirement for monitoring levels in food or the environment.

The detection limit is the concentration of analyte that can be distinguished from zero with a stated degree of confidence. The limit of quantitation is the smallest concentration of analyte that can be measured in samples and yield predicted concentrations with an acceptable level of precision and accuracy. The quantitative range is the lower and upper limits of analyte concentration over which the method yields quantitative results within the stated performance limits. All analytical methods are required to state their limits of detection and quantitation. Several techniques are available for estimating detection and quantitation limits (17,18); regardless of the method chosen, the validation report must define how the limits were derived. The EPA/OPP requests that relevant performance data supporting the determination of the LOQ and/or detection limit be provided in addition to the statistical evaluations (5).

For validation of immunochemical methods, demonstration of the specificity of the method is emphasized much more than for traditional methods. Any information generated during the validation which addresses the selectivity and specificity of the method should be included. The method must be characterized with respect to the potential for interference with other pesticides, metabolites or degradates. Interference due to the matrix or other potential non-specific sources (e.g., solvents, labware) should also be addressed.

Ruggedness testing is strongly recommended for all environmental chemistry methods. Experiments should be conducted to show that the method is amenable to normal variations that might be encountered during normal use (19). These types of tests might include: 1) repeated analysis of a sample or samples on several days (by different analysts, under differing assay or environmental conditions, or with different reagent lots); and 2) measurement of accuracy and precision in fortified samples using control material from several sources (e.g., different soil types). Inter-laboratory validation is required for soil, water, and food chemistry methods (11,13); however, inter-laboratory studies should be conducted if warranted by study, experimental design or by other applications of the method (e.g., product stewardship).

A correlation experiment should be performed when validating an immunochemical method which is intended to complement a traditional analytical method. This experiment is not part of the FIFRA requirements for analytical methods but is intended to establish credibility of the method and raise the confidence level of technical reviewers. The correlation study should be statistically designed to show the goodness of fit between results obtained from immunoassay versus traditional method. A typical correlation study would consist of preparing a series of fortified samples using control or blank matrix then analyzing each fortified sample using both the immunochemical and non-immunochemical method. Once the results are obtained, the correlation between the methods is calculated. In addition, any false positives or false negatives generated during the experiment should be noted.

Use of Immunochemical Methods and Confirmation of Results

The potential uses of immunoassay include generating registration data as well as performing post-registration analyses (e.g. monitoring, enforcement, exposure). Prior to initiating a study, the study director should first evaluate the appropriateness of immunochemical method(s) for the particular study (Table 4). If an immunochemical method is appropriate, the next step is to determine if the performance characteristics of the immunochemical method (e.g. sensitivity, specificity, reproducibility) will meet the needs the particular study. The method should have met all the validation criteria described above. Prior to initiating the study, the protocol should be reviewed with the EPA/OPP.

Table 4. Appropriate and Inappropriate Uses of Immunochemical Methods.

Examples of Appropriate Uses:

- Quantitation/screening of residues in environmental fate field studies.
- Quantitation/screening of residues in magnitude of residue studies.
- Required sensitivity not attainable by other methods.
- Economic reasons (e.g., speed of analysis, cost of performing conventional methods).
- Alternative method inappropriate (e.g., analyses for proteins, biopesticides, genetically engineered pest control products).
- Desire to perform analyses in the field.
- Traditional methods are too cumbersome at the desired LOQ.
- Immunochemical method adds value to traditional method.

Examples of Inappropriate Uses:

- Small studies (<10 samples) / Alternative methods available.
- Matrix not amenable to the immunochemical method.
- Specificity of the method is inadequate (e.g., false positives from metabolites).
- Immunochemical method cannot attain the required sensitivity.
- Purpose of the study is to identify new metabolites or degradates.

The degree to which positives and negative samples from immunochemical measurements require confirmation by alternative analytical techniques is dependent on several factors including the type of study, how the immunochemical method was employed, the performance characteristics of the both the immunochemical and alternative method, and the intended use for the data. As mentioned previously, confirmatory analyses may not be required or possible in certain situations.

Questions which should be evaluated when designing a confirmation scheme for an immunochemical method include:

- What is the "level of concern" for the study?
- Is the data being generated for a pre-registration study or for post-registration monitoring or enforcement?
- Is the method being used as a screening assay or for quantitative results?

- Is the potential for interference from cross-reacting compounds high or low?
- What is the detection and quantitation limit of the confirmatory method and how does it compare to the limit(s) for the immunochemical method?
- What degree of confidence is required at the detection limit (i.e., what is the acceptable level of false positives or false negatives)? Is sufficient data available to demonstrate that this level is attainable (this criteria should be evaluated for both the immunochemical method and the confirmatory method)?
- What is the size of the study (tens, hundreds or thousands of samples)?
- What proportion of the samples are expected to be positive?

Once the criteria for a particular study have been established, a confirmation scheme can be developed. The magnitude of the confirmation effort may vary considerably depending on the answers to the above questions. It is beyond the scope of this article to present all the possible confirmation schemes. Instead we will focus on the general approach which should be followed when developing a confirmation scheme for soil and water analyses from field studies or during monitoring and enforcement applications.

Registration studies which might employ immunochemical methods include:
1) Environmental fate field studies (e.g., aquatic and terrestrial field dissipation, groundwater studies, runoff studies); 2) Plant residue studies (e.g., magnitude of residue studies, processed fraction studies); 3) Measurement of the active ingredient expressed by or contained in genetically engineered plants/microbes or biopesticides and 4) Hazard evaluation and exposure studies. Typically, registration studies are conducted within a defined environment (e.g., pre-screened to ensure lack of contamination) or with a control or blank matrix. Usually only a single compound is applied and few, if any, interferences from other pesticides would be expected to occur. A relatively high proportion of samples may contain residues. A subset of samples with positive and negative results by immunochemical method should be confirmed by a non-immunochemical method. Depending upon the characteristics of the immunochemical method, it may be necessary to confirm the identity of all positive samples (i.e., when parent compound and metabolite cross react in the immunochemical method).

In cases where an immunochemical method is used for post-registration monitoring (water) or enforcement applications (food), all positives must be confirmed. In these applications, the history of the sample is often not well defined and the matrix may be highly variable. Sources for the samples usually have not been characterized prior to collecting the sample, therefore, the potential for interference from cross-reacting compounds cannot be ruled out without confirmation. Positives are typically samples which exceed established tolerances or pre-determined levels of concern. A relatively low proportion (< 5%) of samples will be expected to contain residues at or above the level of concern.

The EPA/OPP has recently proposed a scheme for confirmation of residues in water samples from field studies in which immunochemical methods are used to screen for positives (5). All of the samples are analyzed using the immunochemical method. A statistically representative number of the positive and negative samples should be re-analyzed using a non-immunochemical procedure to confirm the concentration of residues. The identity of the analytes should also be confirmed in a subset of the samples which contain residue in concentrations of toxicological or environmental concern. The degree of confirmatory analyses should be reduced as the study proceeds, assuming good agreement exists between the methods. Criteria must be established to determine what constitutes "good agreement" and how "agreement" between methods will be measured. Examples of criteria for evaluating "agreement" among positive samples might include:
1) The values measured by both methods are within 20 % of each other 90% of the time;

2) The distribution of measurements is not significantly different at a 95% confidence interval. When possible, attempts should be made to re-evaluate the samples which fall outside the acceptable range and to determine the source of the discrepancy. Throughout the study a proportion of the negatives (e.g., 5-10%) should be reanalyzed by an alternative method to guard against false negatives. Once demonstration that the immunochemical method is comparable to the alternative method, the level of confirmation might be reduced to "baseline" levels (e.g. 5% of the positive and negative samples). A hypothetical confirmation scheme for use of an immunochemical method complemented by another analytical method is shown in Table 5.

A similar approach should be applicable to soil samples from field studies and to plant residue studies when an immunochemical screen is employed. Plant residue studies often have fewer samples (tens or hundreds) that environmental fate field studies (several hundred to a few thousand). The proportion of positive and negative samples in plant residue studies confirmed by alternative methods would be similar to that for field studies; however, the size of the experiment is smaller and the absolute number of confirmatory analyses will be reduced.

When an immunochemical method is employed as the principal screening method in water monitoring studies, the concentration and identification of the residue should be confirmed in all positive samples. To ensure that false negatives are not occurring, a subset of the negative samples should be re-analyzed using a traditional technique.

Immunochemical methods used in support of studies conducted on microbial pest control agents, genetically engineered pest control agents or other biorational products are often not amenable to analysis by techniques such as GC or LC. In such cases, where an alternative but also untraditional method is available for detection or quantitation of the analyte, confirmation of positive results or correlation of immunochemical method results to the alternative method is recommended as part of the immunochemical method development or validation process. For many products a confirmatory or alternative method is not practicle to develop or is unavailable. Currently, confirmation of positive and negative samples for this class of products is not required for registration studies (15,20). Often, these products are exempt from the requirements for tolerance level enforcement in commodities, therefore no monitoring or enforcement method would be required (20,21). Application of immunochemical methods for studies supporting the registration and monitoring or enforcement (if needed) for these types of products should be evaluated on a case-by-case basis and reviewed with the proper regulatory personnel prior to initiation of any studies.

Conclusions

The increasing quality and availability of immunochemical methods has led to an increased desire to use these methods in support of pesticide registration data packages and for post-registration monitoring and enforcement applications. The reduce cost, ease of use and relative speed of analysis make immunochemical methods a powerful tool. The technology should eventually lead to more thorough studies which can be completed faster and with reduced analytical costs. Recent notices and communications from EPA/OPP encourage the use of immunochemistry complemented analytical methods. An ongoing dialog between EPA/OPP and the users and developers of this technology will be essential to ensure proper implementation of immunochemical methods. Development of guidelines that incorporate the concepts similar to those reviewed in this document will promote consistent validation, data reporting, and application of immunochemical methods by developers and users of this technology. Consistency in these areas will aid in the acceptance of the technology and may facilitate more rapid review of immunoassay data by the agency.

Table 5.   **Hypothetical confirmation scheme for immunochemical complemented methods for a pre-registration environmental fate field study.**

Study #1

Sampling time (days)	Number of Samples	Expected # of positives	% Positives Confirmed	% Negatives Confirmed
1	100	70	30	5
3	100	60		5
5	100	50	20	5
7	100	40		5
14	100	30	15	5
21	100	25		5
50	100	20	10	5
100	100	15		5
150	100	10	10	5
200	100	5		5
250	100	5	10	5
300	100	2		5
350	100	0		5

Study #2 and all other Studies:

Sampling time (days)	Number of Samples	Expected # of positives	% Positives Confirmed	% Negatives Confirmed
1	100	70	5-10	5
3	100	60		5
5	100	50	5-10	5
7	100	40		5
14	100	30	5-10	5
21	100	25		5
50	100	20	5-10	5
100	100	15		5
150	100	10	5-10	5
200	100	5		5
250	100	5	5-10	5
300	100	2		5
350	100	0		5

Immunoassay offers many benefits to both registrants and regulators. However, it is important to recognize that there are limitations to the technology. Immunochemical methods are not appropriate for all studies. In addition, most current immunochemical methods for classical chemical products are intended to complement traditional techniques such as GC, LC, or GC/MSD or newer technologies like capillary electrophoresis.

This document should be considered an interim guideline. It is simply another step in the ongoing dialog between registrants and regulators regarding the proper use of immunoassay. Many of the opinions and suggestions presented here are those of members of the Analytical Environmental Immunochemical Consortium and not of the EPA/OPP. However, we hopefully have presented suggestions and expressed ideas that are consistent with EPA/OPP policy.

Literature Cited:

1.   Federal Insecticide, Fungicide and Rodenticide Act (FIFRA). Accelerated Reregistration Phase 3 Technical Guidance Prepared by OPTS/EPA, Washington, DC. (1989).
2.   Hock, B. *Acta Hydrochim. Hydrobiol.* **1993**, *21*, 71-83.
3.   Sherry, J.P. *Critical Reviews in Analytical Chemistry.* 1992, *23*, 217-300.
4.   Kimm V.J. OPPTS' Response to EMSL-LV Immunochemistry Summit Meeting Document, OPTS/EPA memorandum, 26 April, 1993.
5.   Marlow, D.A. *Proposed Guideline for EPA/OPP Acceptance of Data from Immunochemistry Methods.* OPTS/EPA memorandum, 20 October, 1993.
6.   Schuda, P.S. *OPP Comments on Proposed AEIC Guidelines for Acceptance of Data From Immunochemistry Methods.* OPTS/EPA memorandum, October 27, 1993.
7.   Jacoby, H.M. *Evaluation of Pesticide Immunoassay Methods* - Report of January 12, 1993. OPTS/EPA memorandum, 11 February, 1993.
8.   *Subdivision N (Environmental Fate) of the Pesticide Assessment Guidelines, prepared by OPTS/EPA; Washington, DC (1982).*
9.   *Subdivision E (Hazard Evaluation) of the Pesticide Assessment Guidelines, prepared by OPTS/EPA; Washington DC (1982)*
10. *Subdivision K (Reentry Protection) of the Pesticide Assessment Guidelines, prepared by OPTS/EPA; Washington, DC (1982).*
11. OPP-00385. Publication of Addenda for Data Reporting; Requirements for Pesticide Assessment Guidelines (N, E, and K). July, 1994. Federal Register and Addenda. Vol. 59 No. 128.
12. *Subdivision O (Residue Chemistry) of the Pesticide Assessment Guidelines, prepared by OPTS/EPA; Washington, DC (1982).*
13. Pesticide Registration Notice 88-5.
14. Nelson, M.J.; Griffith, F.D. Jr. *Health Effects Division Standard Evaluation Procedure: Analytical Method(s).* OPTS/EPA; (1989).
15. *Subdivision M (Microbial and Biochemical Pest Control Agents) of the Pesticide Testing Guidelines, prepared by OPTS/EPA; (1989).*
16. *Code of Federal Regulations.* Food and Drugs, Title 21, Part 211, "Current GoodManufacturing Practices for Finished Pharmaceuticals." (U.S. Government Printing Office, Washington, D.C.)
17. Keith, L.H.; Crummett, W.; Deegan, J.Jr.; Libby, R.A.; Taylor, J.K.; Wentler,G. *Anal. Chem.* **1983**, *55*, 2210-2218.
18. Rodbard, D. *Anal. Biochem.*, **1978**, *90*, 1-12.
19. American Society for Testing and Materials (ASTM). Standard Guide for Conducting Ruggedness Tests. ASTM Designation: E 1169-89, 1989.
20. OPP-00343. EPA/OPP Draft Proposal to Clarify the Regulatory Status of Plant pesticides. November, 1992. SAP hearing December 18, 1992.
21. Biotechnology Science Advisory Committee of the EPA (BSAC); SAP hearing regarding Plant Pesticides, July 1993.

RECEIVED October 26, 1994

Chapter 21

Quality Standards for Immunoassay Kits

James Rittenburg and Joseph Dautlick

Analytical Environmental Immunochemical Consortium,
1427 West 86 Street, #102, Indianapolis, IN 46260

The environmental immunoassay kit industry is growing rapidly with both new kits and new companies entering the market. While the industry is still young, it is an opportune time to develop industry standards for immunoassay kits. The Analytical Environmental Immunochemical Consortium (AEIC) is comprised of agrichemical and immunochemical companies that use or provide immunochemical methods and associated equipment for environmental chemical analysis. One of the goals of the AEIC is to establish performance standards. This paper discusses current programs and potential future activities of the AEIC aimed towards developing quality standards for immunoassay kits.

The rapid increase in the availability of immunoassay kits for detection and quantitation of analytes such as pesticides, environmental contaminants, food and waterborne pathogens, and toxins is now enabling a diverse user group to perform analyses within a variety of laboratory and field environments. One of the objectives of the Analytical Environmental Immunochemical Consortium (AEIC) is to help assure that high quality performance and appropriate interpretation of results are obtained from immunoassay kits used in various applications by operators with varying degrees of experience. To help achieve this goal, the AEIC is developing recommended standards for manufacturers of immunoassay kits. As a model, the AEIC is using the medical diagnostic immunoassay kit industry which has a 25 year history. Initial recommended standards under development include establishing standardized package inserts, sources of immunoassay kit calibrators and quality control samples, and quality control guidelines for users to employ for monitoring kit performance. The AEIC is also developing standardized definitions to enable clear communication within the immunoassay kit industry for both the manufacturers and users.

This paper discusses current programs and potential future activities of the AEIC directed towards establishing uniform quality standards for immunoassay kits. Commercially available immunoassay kits generally contain significant amounts of quality, performance, and general background information that address many of the topics presented in this paper. However, since each kit manufacturer is currently responsible for determining what information should be provided to the user, it is

0097–6156/95/0586–0301$12.00/0
© 1995 American Chemical Society

inevitable that kit performance and quality are measured and presented in different ways by individual companies. Through guidance documents developed by AEIC work groups and adopted by consensus vote of the members, the Consortium hopes to establish a framework of voluntary quality standards aimed at maximizing the utility of immunoassay kits to the end user and ensuring that the kits provide data whose quality is appropriate for given applications.

ESTABLISHING CONSISTENT TERMINOLOGY

The AEIC initially focused on establishing common definitions for terms that are frequently used to describe immunoassay kits and their associated performance characteristics. The list of terms currently being defined by the AEIC is shown in Table 1.

Table 1. Initial list of terms being defined by AEIC

Sensitivity	False Positive
Stability	False Negative
Bias	Method
Accuracy	Kit
Precision	Assay
Reproducibility	Correlation
Specificity	Ruggedness
Interferences	

To effectively describe and communicate kit quality and performance characteristics it is critical that the community of kit manufacturers and users associate the same meaning with words frequently used to describe kit performance. Progress towards this end is shown in Appendix 1 which lists the definitions so far adopted by the Consortium in relation to application of immunochemical methods to environmental analysis. Many of the terms have multiple meanings depending upon the discipline or context in which they are used. The definitions being applied to these terms by the AEIC are intended to standardize their meaning within the context of environmental analysis. For example, since immunoassay kits that are applied to environmental analysis generally equate the term "sensitivity" with some type of analytical detection limit, the AEIC has defined sensitivity in terms of either a limit of detection or a limit of quantitation (see Appendix 1). However, the term sensitivity also frequently refers to the change in analytical response per unit change in analyte concentration (curve slope), or in a clinical context, to the percent test positivity in a diseased population. Although the issue of re-defining or confirming definitions of commonly used words may seem mundane, this simple step will go a long way towards preventing confusion and misunderstanding about the performance characteristics and appropriateness of kits for specific applications. This is also an important step in providing kit users with a common reference point from which to compare and contrast kit information provided by different suppliers.

STANDARDIZING KIT PACKAGE INSERTS

Commercial immunoassay kits are generally supplied with an information package sheet, commonly referred to as the package insert, that describes various aspects of the kit such as kit contents, test procedures, and expected performance characteristics.

The kit package insert provides information to help the user determine appropriate uses for the kit and its limitations. Although the majority of information to be described is currently provided with most commercial kits, standardizing the headings and topics covered in the kit package insert will establish a baseline of information to be provided for each immunoassay kit. The headings described below are currently in review by the AEIC as the basis of a standardized package insert.

Package Insert Headings:

1.   Intended Use
2.   Test Principles
3.   Materials Required to Conduct Test
4.   Preparations for Testing
5.   Test Procedures
6.   Results
7.   Technical Service
8.   References

Content Contained Within Each Heading:

1.   Intended Use

This section includes a statement specifying the analyte(s) and matrix(ces) for which the test kit was designed as well as background information about the analyte, its use, and how it gets into the matrix. Alternative methods of analysis for the analyte should be cited when possible. Any physio-chemical characteristics such as solubility, volatility or temperature sensitivity that are critical to the use of the test should be specified. It should be clear to the user that applications of the test kit beyond the intended use described may produce results of questionable quality. It may also state where or how users can obtain information on additional applications (e.g. application notes produced by the company).

2.   Test Principles

Includes a brief description of the test principles describing the type of immunoassay provided in the test kit. Pertinent reference citations are helpful should the user need to obtain more detailed technical information about the test.

3.   Material Required to Conduct Test

Materials required to conduct the test are described and listed in two categories:

A.   Test Kit Contents - provides a brief description of each reagent including the chemical nature and content of each container, as well as a description of any instruments or associated supplies provided with the kit. Include a definitive statement describing the correct storage conditions for the test kit.

B.   Material Required But Not Provided - includes a brief description of chemicals, instruments or other supplies that are required but not provided in the test kit and suitable sources, such that the user has clear guidance for obtaining these items.

4.   **Preparation for Testing**

     A.   <u>Kit Preparation</u> - provides a description of the test kit preparations that are required prior to testing.   This may include descriptions of any required reagent, instrument, or other preparations that are recommended.

     B.   <u>Sample Preparation</u> - provides description of how to collect and prepare the sample for analysis using the test kit. If these procedures are based on existing methodology then cite relevant references. Includes information about sample holding times, stability constraints, or extraction efficiencies that the user should be aware of.

     C.   <u>Precautions</u> - describes any precautions to be taken when using the test kit.   Describe any dangerous or noxious reactions that could occur in using the kit.   Describes proper protective equipment to be employed while performing analyses with the kit and includes statement about disposal of waste from the kit.

     D.   <u>Limitations</u> - Mentions limitations of the test kit that could impact the quality of the test results.   Information such as kit expiration dates, temperature limitations for conducting the test, critical timing steps, potential interferences, and lot to lot reagent compatibility are examples of potential limitations for use of the kit.

5.   **Test Procedures**

Describes detailed stepwise procedures for performing the test.   Diagrams, flow charts and illustrations are often helpful.

6.   **Results**

Summarizes performance characteristics of the test kit including characteristics such as sensitivity, specificity, precision, and accuracy.   For quantitative tests an illustration of the dose response curve may be helpful.   Interpretation guidelines are often provided and may include items such as curve fitting or other data analysis recommendations and kit quality information that can be inferred from the test results.

7.   **Technical Service**

Information such as telephone, fax or electronic mail numbers should be provided for technical services along with days and hours of service that are available.

8.   **References**

Includes pertinent references describing items such as alternate test procedures, sampling, sample preparation, and test principles.

POTENTIAL SOURCES OF KIT CALIBRATORS AND STANDARDS

The availability of high purity chemical standards is a critical requirement during both the development and commercialization of immunoassay test kits.   Ideally, collaborative or cooperative relationships can be established between the immunoassay kit developer and the manufacturer of the chemical compound(s) that are to be analyzed with the kit. This type of relationship is especially important when the compounds of interest are relatively new and do not have detailed published information about their synthesis, environmental fate, and metabolism. Often times, only the chemical manufacturer has full information about, and access to, impurities, breakdown products, and metabolites that should be examined during the development of the immunoassay kit. The chemical manufacturer also generally has access to "incurred" (real field samples) samples and analytical data generated by the "standard" methods of analysis.

The AEIC may provide an interface between immunoassay kit developers and chemical manufacturers to foster communication and to help ensure that immunoassay kits utilize high purity analytical standards, and that performance characteristics such as sensitivity, specificity, and accuracy are determined using relevant chemical compounds and appropriate field samples. In addition to the chemical manufacturer, other sources of chemical standards may be identified such as reputable commercial chemical suppliers or possibly materials deposited with the EPA.

KIT AND DATA QUALITY INDICATORS

A variety of quality control information usually accompanies the results obtained with an immunoassay test kit. For most environmental chemical analysis some type of competitive inhibition immunoassay is employed. These assays typically generate a sigmoidal type dose response curve (Figure 1) in which the maximal response is obtained for the kit negative control. The level or range of response expected for the negative control is often specified with the test kit, and confirmation of this response by the user following sample analysis provides an indication that the kit reagents were fully active and that the test protocol was performed properly. Precision of replicate standards or samples should lie within expected limits for the test kit indicating both satisfactory kit quality and user technique. Other characteristics of the dose response curve such as slope, 50% inhibition concentration, and range of quantitation provide valuable information about the performance of the test kit (1). Obtaining results within expected limits for these parameters assure the user that the kit reagents and standards performed properly and that the kit procedures were carried out satisfactorily.

Although the parameters mentioned above will provide valuable information about reagent quality and user performance, they do not provide direct quality information about data generated for field samples. Establishment of "operating standards" for use with immunochemical methods is a high priority topic within the AEIC. Operating standards would include guidelines such as recommended frequency of sample duplicates, frequency of laboratory confirmation for positives and negatives, matrix spike recoveries, frequency of instrument calibration if relevant, and level of documentation. In addition, certain levels of training may be required along with documentation of analyst competency if the data from the test kits will be used to meet certain data quality levels. The accompanying paper entitled "Guidelines to the Validation and Use of Immunochemical Methods for Generating Data in Support of Pesticide Registration" discusses some approaches to method validation and confirmation of results.

SUMMARY

The AEIC is working towards establishing voluntary performance standards for the use of immunochemical methods for environmental analysis. Initial efforts are being directed towards establishing consistent definitions for commonly used terms, developing standardized package insert information, establishing sources of kit calibrators and quality control samples, and providing guidelines for user QC.

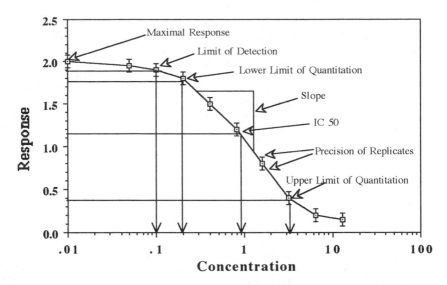

Figure 1.   Typical dose response curve for competitive inhibition immunoassay.   Various curve characteristics are shown that can be used as kit and data quality indicators.

Literature Cited

1.   Vanderlaan, M.; Stanker, L.; Watkins, B. In *Immunoassays for trace chemical analysis;* Vanderlaan, Stanker, Watkins, and Roberts.;   ACS Symposium Series 451, ACS: Washington, D.C. 1991, pp 2-13.

Appendix 1
AEIC DEFINITIONS

Accuracy
The closeness of individual predicted concentrations to the true concentration in a sample.   Note: this definition combines the idea of bias ("systematic error") and precision ("random" error).

Assay
Qualitative or quantitative analysis of a substance.

Bias
a.   Recovery Bias (absolute) - Bias with respect to the actual concentration of analyte - the degree to which the predicted concentration of an analyte differs, on the average, from the actual (or true concentration).

b.   Method Bias (relative) - Bias with respect to another method - the degree to which the predicted concentration in a sample differs, on the average, from the predicted concentration determined by another method.

False Positive
The percent test positivity in a population of true negatives.

False Negative
The percent test negativity in a population of true positives.

Interferences
Effects on the analytical performance of an immunoassay caused by substances other than the analyte(s) of interest.

Kit
A test kit is a packaged system of the key or principal components to detect or measure a specific analyte(s) in a given matrix(ces) within a laboratory or non-laboratory environment. The key components include antibodies, enzyme conjugates, etc. that may only be readily prepared by the purveyor of the kit. Test kits include directions for use and are often self-contained, complete, analytical systems, but they also may require supporting supplies and equipment.

Method
The entire system of procedures that describe how a final measure of analyte concentration is obtained for a target matrix.

Precision
The extent to which replicate analyses of a sample agree with each other. Usually expressed as the standard deviation or percent coefficient of variation (% cv) of a population of values :   % cv = (Standard Deviation/Mean) x 100

Reproducibility
The ability of the analytical method to yield the same result within analyses, between analyses, and between operators.

Sensitivity
a.   Detection Sensitivity (Limit of Detection) - The smallest concentration of analyte that can be statistically significantly distinguished from zero for a given sample matrix with a stated degree of confidence.

b.   Limit of Quantitation - The smallest concentration of analyte that can be measured in samples and yield a predicted concentration with stated relative precision or accuracy (or both).

Stability
a.   Kit Stability (Shelf Life) - The length of time that a complete kit can be stored under given conditions and still yield results that are within the stated performance parameters.

b.   Sample Stability - The length of time that a target analyte level remains stable in a specific matrix under a given set of storage conditions.

RECEIVED January 9, 1995

Chapter 22

Immunochemical Approach for Pesticide Waste Treatment Monitoring of s-Triazines

Mark T. Muldoon[1] and Judd O. Nelson

Natural Resources Institute, Agricultural Research Service,
U.S. Department of Agriculture, Beltsville, MD 20705 and Department
of Entomology, University of Maryland, College Park, MD 20742

Enzyme-linked immunosorbent assay (ELISA) was developed for
the analysis of pesticide waste and rinsate and disposal monitoring.
Three s-triazine-specific monoclonal antibodies that possessed
different specificities were utilized to quantitate individual and total
s-triazine analytes in mixtures. Results from the analysis of field
samples were validated by HPLC. An immunoassay was developed
for chlorodiamino-s-triazine (CAAT), an important degradation
product of chloro-s-triazine herbicides. Antibody recognition of
substituted s-triazines decreased as a function of increased amino
side chain substitution. The assays were sensitive in the low
micromolar range. An s-triazine herbicide class-specific ELISA
was used in conjunction with an ELISA for CAAT for measuring s-
triazine herbicide ozonation followed by microbiological treatment.
The ELISAs were shown to be very accurate and precise for
measuring the concentrations of both atrazine and CAAT. The
information obtained by the two ELISAs could be used for on-site
control of the two stage treatment process and should save time and
expense in s-triazine disposal monitoring applications.

The widespread use and misuse of chemical pesticides worldwide has resulted in
their occurence throughout the biosphere. Pesticides enter the various
environmental compartments through normal use, overapplication, accidents
(including back-siphoning into water supplies), runoff from mixing-loading areas,
and faulty waste disposal (1, 2). Spills and faulty waste disposal may be
considered point sources of contamination since they are usually in a confined area

---

[1]Current address: Food Animal Protection Research Laboratory, Agricultural Research
Service, U.S. Department of Agriculture, College Station, TX 77845

and are often high concentration environmental exposures. This source of exposure may be most readily controlled through safer handling practices and the development and implementation of methods to properly dispose of unusable materials (3).

Some agricultural pesticide spraying operations generate large volumes of pesticide-containing materials consisting of excess pesticide product, leftover tank mixtures, and equipment rinsates. Typically, pesticide concentrations in waste can range from 1.0-10,000 ppm (4). In addition to pesticidal constituents, the material usually contains formulating agents, fertilizers, adjuvants, and machinery wash-off debris. Pesticide wastes are often be generated at very remote areas and appropriate on-site management of these materials depends on the particular situation involved. Management options include reuse, recycling as subsequent make-up water, or if necessary, disposal (5).

On-site disposal methods have been developed for these high-volume wastes. The methods include physical (evaporation, adsorption, filtration), chemical (hydrolysis, oxidation, incineration), biological (composting, landfarming, enzymatic, bioreactors) and combinations of various methods (6, 7). Traditionally, analytes have been monitored using conventional analytical methods such as gas-liquid chromatography (GLC) and high performance liquid chromatography (HPLC) in order to ensure the effectiveness of the treatment and to determine end-points for multiple step processes. These methods are well established, however, they have the disadvantages of being expensive, time-consuming, and are not readily adaptable to in-field analyses.

Hammock and Mumma (8) published a farsighted review on the potential of immunochemical techniques for environmental analytical chemistry. Since then, immunoassays, in particular enzyme-linked immunosorbent assays (ELISAs), have been developed for many environmentally-important analytes (9). They are sensitive, simple to perform, and can be made field adaptable (10). ELISA kits are commercially available for many different environmental contaminants. Immunoassays have been adapted primarily as screening methods for analytical situations where there is a large sample load such as in groundwater monitoring programs (11).

Immunoassays are particularly attractive for use in pesticide waste management applications since they can be performed with minimal operator training, are rapid, require little if any sample preparation, and are field adaptable. The utilization of rapid tests to determine the presence, absence, or concentrations of particular components in a waste material would allow for better management of these materials. In addition, the use of immunoassays for on-site waste disposal monitoring should save time and expense. Pesticide waste analysis introduces some very unique challenges for ELISA analysis which are not encountered in other applications. As described above, pesticide wastes and rinsates are usually complex mixtures of pesticide active ingredients in addition to a number of non-pesticidal matrix components. Therefore, it was necessary to develop and validate immunoassay methods specifically for analysis in this sample matrix.

This study was part of a larger project at USDA to develop strategies for the disposal and remediation of pesticide waste materials. A disposal method developed at USDA used a combined ozonation and microbial mineralization process to oxidize recalcitrant pesticide substrates to more biolabile intermediates which could be readily degraded by indigenous soil and sludge microorganisms (12-15). The s-triazines were shown to be among the most recalcitrant substrates studied and were considered to be useful indicators of treatment effectiveness. Atrazine ozonation (Figure 1) proceeded by either direct N-dealkylation or the oxidation to either or both of the N-alkyl side chains, loss of this acetoamido moiety, and the accumulation of chlorodiamino-s-triazine (CAAT) (14). In the second stage of the process, CAAT was mineralized by soil or sludge microorganisms (12, 15). The process was monitored for the loss of atrazine and the accumulation of CAAT by ozonation, followed by the subsequent biodegradation of this intermediate. The analytical method used for routine analysis was HPLC.

The purpose of the current study was to develop immunoassay techniques for monitoring this disposal process. We utilized three monoclonal antibodies developed by Karu et al. (16), which showed distinct within-class crossreactivities toward various s-triazines herbicides, for discriminating and quantifying individual s-triazines in pesticide waste mixtures (17). ELISAs were developed which were selective for the detection of CAAT (18). The s-triazine herbicide assay which showed the broadest recognition of the parent s-triazine herbicides was chosen for use as a class-specific ELISA for monitoring the loss of atrazine in the disposal process. This assay was used in conjuction with an ELISA for CAAT to monitor the complete disposal process for atrazine (19). This paper summarizes the work pertaining to atrazine disposal monitoring by ELISA.

Materials and Methods

Pesticide Waste and Rinsate Samples. Pesticide waste and rinsate samples were obtained from collection facilities at the Beltsville Agricultural Research Center, USDA, ARS, Beltsville, MD during the spring 1991 growing season. The chemical composition of the samples was described in Muldoon et al. (17).

High Performance Liquid Chromatography. HPLC measurements were made using a Waters 712 WISP automatic sample injector, two Waters Model 510 HPLC pumps, a Waters Model 490 UV detector (210, 220, and 230 nm monitored), and a NEC APC-IV controller with Maxima 820 software. The column was a Waters NOVAPAK 4 μm C-18 in a 8 mm x 10 cm radial compression module. The solvent system used a 15 min gradient (Waters curve 10) of 0 to 75% acetonitrile/phosphoric acid (pH 2) at a flow rate of 2.0 mL/min. The final condition was maintained for 5 min. Analyte concentrations were calculated based on standards curves for each of the individual compounds using authentic analytical standards.

Hapten Synthesis and Hapten-Protein Conjugation. Carboxylic acid s-triazine haptens were synthesized for use in the development of s-triazine herbicide ELISAs and antibody production and ELISA development for the analysis of CAAT. The procedures used were adaptations of those described by Goodrow et al. (20). Briefly, CEPrT (Figure 2) was synthesized by sequentially substituting

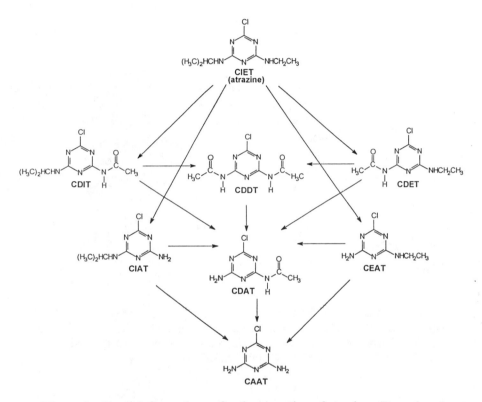

Figure 1. Degradative pathway for the ozonation of atrazine. (Reproduced with permission from reference 14. Copyright 1992 American Chemical Society).

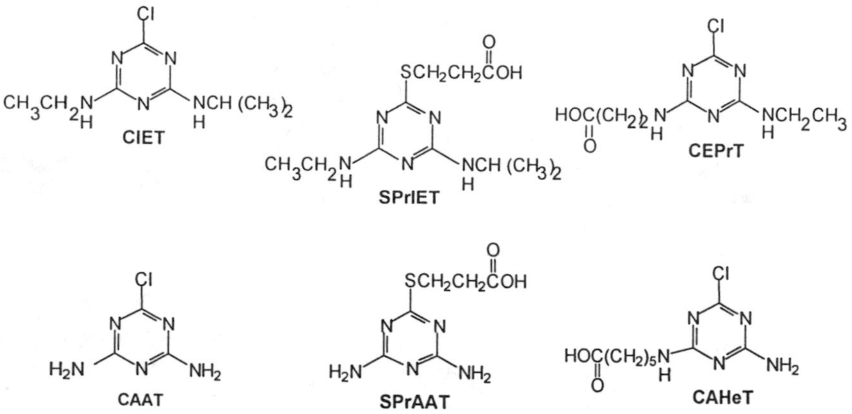

Figure 2. Structures of the target analytes, immunizing haptens, and ELISA haptens used in this study. The immunizing and ELISA haptens used for the analysis of atrazine (CIET) and chlorodiamino-*s*-triazine (CAAT) were SPrIET and CEPrT, and SPrAAT and CAHeT, respectively.

two chlorines of cyanuric chloride, one with an ethylamino and the other with a aminopropanoic acid. CAHeT was synthesized by the substitution of a single chlorine of dichloroamino-*s*-triazine by an aminohexanoic acid. The single chlorine of CAAT was substituted with thiopropanoic acid resulting in the formation of SPrAAT. All of the structures were verified by infrared, mass, and NMR spectral methods. Details of the various methods and spectral data can be found in Goodrow et al. (20) and Muldoon et al. (18).

Hapten-protein conjugates were synthesized by via N-hydroxysuccinimide activated esters of the various carboxylic acid haptens. These were reacted with protein in aqueous solution to form the conjugates (21). SPrAAT was conjugated to keyhole limpet hemocyanin (KLH) for use as an immunogen for the production of antibodies which recognize CAAT (18). The heterologous haptens CEPrT and CAHeT were conjugated to alkaline phosphatase (AP) for use in the *s*-triazine herbicide ELISAs and the CAAT ELISA, respectively.

Antibodies. Mouse monoclonal *s*-triazine-specific antibodies AM7B2, AM1B5, and SA5A1 (primary antibodies), were donated by Dr. A.E. Karu, University of California at Berkeley. Monoclonal antibody production, screening, assay development, and application were previously described (16). The hybridoma cell culture supernatants and ascites fluid preparation (AM7B2) were used without further purification.

Immune polyclonal ascites fluid was produced using KLH-SPrAAT as an immunogen by an adaptation of a previously described method (22). Antibody screening and ELISA development for the analysis of CAAT has been (18). The preparations used here were designated PAb 1 and PAb 5 (primary antibodies).

ELISA Procedure. Antibody screening and ELISA format optimization for the *s*-triazine herbicide ELISAs and the CAAT ELISA have been described in detail elsewhere (17, 18). The ELISA format was adapted from Karu et al. (16). Dilutions of the various immunoreagents used in each procedure were determined by 3-dimensional checkerboard titration. Briefly, microtiter plates were coated with goat anti-mouse IgG (trapping antibody) diluted in coating buffer (15 mM sodium carbonate, pH 9.6), incubated 18 hrs at 4°C, and then washed with phosphate-buffered saline (pH 7.5) containing Tween 20 and sodium azide (PBSTA). One hundred microliters of a predetermined amount of primary antibody diluted in 0.5 mg/mL BSA in PBSTA were applied to the plate, incubated for 60 min, and frozen with the liquid remaining in the wells. The plate was thawed and washed when needed.

For the *s*-triazine herbicide ELISA, 40 μL of sample was mixed with 200 μL of a predetermined dilution of CEPrT-AP in a separate uncoated well. Fifty microliter aliquots were applied to replicate wells of the antibody-coated plate and incubated 30 min; the plate was then washed. Enzyme substrate (*p*-nitrophenyl phosphate) was added and the plate OD measurements (405 nm) were made at 30 min using a Molecular Devices ThermoMax microplate reader (Menlo Park, CA) controlled with SoftMax software.

For the CAAT ELISA, 100 μL of sample was mixed with 100μL of a predetermined dilution of CAHeT-AP in a separate uncoated well. Fifty microliter aliquots were applied to replicate wells of the antibody-coated plate and incubated 30 min; the plate was then washed. Enzyme substrate was added and the plate OD measurements were made at 60 min.

ELISA Characterization. ELISAs were characterized for cross reactivity toward various selected s-triazines. Eleven concentrations of each compound plus a zero dose control were assayed in replicate. Cross reactivites were expressed as IC_{50} values (concentration of analyte which produces a 50% decrease in the maximum normalized response) and were interpreted relative toward atrazine for the s-triazine herbicide ELISAs or CAAT for the CAAT ELISA according to the formula:

$$\% \text{ Reactivity} = (IC_{50} \text{ atrazine or CAAT} / IC_{50} \text{ analog}) * 100 \qquad (1)$$

Reactivity coefficients for pertinent analytes (IC_{50} atrazine or CAAT / IC_{50} analog) were used for calculating the expected summed response of the immunoassays to analyte mixtures (17).

Multianalyte ELISA Analysis of Pesticide Waste and Rinsate. The experimental approach was described in Muldoon et al. (17). It is based on the premise that the observed response of antibody binding to ligands present in a sample as measured by immunoassay (eg. ELISA), is a "summed response" to all the reactive ligands. This summed response is modified by each reactive ligand's "reactivity coefficient" toward the antibody. Therefore the observed ELISA response would follow the equation:

$$\text{ELISA Response} = A(X_A) + B(X_B) + C(X_C) + ... + Z(X_Z), \qquad (2)$$

where A, B, C, and Z are concentrations of the different analytes, and X_A, X_B, X_C, and X_Z, are the reactivity coefficients of the analytes A, B, C, and Z, respectively for that particular antibody. The ELISA response is expressed in the units used for the standard curve for one analyte (eg. atrazine, reactivity coefficient = 1.00), therefore, reactivity coefficients for the other components would be relative to the analyte used in the standard curve (see previous section). By using one antibody (one equation) for each cross-reactive analyte in the mixture, it is possible to solve simultaneous equations to derive quantities of each analyte. The results from the analysis of a sample containing three cross-reactive analytes using three different antibody ELISAs were written in equation form as follows:

$$\text{ELISA Response Ab 1} = A(X_{A1}) + B(X_{B1}) + C(X_{C1}) \qquad (3)$$

$$\text{ELISA Response Ab 2} = A(X_{A2}) + B(X_{B2}) + C(X_{C2}) \qquad (4)$$

$$\text{ELISA Response Ab 3} = A(X_{A3}) + B(X_{B3}) + C(X_{C3}) \qquad (5),$$

where, in equation 3, ELISA Response Ab 1 is the amount determined by ELISA using Ab 1, expressed in units of the standard curve (ie., μM atrazine); A, B, and C are the unknown concentrations of the analytes A, B, and C, and X_{A1}, X_{B1}, and X_{C1} are the known reactivity coefficients for antibody 1 for the analytes A, B, and C. Respective designations are also given to equations 4 and 5. The three equations were solved simultaneously for the unknown concentrations of analytes A, B, and C by matrix inversion.

Pesticide waste and rinsate samples were fortified with additional amounts of atrazine, simazine, or cyanazine in order to establish a larger analyte concentration range for analysis. Samples were diluted in acetonitrile for HPLC analysis and further diluted in PBSTA and analyzed by ELISA using antibodies AM7B2.1, AM1B5.1, and SA5A1.1.

Concentrations of s-triazine were initially calculated as μM atrazine equivalents (based on the standard curve for atrazine) using the two lowest sample dilutions which gave an OD value within the working range of the assay. Individual s-triazines in each sample were quantified using the analyte reactivity coefficients for each antibody and solving three simultaneous equations (one per antibody) with three unknowns (one per analyte) by matrix inversion. Individual single antibody ELISAs were evaluated by geometric mean regression (23) of the amount found by ELISA on the expected response to total s-triazine determined by HPLC utilizing the individual antibody/analyte reactivity coefficients. Estimation of individual and total s-triazines in the samples were evaluated by geometric mean regression of the amount found by ELISA after solving simultaneous equations, on the amount determined by HPLC.

Ozonation Experiments. Bench scale (250 mL) and pilot scale (208 L) ozonation experiments were conducted on 100 mg/L atrazine solutions of Aatrex Nine-O. Ozone was generated using a PCI Model GL-1B (PCI Ozone Corporation, West Caldwell NJ) with oxygen feed. Ozone was delivered at a rate of 1.0 L/min at 1.0 and 3.0 % w/w (O_3/O_2) for the bench scale and pilot scale reactions, respectively. Samples were purged with nitrogen to remove residual ozone prior to analysis and analyzed by HPLC either undiluted or diluted 1:2 with acetonitrile (for atrazine, deethylatrazine, and deisopropylatrazine analysis of early samples). Samples were diluted in PBSTA and analyzed by the ELISAs.

For the bench scale reaction, the pH was adjusted to 10.5 with the addition of 1 N NaOH and maintained at pH 9.5 to 10.5 throughout the reaction. Ozonation was monitored by HPLC and was carried out until atrazine was converted to CAAT (150 min).

The pH was not adjusted for the pilot scale reaction. Ozonation was monitored by HPLC and was continued until all atrazine was converted to CAAT and the acetoamido intermediate CDAT (18.5 hr). The acetoamido product was hydrolyzed to CAAT by fortifying the solution to 2 mM KOH resulting in a pH value of 10.7. After 3 hr, the pH was adjusted to pH 6.6 by fortification to 5 mM KH_2PO_4.

Biodegradation of Ozonated Aatrex. Biodegradation of ozonated Aatrex was conducted on a bench scale (150 mL) to demonstrate the effectiveness of the CAAT ELISA for monitoring this process. Ozonated Aatrex was fortified to 10 mM phosphate buffer (pH = 7.0), 0.1 % w/v Tru-Sweet high fructose corn syrup (American Fructose-Decatur, Decatur, AL), 0.5 mM $MgCO_3$, 50 μM $CaCO_3$, 50 μM $MnSO_4$, and 5 μM $FeCl_3$. This was inoculated with 20 mL of a culture of *Klebsiella terragena* strain DRS-1-S (Klett 660 = 30 units) (15). Samples were removed from the flask and centrifuged to remove cellular material. An aliquot was injected immediatley on HPLC and the remainder frozen. Biodegradation was carried out until CAAT was no longer detected by HPLC (< 50 ppb). Samples were thawed and diluted in PBSTA prior to ELISA analysis.

Results

Hapten Chemistry. Figure 2 shows the structures of the two target analytes, the haptens used for the generation of analyte-specific antibodies, and the haptens used in the heterologous ELISAs. The 2-chloro positions of atrazine (20) and CAAT (18) were substituted with a thiopropanoic acid bridging group and resulted in structures which were used as immunizing haptens. Heterologous haptens (different than that used for animal immunization) were synthesized for use in the ELISA procedures and were extremely valuable in improving assay sensitivity for both s-triazine herbicide ELISAs (24, 17) and CAAT ELISAs (18). The use of heterologous haptens in ELISAs was previously shown to be important for improving assay sensitivities for other analytes (25, 26).

ELISA Development and Characterization. The application of immunoassays for monitoring treatment required characterization of the various assays using all of the pertinent structures which may be present during the course of the process. In addition to these structures, other s-triazines were tested in order to fully characterize and compare antibody crossreactivities. Table I shows the reactivity profiles for the monoclonal antibodies AM7B2, AM1B5, and SA5A1, and the polyclonal antibodies PAb 1 and PAb 5. Although s-triazine herbicide-sensitive assays showed varied reactivities toward the parent structures, analyte recognition was greatest for propazine in all cases. The addition of oxygen to the alkyl side chains greatly diminished monoclonal antibody recognition and this was an important consideration for the application of these assays in monitoring s-triazine herbicide ozonation. Loss of the alkylamino groups resulted in diminished recognition for the s-triazine herbicide-sensitive assays. Monoclonal antibody AM7B2 showed the broadest range of sensitivity for the parent herbicides atrazine, simazine, and cyanazine. This assay was chosen for use in disposal monitoring since it was a broad-spectrum s-triazine herbicide assay and demonstrated insensitivity toward the reaction products.

The ELISAs using polyclonal antibodies PAb 1 and PAb 5 showed high selectivity toward the environmental degradation and ozonation product CAAT. These antibodies did not recognize parent herbicides, other dialkylamino side chain substituted s-triazines, nor the monodealkylated product deethylatrazine (CIAT). The antibodies did recognize the monodealkylated product deisopropylatrazine (CEAT), however this is a minor intermediate in atrazine ozonation (14). The ELISA using PAb 1 was chosen over PAb 5 for use in disposal monitoring since it was less sensitive toward the ozonation intermediate 2-chloro-4-acetoamido-6-amino-*s*-triazine (CDAT).

Multianalyte ELISA Analysis of Pesticide Waste and Rinsate. The monoclonal antibodies AM7B2, AM1B5, and SA5A1 possessed different reactivity profiles for the parent *s*-triazine herbicides atrazine, simazine, and cyanazine (Table I). These differences were utilized in a multiple regression method for estimating individual *s*-triazines in mixtures. Table II shows the geometric mean regression data from the analysis of actual pesticide waste mixtures. The most accurate individual assay employed antibody AM1B5, which was the most selective antibody for atrazine. The less selective assays (AM7B2 and SA5A1) gave lower slopes of regression (underestimates of analytes by ELISA) which may have been caused by a potent interfering material present in the waste samples. The samples were analyzed without sample clean-up. Samples required a minimum 100-fold dilution for ELISA analysis (versus 2-fold for HPLC) which may have magnified any subsampling error initially present. The variability of the ELISA data was greatest at high analyte concentrations (17). Atrazine estimation was the most accurate and precise among the analytes studied and probably resulted from the high selectivity of the ELISA which utilized antibody AM1B5. The estimations of the other *s*-triazines were dependant on less selective assays complicates these estimates. Total *s*-triazine estimation was highly correlated to HPLC data and should be valuable for estimating theoretical yields in disposal processes.

Karu et al. (27) presented a summary of the various statistical methods available for analyzing multianalyte ELISA data. The multiple regression method was used because of its simplicity and ability (theoretically) to be applied to analyte mixtures.

In a related study, the effects of selected agricultural waste components on the ELISA which utilized monoclonal antibody AM7B2 were examined. This ELISA was found to be particulary sensitive to magnesium, ionic surfactants, and some commercial formulated surfactants (28). A simple solid-phase extraction technique (C_{18}) was used which improved assay precision but had little effect on improving the slopes for regression (approx. 0.90) of the amount detected by ELISA on the amount added.

Atrazine Ozonation Monitoring. The ELISAs used for ozonation monitoring utilized either monoclonal antibody AM7B2 or polyclonal antibody PAb 1. Results from these assays were compared to results obtained from a multiresidue HPLC method. Samples were taken from either a bench scale ozonation reaction (250 mL) carried out at pH 10 or a pilot scale reaction (208 L) carried out without pH control. Figure 3 shows the reaction product profile determined by HPLC analysis compared to the results from ELISA for the bench scale reaction. For the *s*-triazine herbicide ELISA, geometric mean regression of the amount found by the ELISA (atrazine equivalents) on the amount found by HPLC had a slope of 1.15 and R = 0.99. For the CAAT ELISA, geometric mean regression

Table I. Percent Reactivities of Various Antibodies to Toward Selected s-Triazines[a].

Common Name	Cook System[b]	AM7B2	AM1B5	SA5A1	PAb 1	PAb 5
atrazine	CIET	100	100	100	0.2	0.2
simazine	CEET	34.3	7.9	86.8	0.2	0.2
cyanazine	CENT	90.4	0.3	6.2	0.2	0.3
propazine	CIIT	227.8	481.8	124.0	n.d.[c]	n.d.
ametryne	SMeIET	14.8	70.6	0.9	n.d.	n.d.
atratone	OMeIET	2.5	6.6	0.4	n.d.	n.d.
hydroxyatrazine	OIET	8.8	1.4	0.0	n.d.	n.d.
deethylatrazine	CIAT	0.5	10.1	13.3	2.2	3.7
deethylsimazine	CEAT	0.5	0.6	28.2	92.8	92.0
chlorodiamino-s-triazine	CAAT	0.0	0.0	0.1	100	100
N-isopropylammeline	OIAT	0.1	0.2	0.0	<0.06	0.2
N-ethylammeline	OEAT	0.2	0.0	0.0	<0.06	0.1
N-ethylammelide	OOET	0.1	0.0	0.0	0.6	0.3
ammeline	OAAT	n.d.	n.d.	n.d.	0.8	0.5
ammelide	OOAT	n.d.	n.d.	n.d.	0.1	<0.04
cyanuric acid	OOOT	n.d.	n.d.	n.d.	n.i.[d]	0.04
melamine	AAAT	n.d.	n.d.	n.d.	1.0	1.2
cyromazine	CyPrAAT	n.d.	n.d.	n.d.	1.0	0.6
diamino-s-triazine	HAAT	0.1	0.0	0.1	1.1	0.6
	CDDT	8.9	26.1	6.4	1.1	1.0
	CDIT	10.6	3.4	6.7	1.1	1.0
	CDET	0.1	0.0	5.7	2.3	2.0
	CDAT	273.6	46.0	4.2	n.d.	15.7
	SPrIET	1.1	5.4	1.8	n.d.	n.d
	ClPrT	0.9	0.2	1.5	n.d.	n.d
	CEPrT	n.d.	n.d.	n.d.	n.d.	n.d
	SPrAAT	n.d.	n.d.	n.d.	140.5	723.8
	SBeAAT	n.d.	n.d.	n.d.	242.1	345.1
	CAHeT	n.d.	n.d.	n.d.	7.7	5.7
	SAAT	n.d.	n.d.	n.d.	35.8	103.6

IC_{50} values (concentration of analyte which produces a 50% decrease in the maximum normalized response) and were interpreted relative toward atrazine for the *s*-triazine herbicide ELISAs (AM7B2, AM1B5, and SA5A1) or CAAT for the CAAT ELISAs (PAb 1 and PAb 5). IC_{50}'s were determined by assaying a zero-dose control (PBSTA) and 11 concentrations of each analog and deriving the value from the 4-parameter logistic curve fitting function. Each concentration was assayed in duplicate or triplicate wells of an antibody coated plate. [a] % Reactivity = (IC_{50} atrazine or CAAT / IC_{50} analog) * 100. [b] T= *s*-triazine ring, H= hydrogen, C= chlorine, A= amino, O= hydroxyl, S= thio, I= isopropylamino, E= ethylamino, D= acetoamido, N= cyanoisopropylamino, CyP= cyclopropylamino, SMe= thiomethyl, OMe= oxymethyl, SPr= thiopropanoic acid, SBe= thiobenzoic acid, He= aminohexanoic acid, Pr= aminopropanoic acid. Refer to Figures 1 and 2 for examples. [c] n.d., not done. [d] n.i., no inhibition. Data was obtained from references 17-19.

Table II. Parameters from the Geometric Mean Regression of ELISA Results on HPLC Results from Analysis of Pesticide Waste and Rinsate (39 samples).

Antibody	Slope[a]	Standard Error[b]	Intercept[c]	R
AM7B2	0.84	0.07	-10.17	0.88
AM1B5	0.98	0.06	-38.55	0.86
SA5A1	0.82	0.07	-13.90	0.86
Analyte				
atrazine	1.00	0.05	-39.72	0.94
simazine	0.66	0.05	-5.00	0.90
cyanazine	1.19	0.08	-98.08	0.92
total s-triazines	0.83	0.07	-7.80	0.85

[a]Slope for geometric regression of results of ELISA on HPLC. [b]Standard error of the slope (μM) from least squares estimates. [c]y-Intercept value (μM). Data was obtained from reference 17.

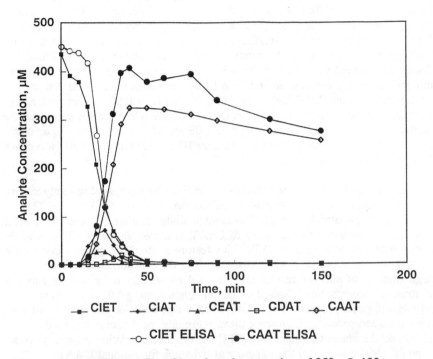

Figure 3. Reaction profile of bench scale ozonation of 250 mL 100 ppm Aatrex carried out under alkaline conditions as measured by HPLC and the ELISAs. (Reproduced with permission from reference 19. Copyright 1994 American Chemical Society).

of the amount found by the ELISA (CAAT equivalents) on the amount found by
HPLC had a slope of 1.23 and R = 0.98. To test for assay selectivities, the
regressions were repeated using only atrazine or CAAT concentrations detected by
HPLC (versus a summed response) and this did not change the correlations
(atrazine ELISA R = 0.99, CAAT ELISA R = 0.97). This indicated that the
ELISAs can be used to measure these two analytes exclusively in the treatment
process due the high analyte selectivities of the antibodies used in the ELISAs.

Similiar results were obtained when the assays were used for monitoring a pilot
scale reaction (19). The pH of the reaction mixture decreased from 6.5 to 3.8
during the course of the reaction and the acetoamido intermediates CDDT and
CDAT were more stable under these conditions. Quantitative conversion to
CAAT was accomplished by alkaline hydrolysis of these products. Again, results
from the ELISA analysis of this process were highly indicative of atrazine and
CAAT concentrations determined by HPLC. Geometric mean regression of the
amount found by the s-triazine herbicide ELISA (atrazine equivalents) on the
amount detected by HPLC had a slope of 1.06 and R = 0.99. Geometric mean
regression of the amount found by the CAAT ELISA (CAAT equivalents) on the
amount detected by HPLC had a slope of 1.08 and R = 0.95. The use of only
atrazine or CAAT concentrations detected by HPLC for regression did not change
the correlations found.

Biodegradation of Ozonated Atrazine. The ELISAs were used to analyze the
biodegradation of ozonated atrazine by *Klebsiella terragena* sp. DRS-1. This
organism was isolated from municipal sewage sludge and was unique in its ability
to utilize the s-triazine ring nitrogen of CAAT as a sole source of nitrogen in the
presence of ammonia nitrogen (15). This feature made this organism particulaly
suitable for use in pesticide waste disposal since these materials often contain
large amounts of nitrogen fertilizers. Figure 4 shows results from the ozonation
of atrazine (bench scale) followed by biodegradation using DRS-1 to show the
complete disposal process. The CAAT ELISA was used for the analysis of the
biodegradation process. Geometric mean regression of the amount found by
ELISA on the amount detected by HPLC for the biodegradation process gave a
slope of 0.903 and R = 0.98. There was evidence for residual CAAT
degradation during the freeze-thaw process as evidenced by low ELISA recovery
in comparison to HPLC. For all of the data depicted in Figure 4 (n = 43), the
geometric mean regression equation for the amount found by the two ELISAs on
the amount detected by HPLC was Y = 1.12 X - 7.13, standard error = 0.038
μM, R = 0.95.

Discussion and Conclusions

This project involved the utilization of existing antibodies in an ELISA for
parent s-triazine herbicides to first characterize the mixed waste samples and the
development of antibodies and ELISAs for CAAT in order to monitor the
complete disposal process by immunoassay. The treatment process considered
here required monitoring for the degradation of parent substrates and, in the case

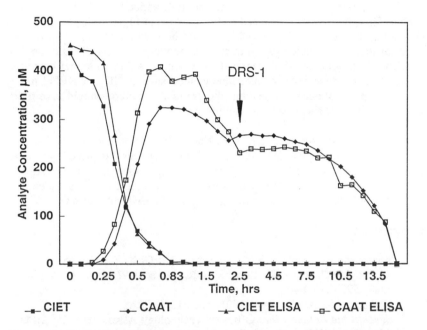

Figure 4. HPLC and ELISA results for atrazine ozonation and CAAT biodegradation monitoring. (Reproduced with permission from reference 19. Copyright 1994 American Chemical Society).

of chloro-*s*-triazine herbicides, the accumulation of CAAT. When this was complete, the material was subjected to biological degradation and monitored for the loss of intermediates.

A simple multiple regression technique was used to analyze the results from multiple crossreactive immunoassays for the *s*-triazine herbicides. It was found to be suitable for estimating individual and *total s*-triazines in mixtures which is important for the initial characterization of actual pesticide wastes prior to their disposal.

Two ELISAs were used for monitoring the disposal of atrazine. Together, they were shown to be very accurate and precise for the quantitation of atrazine and CAAT. This was possible due to the utilization of antibodies which were highly selective for either parent herbicides or CAAT. From Figure 4 it can be seen that information from either HPLC or ELISA could be used for process control, ie. the termination of ozonation (when atrazine is converted to CAAT) and the initiation and termination of biodegradation (when CAAT is no longer detected). Coupled with a conventional multiresidue method such as GLC or HPLC for initial and final waste characterization, the use of this method should save time and expense in *s*-triazine treatment monitoring.

Acknowledgments. We thank Dr. A.E. Karu, Hybridoma Facility, UC Berkeley, for generously donating the monoclonal antibodies used in this project and Drs. Cathleen J. Hapeman and Jeffrey S. Karns, Natural Resources Institute, ARS, USDA, Beltsville, MD, for their assistance. Contribution No. 8836, Scientific Article No. A6620, of the Maryland Agricultural Experiment Station.

Literature Cited.

1.   Aharonson, N.; Cohen, S.Z.; Drescher, N.; Gish, T.J.; Gorbach, S.; Kearney, P.C.; Otto, S.; Roberts, T.R.; Vonk, J.W. *Pure Appl. Chem.* **1987**, 59, 1419-1446.
2.   Graham, J.A. *Anal. Chem.* **1991**, 63, 613A-622A.
3.   *Beneath the Bottom Line;* Office of Technology Assessment. US GPO Washington, DC. 1990. pp 79.
4.   Seiber, J.N. *Proceedings: National Workshop on Pesticide Waste Disposal.* EPA/600/9-87/001, 1987; pp 11-19.
5.   Dwinnell, S.E. *National Symposium on Pesticide and Fertilizer Containment: Design and Management.* Midwest Plan Service. Ames, IA., 1992; pp 5-12.
6.   Ferguson T.D. *Proceedings of International Workshop on Research in Pesticide Treatment/Disposal/Waste Minimization.* EPA/600/9-91/047, 1991, 193 pp.
7.   Seiber, J.N. In *Pesticide Waste Management. Technology and Regulation.* Bourke, J.B., Felsot, A.S., Gilding, T.J., Jenson, J.K., Seiber, J.N., Eds.; ACS Symposium Series 510; American Chemical Society: Washington, D.C. 1992: pp 138-147.

8.   Hammock, B.D.; Mumma, R.O. In *Recent Advances in Pesticide Analytical Methodology;* Harvey, J., Jr., Zweig, G. Eds.; ACS Symposium Series 136; American Chemical Society: Washington, D.C., 1980; pp 321-352.

9.   Sherry, J.P. *Crit. Rev. Anal. Chem.* **1992**, 23, 217-300.

10.   Bushway, R.J.; Perkins, B.; Savage, S.A.; Lekousi, S.J.; Ferguson, B.S. *Bull. Environ. Contam. Toxicol.* **1988**, 40, 647-654.

11.   Thurman, E.M.; Meyer, M.; Pomes, M.; Perry, C.A.; Schwab, A.P. *Anal Chem.* **1990**, 62, 2043-2048.

12.   Kearney, P.C.; Muldoon, M.T.; Somich, C.J.; Ruth, J.M.; Voaden, D.J. *J. Agric. Food Chem.* **1988**, 36, 1301-1306.

13.   Somich, C.J.; Kearney, P.C.; Muldoon, M.T.; Elsasser, S. *J. Agric. Food Chem.* **1988**, 36, 1322-1326.

14.   Hapeman-Somich, C.J.; Gui-Ming, Z.; Lusby, W.R.; Muldoon, M.T.; Waters, R. *J. Agric. Food Chem.* **1992**, 40, 2294-2298.

15.   Leeson, A.; Hapeman, C.J.; Shelton, D.R. **1993**, *J. Agric. Food Chem.* 41, 983-987.

16.   Karu, A.E.; Harrison, R.O.; Schmidt, D.J.; Clarkson, C.E.; Grassman, J.; Goodrow, M.H.; Lucas, A; Hammock, B.D.; White, R.J.; Van Emon, J.M. In *Immunoassays for Trace Chemical Analysis: Monitoring Toxic Chemicals in Humans, Foods, and Environment;* Vanderlaan, M., Stanker, L.H., Watkins, B.E., Roberts, D.W., Eds.; ACS Symposium Series 451; American Chemical Society: Washington, D.C. 1991: pp 59-77.

17.   Muldoon, M.T.; Fries, G.F.; Nelson, J.O. *J. Agric. Food Chem.* **1993**, 41, 322-328.

18.   Muldoon, M.T.; Huang, R.N.; Hapeman, C.J.; Fries, G.F.; Ma, M.; Nelson, J.O. *J. Agric. Food Chem.* **1994**, 42, 747-755.

19.   Muldoon, M.T. and Nelson, J.O. *J. Agric. Food Chem.* **1994**, 42, 1686-1692.

20.   Goodrow, M.H.; Harrison, R.O.; Hammock, B.D. *J. Agric. Food Chem.* **1990**, 38, 990-996.

21.   Langone, J.J.; Van Vunakis, H; *Res. Comm. Chem. Pathol. Pharmacol.* **1975**, 10, 163-171.

22.   Lacy, M.L.; Voss, E.W. *J. Immunol. Meth.* **1986**, 87, 169-177.

23.   Sokal, R.P.; Rohlf, F.J. In *Biometry*; Freeman: New York, 1981. 2nd Ed. Chapter 14.

24.   Harrison, R.O.; Goodrow, M.H.; Hammock, B.D. *J. Agric. Food Chem.* **1991**, 39, 122-128.

25.   Wie, S.I.; Hammock, B.D. *J. Agric. Food Chem.* **1984**, 32, 1294-1301.

26.   Harrison, R.O.; Brimfield, A.A.; Nelson, J.O. *J. Agric. Food Chem.* **1989**, 37, 958-964.

27.   Karu, A.E.; Lin, T.H.; Breiman, L.; Muldoon, M.T.; Hsu, J. *Food Agric. Immunol.* **1994**, submitted.

28.   Muldoon, M.T. and Nelson, J.O. *Food Agric. Immunol.* **1994**, accepted.

RECEIVED January 9, 1995

Chapter 23

Evaluation and Application of Immunochemical Methods for Mycotoxins in Food

Mary W. Trucksess[1] and Donald E. Koeltzow[2]

[1]Division of Natural Products, Center for Food Safety and Applied Nutrition, U.S. Food and Drug Administration, Washington, DC 20204
[2]Federal Grains Inspection Service, U.S. Department of Agriculture, Kansas City, KS 64153

Immunoassays have been developed for several mycotoxins including aflatoxins, deoxynivalenol, zearalenone, and fumonisins. These assays can determine such analytes in a variety of matrices and provide rapid analyses of a large number of test samples. Commercial immunochemical test kits are often evaluated on the basis of sensitivity, specificity, reproducibility, cost, time stability, and ease of use. Laboratory quality assurance checks, such as standard curves and positive controls, are essential. It is also important to differentiate interferences due to matrix effects from high levels of analyte. The criteria for interpretation of results from yes/no tests and quantitative tests are presented, as well as results of collaborative studies of immunochemical methods for aflatoxins and zearalenone in grains and grain products and surveillance findings obtained by using these methods for aflatoxins, deoxynivalenol, zearalenone, and fumonisins. Some of the criteria used to evaluate mycotoxin immunoassay procedures and experience in using them as surveillance tools can serve as models for similar approaches to the determination of pesticide residues in foods.

Mycotoxins are toxic secondary metabolites produced by fungi in agricultural commodities. The fungi can invade grains growing in the field and during storage. Their ability to parasitize the plants depends on many factors, such as density of inoculum, plant species, prevailing temperature, tissue damage, insect activity, moisture, and harvesting practices. Many fungi produce mycotoxins, some of which are toxic to humans and animals. The mycotoxins found in significant quantities in naturally contaminated foods and feeds include the aflatoxins, deoxynivalenol (DON), zearalenone, and the fumonisins. These toxins can be present in corn, peanuts, cottonseed, tree nuts, cereal grains (wheat, barley, rice, oats), and many other commodities. Aflatoxins, a group of structurally related mycotoxins identified in the early 1960s, are well known for their acute and chronic toxicity in animals and humans (*1*). The major aflatoxins of concern are B_1, B_2, G_1, G_2, and M_1. DON is also known as vomitoxin from its effect on swine (*2*). Zearalenone is an estrogenic metabolite; it causes feed refusal and hyperestrogenism in swine (*3*).

The fumonisins, a new class of mycotoxins, were isolated and characterized in 1988. They are produced primarily by *Fusarium moniliforme* and *F. proliferatum*. *F. moniliforme* is the most common mold found on corn in the United States. Currently, seven fumonisins have been identified: FB_1, FB_2, FB_3, FB_4, FA_1, FA_2, and FC_1 (*4,5*). FB_1 causes brain damage in horses and lung edema in pigs. Fumonisins have been suggested as the possible cause of human esophageal cancer in South Africa (*6*).

Analysis for Mycotoxins

Analysis for mycotoxins is essential to minimize the consumption of contaminated food and feed. The problem is not simple. Determining the concentrations of toxins in grains at the parts-per-billion levels required for the most important mycotoxins is difficult. A systematic approach is necessary. The approach generally followed consists of obtaining a relatively large sample, reducing it in bulk and particle size to a manageable quantity, and finally performing the analysis. Sampling commodities for aflatoxin contamination follows the U.S. Department of Agriculture (USDA) recommendations. Laboratory samples of at least 5-25 kg of corn, milo, and other grains are collected (*7*). To prepare a representative test portion for analysis, the laboratory sample is ground and mixed, so that the concentration of toxin in the test portion is the same as in the original laboratory sample collected. For pelletized feed a 1 kg laboratory sample is adequate, because the mycotoxins in the individual contaminated ingredients have presumably been uniformly distributed during feed manufacture. Similarly, a smaller laboratory sample is adequate for processed, comminuted, and mixed foods.

Selection of Methods. For analysis we use authentic toxin standards and available methods, selecting appropriate methods for particular needs. The following criteria should be considered in selecting a method: number of analyses, time, location, cost of equipment, safety, waste disposal, and availability of experienced analysts. The new antibody-based immunochemical methods are simple, specific, and sensitive. Because mycotoxins are low molecular weight compounds, they do not independently induce an immune response when injected into laboratory animals; in most cases, they must first be derivatized and then conjugated to a carrier protein. The preparation of a suitable immunogen is an important step in the production of a specific antibody to a mycotoxin and in the development of an immunoassay.

Immunochemical Methods

Three major immunochemical techniques have been developed for mycotoxin determination: radioimmunoassay (RIA), enzyme-linked immunosorbent assay (ELISA), and immunoaffinity column assay (8). The first two methods are based on competition between a free mycotoxin and a labeled mycotoxin for an antibody binding site. In the immunoaffinity column assay the antibodies are bound covalently to beaded agarose. The affinity column is used to bind the analyte to the antibody-agarose packing and serves as a concentration tool for the analyte. After elution, the analyte is subjected to further procedures. At present the ELISA and the immunoaffinity column assay techniques are more commonly used than the RIA. Solid foods are typically extracted with aqueous methanol and diluted with water before analysis to maintain the native protein structure of the antibody and enzyme conjugate.

Commercial Immunoassay Kits. Many commercial immunoassay kits for mycotoxins in agricultural products are being marketed. These kits provide the reagents, materials, and instructions necessary to perform the tests. A list of some of the manufacturers and types of immunoassay kits is found in Table I. The kits are intended for rapid qualitative identification or quantitative determination. This listing includes only those applicable to mycotoxins for which our laboratory has received descriptive information, and is not intended to be all-inclusive.

Criteria for Evaluation of Commercial Immunoassay Kits

At present there are no standard criteria for evaluating commercial immunoassay kits. Several organizations have been actively engaged in developing evaluation guidelines: Association of Official Analytical Chemists International (AOACI), U.S. Environmental Protection Agency (EPA), U.S. Department of Agriculture (USDA), and U.S. Food and Drug Administration (FDA). Table II summarizes some of the criteria used by the USDA-FGIS (Federal Grains Inspection Service) to evaluate immunoassays for aflatoxin.

Initial Evaluation. In FDA laboratories, requested information obtained from manufacturers is evaluated with emphasis on sensitivity, applicability, stability, clarity of instructions, quality control, comparison with reference method (if possible), cost, and equipment. An initial choice is made after this review is completed. Kits are purchased and critically evaluated.

A slightly different approach is used in FGIS laboratories. A notice is published in the Federal Register requesting that all manufacturers of test kits capable of detecting the analyte of interest submit information on test kit capabilities. These data are reviewed according to criteria similar to those described for FDA laboratories. Test kits representing each of the different analytical methods available (well, cup, column, etc.) are examined in the laboratory to provide experience in using the various test methodologies. Kits that meet the basic FGIS timeliness and safety requirements shown in Table II are purchased and collaboratively tested in several FGIS field inspection laboratories.

Table I. Commercially Available Mycotoxin Immunoassay Kits

Mycotoxin	Company	Type
Aflatoxins	Environ. Diag. Sys.	card
	Idexx	well, probe
	Idetak	well
	Inter. Diag. Sys.	cup
	Neogen	well
	Vicam	column
Deoxynivalenol	Environ. Diag. Sys.	card
	Neogen	well
Fumonisins	Neogen	well
	Vicam	column
Zearalenone	Environ. Diag. Sys.	card
	Neogen	well

**Table II. Design Criteria and Test Performance Specifications for
Quantitative Aflatoxin Test Kits**

1.   Time required for completion: 30 min.
2.   Capability of analyzing for B_1, B_2, G_1, and G_2.
3.   Applicability: corn, corn meal, etc.
4.   Acceptable accuracy, precision limits.

Aflatoxins added ng/g	Accuracy ng/g	Precision ng/g
0	≤ 7.0	4.0
10	± 8.0	6.0
20	± 10.0	8.0
30	± 12.0	10.0
320	± 60.0	32.0

5.   Should not include toxic solvents and reagents.
6.   Comparative accuracy of test kits on corn samples naturally
     contaminated with aflatoxins at about 20 and 100 ng/g.
7.   The limit of detection of test kits, ≤ 5 ng/g.
8.   Insensitivity to temperature change, 18-30°C.
9.   Stability data to support expiration date.
10. Free of matrix interference.

In-Depth Evaluation. FDA laboratories use the same systematic approach to evaluate immunochemical and traditional analytical methods. This approach includes estimation of accuracy, precision, sensitivity, and specificity for quantitative methods, whereas only sensitivity and specificity apply to qualitative methods. Two sets of test samples are used: the control and the naturally contaminated commodity. These test samples are analyzed by using a reference method, i.e., a method which has been evaluated by our laboratory as well as in a method performance collaborative study. The control test sample must be free of the mycotoxin of interest or contain a level at or below the limit of detection; the naturally contaminated test sample contains the mycotoxin at a level close to the target, specification, or action level. A third set of test samples, the spiked test samples, are prepared by adding a known amount of analyte to the control. The spiking levels are the target level, half the target level, and one-and-a-half to two times the target level.

In FGIS laboratories, data obtained in the FGIS collaborative studies for each grain matrix of interest are used to establish minimum performance specifications that are possible with the technology being evaluated. Items 4 and 7 in Table II are examples of performance specifications for quantitative aflatoxin test kits in analysis of corn. Again, a public announcement is made that informs the industry of FGIS's intentions to initiate official testing of the analyte of interest. These requirements include the analysis of both spiked and naturally contaminated test samples. For quantitative test kits, the levels of mycotoxins are evenly distributed across the test range of interest. For qualitative test kits (those that provide a yes/no or positive/negative result), target levels are similar to those used by the FDA. Performance claims are spot-checked in FGIS laboratories. Those test kits that meet all performance requirements and verification tests are approved for use in the official grain inspection system.

Quantitative Immunoaffinity Column Methods. Performance of the method is assessed on the basis of results of analyses of the three sets of test samples. The first quantitative method evaluated by the FDA was the immunoaffinity column assay of aflatoxins in corn, peanuts, and peanut butter (9). Results of the study indicated that the accuracy and precision of the method were suitable: recovery of the added aflatoxin was equal to 97-131% with a within-laboratory relative standard deviation of <20%. The high relative standard deviation (70%) for the control level was due to the extremely small amount of aflatoxin (<2 ng/g) in peanuts and peanut butter. At the limit of detection the standard deviation would be expected to be higher. The sensitivity and specificity were demonstrated by confirming that the control contained aflatoxins at <2 ng/g. The method also showed good correlation with the reference method.

This method was further evaluated by an international collaborative study. All 11 participating laboratories from seven countries produced acceptable results (10). The method was used to analyze 336 samples of commercial peanuts and peanut butter for aflatoxins. Only two samples of peanuts were found to contain aflatoxin above the action level of 20 ng/g. The final test of acceptability of any method is satisfactory performance when the method is used by typical analysts in actual practice.

The immunoaffinity column method for fumonisin B_1 (FB_1) has been evaluated in a similar manner. A collaborative study will be conducted in the near future. The method was used to conduct a survey of FB_1 in canned corn and frozen sweet corn (11). Results indicated that FB_1 was present in sweet corn products destined for human consumption. For the 1993 crop year, a moderate number (36%) of the sweet corn samples tested were found to be contaminated by a low level (4-350 ng/g) of FB_1.

Direct Competitive ELISA Methods. These methods are susceptible to matrix interferences. Most interfering substances act either by affecting antibody recognition of analyte or by modifying the activity of the enzyme label. It is important to check for matrix interference. Two approaches were used to validate the ability of the Veratox test kit by Neogen to measure deoxynivalenol in grains and grain products. First, the standard curve of DON in water was found to be similar to a standard curve of DON in blank-matrix extract. Second, the mass of analyte found in an assay was plotted against the volume of extract analyzed. The graph showed good linearity between the amount of deoxynivalenol versus the volume of extract analyzed. This method was used to analyze more than 630 samples of wheat and barley collected from midwestern areas in 1993 (12). The average DON contamination in the 483 wheat samples was 1.2 μg/g, ranging from 0 to 18 μg/g. The average DON contamination in the 147 barley samples was 2.7 μg/g, ranging from 0 to 26 μg/g. This was not unexpected; the unusually wet weather in many areas of the midwestern United States in 1993 provided favorable conditions for proliferation of *F. graminearum,* which produces DON.

One recent event illustrates the importance of good analytical practice. In 1992 a load of raisins exported from the United States to Greece was analyzed for aflatoxins by an ELISA method developed for grains and was found to contain 60 ng aflatoxins/g. The raisins had been analyzed before shipment by a liquid chromatographic (LC) method which indicated that no aflatoxin was present. To settle the dispute, the shipment was reanalyzed in England by another LC method which confirmed the original finding of no aflatoxin contamination. All commercial kits always need to be checked for performance by using the specific commodity. It is important to do a comparative study to demonstrate what the new assay will measure with respect to a reference method. Since aflatoxins are not commonly found in raisins, the use of a mass spectrometric technique to confirm the identity of the isolated toxins was absolutely necessary. As a rule, the use of a chemical derivatization method to confirm the identity of aflatoxins in corn, peanuts, cottonseeds, and pistachio nuts is recommended, even though aflatoxins are often found in these commodities.

Qualitative Immunochemical Methods. Performance of the qualitative methods, the yes/no or positive/negative tests, is commonly assessed by the sensitivity and specificity of the tests in correctly classifying test samples as either positive or negative at a certain target level. Sensitivity is defined as the ability to identify positive materials as positive at some target level. Specificity is defined as the ability to identify negative materials as negative below the target level. Usually at least 15 test samples are used at each spiking level. If there is no overlap in test results between positive and negative, the test can identify all test samples correctly, i.e., distinguish the +/- categories correctly. However, if the test results for the +/- categories overlap somewhat, the test does not distinguish them perfectly.

Operating Characteristic Curves. A perfect qualitative test exhibits no incorrect results (identifying positive test samples as negative at some target level and identifying negative test samples below the target level as positive). However, tests are rarely perfect. Each test has a particular response pattern that is a function of concentration. Statistically this can be determined by the use of operating characteristic (OC) curves. The OC curve plots the positive rate or percentage positive [true positive/(true positives + false negative)] as a function of concentration. The false positive rate [false positive/(true negatives + false positive)] is also a function of concentration. The false positive rate is usually highest as the concentration approaches the target level. Good performance of a test is characterized by a high true positive rate (>90%) at the target level of 20 ng/g of aflatoxin. A collaborative study was conducted to validate a yes/no test for zearalenone in corn, wheat, and feed at 500 ng/g (*13*). The OC curve shows a low positive rate (75%) at a target level of 500 ng/g. Therefore, the assay failed at the target level. At 800 ng/g, the positive rate was 96%. Subsequently the assay was recommended for adoption as a screening method for zearalenone at ≥800 ng/g in corn, wheat, and feed.

Future of Immunochemical Methods

Immunoassays are simple, specific, and rapid and can be performed with minimum training. They are gaining acceptance and the confidence of analytical chemists. They are competing successfully with traditional analytical methods because they are evaluated by the same criteria and follow the same quality assurance plan, including preparation of standard curves and checking for recovery and repeatability of results in analyses of spiked test samples. Immunoassays are being used more and more as analytical tools for monitoring the presence or absence of particular hazardous residues in agricultural commodities. They also are of great value when used in combination with existing thin layer, gas, and liquid chromatography methods. When the result obtained for a new commodity by immunoassay is positive or above the target level, it should be confirmed by other reference methods.

Literature Cited

(1)   Cullen, J.M.; Newberne, P.M. in *The Toxicology of Aflatoxins*; Eaton, D.L.; Groopman, J.D. Eds.; Academic Press, Inc.: San Diego, CA, 1993, pp 3-26.
(2)   Scott, P.M. in *Trichothecene Mycotoxicosis: Pathophysiologic Effects*, Beasley, V.R., Ed.; CRC Press: Boca Raton, FL, 1989, pp 1-26.

(3)   Tanaka, T.; Hasegawa, A.; Yamamoto, S.; Lee, U.-S.; Suguira, Y.; Ueno, Y. *J. Agric. Food Chem.* **1988**, *36*, 979.

(4)   Cawood, M.E.; Gelderblom, W.C.A.; Vleggaar, R.; Behrend, Y.; Thiel, P.G.; Marasas, W.F.O. *J. Agric. Food Chem.* **1991**, *39*, 1958.

(5)   Branham, B.E.; Plattner, R.D. *J. Nat. Prod.* **1993**, *56*, 1630.

(6)   Sydenham, E.W.; Thiel, P.G.; Marasas, W.F.O.; Shephard, G.S.; Van Schalkwijk, D.J.; Koch, K.R. *J. Agric. Food Chem.* **1990**, *38*, 1900.

(7)   Food and Drug Administration, *Inspection Operations Manual*, Sample Schedule, Chart 6, Washington, DC, 1983.

(8)   Chu, F.S. *Vet. Hum. Toxicol.* **1990**, *32*, 42.

(9)   Trucksess, M.W.; Page, S.W. in *Biodeterioration Research*; Llewellyn, G.C.; O'Rear, E.O., Eds., 3, Plenum Press: New York, NY, 1990, 161.

(10)   Trucksess, M.W.; Stack, M.E.; Page, S.W.; Albert, R.H. *J. Assoc. Off. Anal. Chem.* **1991**, *74*, 81.

(11)   Trucksess, M.W. *J. AOAC Int.*, in press.

(12)   Trucksess, M.W. *J. AOAC Int.*, in press.

(13)   Bennett, G.A. *J. AOAC Int.*, in press.

RECEIVED December 28, 1994

Chapter 24

Immunodetection of Ecosystem Contaminants
Research, Application, and Acceptance in Canada

James P. Sherry

Ecosystem Conservation Branch, National Water Research Institute,
Canada Center for Inland Waters, 867 Lakeshore Road, P.O. Box 5050,
Burlington, Ontario L7R 4A6, Canada

Immunoassay (IA) techniques for ecosystem contaminants are poised to play a key role in Canadian environmental programmes. IAs shall probably first become established in routine laboratories where they would be ideal for screening out negative samples from large sample sets. Several Canadian groups have worked on the development of IAs for pesticides and halogenated hydrocarbons such as dioxins and PCBs. The availability of IA kits from commercial sources has made the technology more widely access-ible. Some favourable Canadian validation studies have been completed and a number of agencies have trial studies in progress. It is expected that several leading laboratories shall soon offer an IA option to their clients. A strategy to promote IA techniques in Canada should include improved communications between interested analysts. Analysts must be made aware of both the strengths and weaknesses of immuno-techniques. The credibility of immuno-techniques shall be established through rigorous validation studies. Quality control and assurance programmes shall allow analysts and scientists to have confidence in their IA data. Immuno-techniques' unique blend of sensitivity, low cost, and small sample needs make them powerful research tools.

Interest in the use of antibody (AB) based techniques for the detection of ecosystem contaminants, has grown markedly in recent years, despite a slow start following Ercegovich's (1) introduction of such techniques in the early 70's. Numerous immunoassays (IAs) and related techniques have been developed for a broad range of pesticides and many contaminants of industrial origin (*2-5*). IAs have a proven track record in the clinical laboratory where they have helped to transform analytical strategies. The development of the early IAs (*6*) coincided with an emergent demand for sensitive analytical methods for a variety of difficult to analyze hormones and pharmaceuticals which resulted from advances in the disciplines of biochemistry and physiology, and increased access to quality health care. There was, and is, constant pressure in clinical laboratories to shorten sample turn-around time without compromising data quality; pressures that are familiar to the environmental analyst in Canada, and elsewhere.

The level of current interest in immuno-techniques for ecosystem contaminants can be gauged from the number of recent publications that describe IAs and other antibody based techniques (for reviews see *3-5, 7*). Several reviewers have concluded that, whereas IAs may not be a panacea for all analytical problems, they should flourish as screening techniques that can free the analyst and his instruments for more demanding and challenging tasks. Many analysts and their managers, however, remain sceptical as to whether immuno-techniques can realize their potential.

With its highly developed agricultural, forestry, mining, and industrial sectors, Canada suffers from many of the environmental problems that afflict other advanced economies. Canadian society, through its elected representatives, is committed to the conservation and protection of its extensive fresh water resources and their associated ecosystems (*8*). Responsibility for the ecological well being of Canada's international and interprovincial boundary waters is shared by the Government of Canada with both the Government of the United States (U.S.) and the appropriate Provincial Governments. Environment Canada, the Department of Fisheries and Oceanography, and the Provincial Ministries of the Environment share responsibility for studying and monitoring the effects of pollutants on the health of other aquatic systems. Local Government agencies work in cooperation with their provincial counterparts to ensure the quality of recreational and drinking waters. The extent and variety of Canada's ecosystem resources means that the various agencies face formidable analytical challenges. For example, technology is needed to determine a wide variety of analytes and their degradation products in large numbers of varied environmental samples.

The high cost of many organic analyses has encouraged efforts to develop alternative techniques that will lower costs and improve efficiency. Commonly in monitoring studies a large proportion of the sample sets are analyte free or contain undetectable or meaningless levels of analyte. A screening strategy that could identify those negative samples and help to prioritize the positive samples would help to reduce costs and improve analytical services.

Immuno-techniques have emerged as a realistic screening option because they are versatile, easy to use, and relatively inexpensive. Most modern IAs for ecosystem contaminants use enzyme tracers to provide the quantification signal. The long shelf lives, ease of distribution, and suitability for field use of enzyme based IAs (EIAs) make them particularly attractive to environmental analysts. Figure 1 outlines an IA based screening. As indicated in the diagram, IAs are intended to complement, not replace, conventional techniques. After all samples

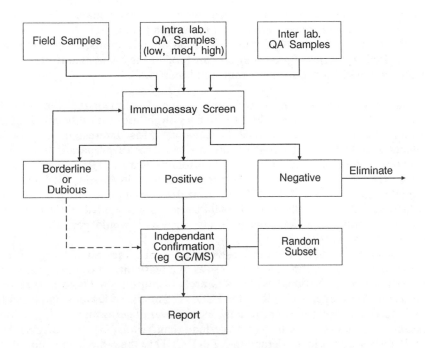

Figure 1.   Outline of an IA screening strategy.   (Reproduced with permission from reference 5.   Copyright 1992 CRC Press).

are processed through the IA, the positive samples and a random selection of negative samples should be confirmed by an independent technique, such as gas chromatography (GC) or high performance liquid chromatography (HPLC), although that requirement can be relaxed in some cases (5). The present chapter suggests that several factors, including maturity of the technology, analytical demand, and escalating costs indicate that IA screening techniques are poised to play an important role in Canadian environmental programmes.

Immunodetection Research in Canada

Through the 80's a small number of Canadian researchers maintained an active interest in environmental IAs. Researchers at Health and Welfare Canada under the leadership of Dr. H. Newsome made a major contribution to IA technology by developing a series of promising IAs for food borne contaminants that can be readily adapted for use with environmental matrices.

Contaminants of Industrial Origin. Table I summarizes the IAs for polychlorinated biphenyls (PCBs) and polychlorinated dibenzo-p-dioxins (PCDDs) that have been studied in Canada. Most interest in IAs for the detection of PCBs has focused on the various Aroclor preparations as target analytes. Newsome and Shields' (9) radioimmunoassay (RIA) was targeted against the higher chlorinated PCBs that predominate in Aroclor 1254 and 1260. Dimethyl sulfoxide (DMSO) (25 %) proved to be superior to the non-ionic surfactant Cutscum for the critical solubilization of the analyte molecules. The assay's performance at low analyte levels suggests it would perform well with environmental matrices.

Dioxins are notoriously expensive to determine: the cost of analysis ranges from $1000 - $1500 per sample. The results of a feasibility study (10) encouraged researchers at the National Water Research Institute (NWRI) to pursue their interest in Albro et al.'s (11) RIA for PCDDs. The use of DMSO simplified the solubilization process and improved the assay's overall performance (12). DMSO was, however, more prone to matrix overload than Triton (13). The original RIA for PCDDs used ^{125}I-iodovaleramido-3,7,8-T$_3$CDD as the radioligand. Polyclonal (PAB) and monoclonal (MAB) versions of the assay have been modified for use with tritiated 2,3,7,8-T$_4$CDD (14, 15), primarily to avoid the tricky iodination reaction. The assay's performance at low analyte levels was improved by lowering the levels of tracer and ABs.

In an effort to improve the sensitivity of the IA for PCDDs, Gerry Reimer of CanTest in British Columbia (16) has developed an interesting time resolved fluorescent IA (TRFIA) for PCDDs. Fluorescent labels can be measured easily, rapidly, and precisely using modern instruments. The ability to resolve the emitted signal over time minimizes the influence of background interferences and improves assay sensitivity. Rare earth chelates are particularly suited to time TRFIA because of their long signal decay times. A preliminary version of the assay was developed for the detection of atrazine. The PCDD assay, which uses biotinylated hapten and an immobilized second antibody to trap the anti-dioxin

Table I. Some Canadian Research on IAs for Contaminants of Industrial Origin

Analyte	Format	Binder	Working Range	DL	I_{50}	Ref.
PCDDs 2,3,7,8-T$_4$CDD	RIA[125I]	PABs	GC$_5$ DMSO: 20 pg - 2 ng	27 pg		12
PCDDs 2,3,7,8-T$_4$CDD	RIA[3H]	PABs	GC$_5$ DMSO: a: 20 pg - 2 ng b: 2.5 pg - 200 pg	21 pg 3.9 pg	350 pg 42 pg	14
PCDDs 2,3,7,8-T$_4$CDD	RIA[3H]	MABs	GC$_5$ DMSO: a: 25 - 1600 pg b: 10 - 100 pg	19 pg 6 pg	200 pg 42 pg	15
PCDDS 2,3,7,8-TCDD	TRFIA	MABs	0.1 - 50 ng / mL in well	70 pg/well	190 pg/well	16
PCBs	RIA[125I]	PABs	Aroclor 1260: 100 pg - 3 ng	100 pg	400 pg	9

SOURCE: Data are from listed references.

MABs, shows promise and is presently being optimised for improved sensitivity - the performance data in Table I are provisional. The strong binding affinity of the avidin-biotin complex helps to enhance the assay signal. It is planned to interface the assay with a supercritical fluid extraction (SFE) system for use with soil and pulp mill effluents.

Herbicides. Table II summarizes the IAs for herbicides that have been developed in Canada. Hall et al. used herbicide derivatives coupled to [³H]-glycine as the radioligands in selective RIAs for picloram and 2,4-dichlorophen-oxyacetic acid (2,4-D) (17). Anti-picloram MABs were later produced and used in an effective EIA (18). Both MAB and PAB based versions of the EIA were more sensitive than the earlier RIA. Neither assay cross reacted appreciably with 2,4-D or with a variety of pyridine herbicides. The Guelph group has also developed an IA for the detection of metolachlor in water and soils (19).

The IA group at Health and Welfare Canada have developed ABs to the phenylurea (carbamide) herbicides (20). The coating antigens were made heterologous with respect to the immunogen so as to weaken the AB-binding reaction and to allow the analyte molecules to compete for binding sites. A combination of serum and coating antigen was selected that yielded workable assays for monolinuron, diuron, and linuron.

Fungicides. Newsome's group have produced a series of IAs for fungicides such as benomyl, metalaxyl, triadimefon, and iprodione (22 - 25). The assays have been mainly used to measure residues in food. The EIA for metalaxyl improved analytical efficiency by a factor of 4.5 when used to screen food samples for metalaxyl. The assay's broad specificity could make it useful as a screening tool.

Application and Validation Studies.

Once an assay has been developed and optimized its ability to accurately and precisely recover the target analyte from a variety of matrices must be validated. The availability of commercial IA kits, has encouraged a minor boom in the number of Canadian laboratories that are actively evaluating the IA option. Table III provides an overview of those validation studies.

The RIA for PCDDs was able to detect about 70 ppt of 2,3,7,8-TCDD in 300 mg equivalents of tissue of a Lake Trout extract. A further improvement in the detection limit (DL) would probably require an increase in the sample size, which would probably require an extra clean-up step, such as chromatography on carbon fibre (10, 13). The assay's DL must be improved if it is to be used in routine applications. The RIA for PCBs was able to detect as little as 20 ppb of Aroclor 1260 in milk and 2 ppb in blood (9). On average the RIA estimates were lower than those of a gas liquid chromatography (GLC) confirmation method, although the data from the two methods were well correlated. The ability of Agri-Diagnostics' BTEX kit (benzene, toluene, ethyl benzene and o-, m-, and p-xylene) to recover spikes of the aromatic components of petroleum from water and soil samples suggest that it could be used to assess petroleum

Table II. Some Canadian Research on IAs for Herbicides

Analyte	Format	Binder	Working Range	DL	I_{50}	Ref.
2,4-D	EIA	PABs	100 ng/mL - 10 μg/mL	100 ng/mL	2 μg/mL	17
2,4-D	RIA (³H-glycine)	PABs	50 ng/mL - 10 μg/mL	50 ng/mL	1 μg/mL	17
Picloram	RIA (³H-glycine)	PABs	50 - 5000 ng/mL	50 ng/mL	760 ng/mL	18
Picloram	EIA	PABs	5 - 5000 ng/mL	5 ng/mL	140 ng/mL	18
Picloram	EIA	MABs	1 - 200 ng/mL	1 ng/mL	10 ng/mL	18
Carbamide	EIA	PABs	Monolinuron: 0.08 - 5 ng/mL Diuron: 0.08 - 5 ng/mL Linuron: 0.5 - 50 ng/mL		0.9 ng/mL 1.6 ng/mL 12 ng/mL	20
Imazmethabenz	EIA	PABs	0.5 - 32 ng/mL		12.6 ng/mL	21
Atrazine	TRFIA	PABs		0.08 ng/mL		29

SOURCE: Data are from listed references.

Table III. Overview of Validation Studies

Agency/Research Group	Analyte	Matrix	Clean-up	Ref.
NWRI	2,3,7,8-T$_4$CDD	fish	gel permeation chromatography, acid/base treatments, alumina chromatography	13
Health & Welfare	PCBs	blood & milk	solvent extraction, chromatography on neutral alumina	9
Environment Canada, Emergencies Science Division	BTEX	water & soil	none	26
NWRI	atrazine	river, lake, surface run-off water	none	27
University of Guelph	atrazine, metolachlor, 2,4-D	precipitation and surface water	none	28
CanTest	atrazine	creek, river, marsh water	none	29
University of Guelph/Agriculture Canada	metolachlor	soil & water	filter water samples extract soil with methanol, dilute with PBS	19
University of Guelph	picloram, 2,4-D	water	none	17, 18

Organization	Analyte	Matrix	Sample preparation	Reference
Quebec Ministry of the Environment	atrazine, total triazines	water	none	30
Agriculture Canada (Harrow Research Station)	atrazine, metolachlor	water	none	31
Ontario Ministry of Agriculture (Guelph Laboratory)	triazines & 2,4-D	surface & well water, groundwater	SPE on some samples	32
Health & Welfare Canada	methyl 2-benzimidazole carbamate	food crops	reflux in ethyl acetate, filter	22-25
	benomyl	foods	reflux in ethyl acetate, filter	
	metalaxyl	foods	homogenize in methanol, filter	
	triadimefon	foods	homogenize in solvent, filter	
	iprodione	foods	homogenize in ethyl acetate, filter	
Agriculture Canada (Food Production & Inspection Branch)	atrazine, cylcodienes	cornmeal & corns	none	33, 34

SOURCE: Data are from listed references.

spills or contaminated sites, such as decommissioned gasoline stations (26). For some water and soil samples, however, the IA tended to underestimate the higher level spikes. That discrepancy may have been caused by loss of the analyte through volatilization, or by inadequate solubilization of the analyte (26).

IMS's tube based assay was used to estimate atrazine levels in a variety of surface waters from various parts of Canada (27). The IA and a gas chromatography-nitrogen phosphorous detector (GC-NPD) technique compared well in their abilities to estimate atrazine levels in a set of 124 samples (R=0.92). The sample load was reduced by 71% at a cut-off threshold of 1 ng/mL. There were 2.4% false negatives and 0.8% false positives in the sample set. An evaluation of the ability of Ohmicron's magnetic particle based IA for the detection of atrazine yielded a correlation coefficient of 0.98 when spiked samples were analyzed by GC and EIA techniques. The EIA tended to overestimate the atrazine levels in a set of 24 field samples, although the results were still satisfactory for screening purposes (R=0.89). Interferences in the creek water samples caused some loss of accuracy for CanTests' TRFIA for atrazine at low spike levels, and the assay had a slight positive bias with the other water samples; overall, however, the TRFIA's ability to recover atrazine was satisfactory. The TRFIA's performance is now being compared to that of an EIA, with the same ABs being used in each assay.

EIA and GC techniques yielded comparable results when used to study the persistence of metolachlor in soil (19). Hall et al.'s IAs for picloram and 2,4-D have also been successfully validated using spiked and unspiked soil and water samples (17, 18). The assays proved robust and, with the exception of the PAB based version of the EIA for picloram, were generally resistant to matrix interferences.

Several key performance parameters such as susceptibility to likely interferences, tendency to generate false results, and the ability to detect atrazine and total triazines in water samples were used in a favourable evaluation of two IA kits for the detection of triazine herbicides (30). Both kits compared well to a conventional gas chromatography/mass spectrometry (GC/MS) method. A double blind study design was used to test the selected kit's ability to recover atrazine, total triazines, and weighted total triazines herbicides from a set of 350 environmental water samples in a follow-up study (30). A GC/MS method was used as a reference. The weighted triazine parameter took into account the ABs cross reactivity profile. The correlation coefficients for the linear regression of the EIA data on the GC-MS data were > 0.95 and the slope values of the regression lines were close to 1 which suggests a high level of accuracy. The Quebec team that undertook the foregoing research has now proceeded to evaluate the ability of IAs to recover PCBs from soils.

Two of IA technology's strong points, their excellent sensitivity and the ability to assay small volume samples, are being exploited in an innovative study of the distribution patterns and dynamics of herbicide residues in surface run-off water, soil pore water, and ground-water (31). IAs' small sample size require-ments allow the researcher to trace the herbicides' movement within non-saturated soils and to study adsorption/desorption kinetics in the field. The IA is being used to verify some laboratory developed fate and transportation models for both atrazine and metolachlor. In this case IA technology is facilitating a study that would be difficult to undertake using conventional techniques.

Clegg (*32*) has assessed the abilities of IA kits from Agri-Diagnostics and Millipore to screen water samples for triazine herbicides and 2,4-D (*32*). Both kits were favourably compared with a GLC reference technique (n>200). In a follow up study that is in progress solvent extraction, solid phase extraction (SPE) on C18 columns, and IA based methods are being used to screen 1300 samples for triazine herbicides. Clegg and Harris (*32*) have also compared the abilities of EIA, RIA, and GLC-techniques to measure 2,4-D in river water and urine. The methods performed well with fortified samples, although the RIA, gave some false positive results with river water samples. Although generally pleased with the IAs' performances Clegg's laboratory has yet to commit to routine use.

Wigfield and Grant (*33*) used Millipore's microtitre plate EIA to detect atrazine in spiked cornmeal and corns. The atrazine spikes were quantitatively recovered from the spiked extracts over the assay's working range (93 - 117%). The authors recommend that the calibration curve be prepared in matrix blanks to correct for slight interference effects. The same group has also evaluated the ability of Millipore's Res-I-Mune kit to detect cyclodiene insecticide residues in some food products (*34*).

Newsome and co-workers have compared the abilities of EIA and liquid chromatography (LC)/GC based methods to recover several fungicides from a variety of foods and agricultural produce. In most cases there was close agreement between the IA and the reference method.

Growing Acceptance?

There are signs of growing interest in environmental IAs among Canadian analysts. For example, Environment Canada's Laboratory Management Committee, apparently in response to demands from clients, has begun to seriously consider the potential of IA screening tests. Environment Canada's National Laboratory for Environmental Testing (NLET), after some trial experiments, is currently offering an IA screening service, based on commercially available assays, to client laboratories on a trial basis. The Laboratory Services Branch (LSB) of the Ontario Ministry of the Environment and Energy recently appointed a committee to assess the usefulness of environmental IAs. The committee's report will form the basis of a decision on whether or not the LSB becomes more actively involved with IA screening strategies (*35*). While favourably disposed to the concept of IA screening tests, the Laboratory Services Branch of Agriculture Canada have yet to use IAs for the routine screening of foods mainly because the methods they evaluated require some clean-up of the sample and the preparation of calibration curves in blank matrix (*36*). Some preliminary performance data for Ohmicron's magnetic particle based EIA for captan suggest that assay is less prone to matrix effects and may be a promising prospect for routine use (*36*). The Quebec Ministry of the Environment is also considering whether or not offer an IA screening option to its clients. IAs may also be of value to municipal laboratories for the analysis of drinking water, river water, and recreational water samples. The Hamilton-Wentworth Regional Laboratory, for example, is considering the use of IA techniques as part of their rapid response to emergencies that may threaten drinking water (37).

Despite those encouraging signs there is still substantial resistance in Canada to the use of IA screening tests. Given the positive nature of most of the published validation studies it is likely that much of that resistance is due to factors other than assay performance (5, 38). One of IA's drawbacks is the difficulty and cost of developing a fully fledged and certified IA. Until recently it was difficult to obtain key assay reagents other than as gifts via the research network. The efforts of private sector companies and their Canadian distributors such as Millipore (Biomann), Agri-diagnostics (Beak), and Ohmicron (Kalyx) have solved the distribution problem at least for those analytes for which assay kits are available.

Some resistance may also stem from a well founded trust in classical methods combined with a scepticism about the ability of biological molecules to match the performance of state of the art high performance liquid chromatography (HPLC) and GC systems. A compounding factor is a tendency to view IAs as a threat to the quest to equip laboratories with the latest and best in GC/MS/HPLC hardware. That attitude is based on the false notion that IAs can replace existing analytical methods, and that the elimination of negative samples from sample sets will lead to a loss of work. On the contrary, the desire to advance knowledge and to improve our understanding of ecosystem contaminants should ensure that savings from the use of screening techniques will likely be used to enhance the quality of environmental studies through more intensive sampling of contaminated sites, broader environmental surveys, or additional research and monitoring programmes.

Some Key Challenges

Several challenges must be met if IAs are to make the transition from research tools to widely used analytical methods.

Challenge 1: Improve Distribution of Key IAs. The many IAs that have been developed will remain of limited interest and value unless the key reagents, or kits made from those reagents, become widely available to the analytical community. The distribution of IA reagents has improved greatly in recent years. In Canada, assay kits have become the preferred route of introduction for many analytical laboratories that lack the expertise and resources to develop their own assays.

Challenge 2: Build Awareness. There is a need to promote an understanding of IA screening strategies within the wider analytical community. Analysts and environmental programme managers also need to become more aware of the benefits and shortcomings of IA screening techniques.

Challenge 3: Establish Clear Performance Criteria. Confidence in individual IAs will be stronger if it is demonstrated that the assay meets clearly defined, and widely accepted performance criteria. Those criteria should describe the technique's DL, selectivity, working range, and precision. The ability of IAs to accurately and precisely recover analyte molecules from environmental matrices must also be validated, preferably in "round robin" studies. Consensus is needed on how those criteria should be defined and measured.

Challenge 4: Implement Flagship Assays. Several Canadian Laboratories have now validated the performances of commercial IA kits, mainly for the triazine herbicides. Those studies have helped to enhance the credibility of IA screening tests. The stage is set for analytical laboratories to meet the challenge and to offer the IA screening option to their clients - at least on a trial basis.

Promoting Immuno-Detection Techniques in Canada

The availability of a suite of IAs for priority analytes would enable laboratories to offer screening tests as part of their overall analytical service, and would probably tip the balance in favour of environmental IAs. The compilation of a prioritized list of target analytes would facilitate the rational development of candidate assays. High priority analytes should meet several criteria. There should be a genuine long-term demand for data on their occurrence, the analytes should be amenable to detection by IA and, preferably, be difficult to determine by conventional means. Attention should be given to the development of assay kits for emerging pesticides such as Pursuit (imazethapyr-ammonium).

    There are a number of ways to promote the concept of IA screening tests. A blend of successful peer reviewed studies, review articles, and workshops is the track proven approach that is favoured by the scientific community. Undoubtedly the advertising campaigns of the major kit manufacturers have improved product visibility, and are helping to erode some barriers. Given the growing acceptance of environmental IAs as a legitimate analytical tool, it is perhaps timely to consider the introduction of the technique into analytical chemistry courses. IA kits are ideal for class experiments and could be used in conjunction with conventional GC-mass selective detector (MSD) based techniques to provide hands on experience in all components of a modern screening strategy.

    Some idea of the potential scope for IA screening tests in Canadian laboratories can be gleaned from the sample loads of the Ontario Ministry of Environment and Energy which during 1993 analyzed a total of 2600 water samples for triazine herbicides and metolachlor and 800 samples for 2,4-D as part of wider analyte scans (*39*). Commercial IA kits are available for each of those analytes. Based on data from Sherry and Borgmann (*27*) for a set of 124 samples that were analyzed for atrazine, an IA screening step would have saved $7564 at an IA replication level of 3x or $9424 at a 1x replication level - the use of positive controls and matrix blanks is assumed. It should be borne in mind however, that failed GC runs can usually be repeated at little cost to the operating laboratory other than instrument time, whereas, a faulty IA run requires the purchase of a replacement kit from an external source. Hopefully, market pressures shall serve to make the cost of IA kits attractive to potential users.

Ultimately the credibility and acceptance of IA screening tests will be based on proven performance rather than fine promises, no matter how well thought out or presented. The widening distribution of IA kits should facilitate the necessary validation studies. In view of the concerns of many conventional analysts about the ability of ABs to discriminate between similar compounds, it is important to verify an IA's ability to select its target compound from among those likely to occur in the matrix of interest. Any need for sample preparation detracts from an IA's attractiveness as a screening test. Assays for the detection of non-lipophilic analytes usually need little or no preparation of aqueous samples. Solid matrices may require more preparation. Soil, sediment, and biota samples are likely to present more problems than water samples and often require some extraction and clean-up. For that reason it is important that IAs be validated for each environmental matrix.

The achilles heel of a screening test is any tendency to produce false negative results. Once eliminated from the data set the falsely identified sample is lost to the study. Assay performance criteria and assay validation studies must guard against such tendencies. False positives samples, while not desirable, are not as big a problem, since all positive samples should be confirmed by means of an independent technique anyway. Berthold Hock of the Technical University in Munich, recommends that the screening threshold be established in the mid-point of the assay calibration curve so as to minimize false negative samples (40). The Food Safety and Inspection Service (FSIS) of the US Department of Agriculture (41) has wisely recommended that particular attention be paid to points slightly above or below the cut-off value in validation experiments. Once approved for routine use it is critical that a kit's performance be assured by the manufacturer in the long term (42), which means there must be adequate supplies of ABs to meet projected needs.

In Canada, it is the practice to accredit or certify a laboratory based on its ability to meet specified performance standards rather than to restrict the analyst's choices to a range of certified methods. Once certified, laboratories are subject to routine audits, and are also expected to follow good analytical practices. The accreditation process is administered by the Canadian Association of Analytical Laboratories (CAEAL): an industry run and government supported organization. A similar programme could be used to licence IA laboratories if, and when, environmental IAs become more widely used.

Approval of IA screening tests by major agencies in the European Community (EC) or the USA would have a positive effect in Canada. It would help assure Canadian analysts and their managers that a decision to launch an IA screening programme would be perceived as part of a global trend rather than a case of unnecessary risk taking. Similarly the success of an IA programme in Canadian laboratories would probably encourage more serious appraisal of the technique in other countries.

Quality Assurance/Quality Control (QA/QC)

Regulatory bodies, those who contract out analytical work, and laboratories wishing to assure clients of the quality and integrity of their data, each have a vested interest in QA programmes. The assurance of data quality is a mission critical goal for both the analytical laboratory and programme manager alike. Because Canadian laboratories use a wide variety of analytical techniques, not all of which have gone through an approval process, it is imperative that the accuracy, precision, and comparability of data sets be assured. Two approaches are used to ensure the quality of that data: a laboratory accreditation process and participation in regular quality assurance studies. QA programmes are usually externally administered and use carefully prepared samples or certified reference materials (CRMs) in round robin style studies (*43*). The data from such studies identify outlier laboratories, which then have an opportunity to correct their problems. The data also allow analysts to compare their performance with those of peer laboratories. Researchers who contract out analyses may purchase CRMs or round robin samples and use them to assess candidate laboratories. Existing QA protocols could be readily modified for use with IA screening tests. For example, reference samples could be designed to test an assay's tendency to record false negative results.

In-house programmes allow the bench analyst and the laboratory management to control analytical quality. At the bench level it would be prudent to maintain records of key assay parameters so that variations in the reference kits, inter-analyst variations, or assay drift can be promptly noted. Low, medium, and high concentration controls should be included with each assay run. It would also be wise for the analyst to occasionally verify the assay's selectivity, particularly if a long term commitment is being made to a kit from a particular manufacturer.

Prospects

Informal contacts made with several agencies that would be expected to benefit from the use of environmental IAs suggest that IAs will soon find a niche as screening tests in Canadian laboratories. There is also much interest in the application of immuno-techniques to problems that are difficult or expensive to tackle using conventional techniques. The unique qualities of immuno-reagents such as the need for small sample volumes, limited clean-up, and the ability to easily detect biological molecules make them ideally suited to many niche applications. At present most sample sets that are submitted to routine laboratories are usually analyzed for a broad scan of related analytes. Although groups of IAs can be used to detect more than one target analyte by running multiple chemistries on a single plate, that option is costly. The development of true multianalyte IAs would be an important breakthrough since it would permit the detection of non-structurally related analytes in a single assay. It may, also make sense, however, if environmental studies were designed to answer clear hypotheses related to single analytes, or groups of closely related analytes. There is a suspicion that many requests for multianalyte data are related less to a real need for that data than to the known ability of the conventional method to detect the extra compounds. IAs are ideally suited to the more focussed approach.

At present in Canada, there are several groups of IA enthusiasts who work largely in isolation from each other, even when they exist within the same government department, or building, in some cases. Improved communications between those groups would allow a productive cross fertilization of ideas and could also yield a strategy for promoting the use of environmental IAs. A Canadian Network of Environmental Immunoassayists could be the vehicle for such interaction. The network could include interested analysts from Government, Industry, Academic and Private laboratories, and could be instrumental in devising acceptance criteria for IAs, guidelines for validation studies, and appropriate quality assurance and control programmes. Members of the network would be an instant resource base for Canadian analysts who are unfamiliar with environmental IAs.

It is important that an assay's performance be capable of meeting the analytical requirements for the target analyte. Experience at NWRI with the RIA for dioxins is a good example of the dampening effect that can occur when a notoriously difficult ultra-trace analyte is matched up with a reasonably sensitive assay. NWRI's original decision to become involved with an IA for dioxins was based on the phenomenal savings that could accrue from such a screening test. In the interim analytical requirements for dioxins were pushed to the fg level, below the reach of present versions of the RIA.

Tube based EIAs are suitable for field use where they can be used to identify interesting collection sites or to rapidly screen samples in the field - which are important advantages. Tube based IAs can also form part of a suite of techniques for use in environmental emergencies where time is usually of the essence. That format is also well suited to the intermittent screening of small groups of samples.

There is interest in Canada, as elsewhere, in the possible use of SFE techniques in conjunction with IAs for the screening of solid phase matrices such as sediments and biota samples. SFE can also be used to extract analyte molecules that have been pre-concentrated on SPE columns. An IA/SFE system may prove useful for the screening of organisms, such as zebra mussels, that can bio-concentrate contaminants from ambient water and sediment. Such measurements would reflect long term exposure of the sentinel organisms to the contaminants. IAs can also play a lead role in investigations of the fate and transportation of herbicides in the environment. The Ontario Ministry of the Environment and Energy, for example, is considering the use of IAs to study the tendencies of herbicides to accumulate in irrigation waters (44).

Consideration should be given to channelling some Canadian research support into several promising immuno-techniques that have recently emerged. Flow-injection IAs have several attractions including rapidity, good sensitivity, and ease of automation (45). Immuno-probes, particulary those that combine optical fibre technology with fluorescent tracers, are also worthy of attention. Once perfected such probes could be used to continuously monitor effluents or ambient waters. The use of the selective AB-antigen (Ag) reaction to extract hapten molecules from complex solutions may develop into a valuable analytical aid (*46*). IA screening strategies shall succeed in Canada if they are seen to clearly benefit analytical, research, or regulatory programmes. If the major routine laboratories continue to delay their involvement, their clients could avail of commercial IA kits, which are generally easy to use, to screen their own samples. Indeed for many studies that may well be the preferred option. Overall, the prospects for immuno-detection techniques in Canada seem bright. Interest among analysts and programme managers, who until recently were largely indifferent, is growing. There is a sense that many Canadian analysts are waiting for a respected laboratory or agency to break the log jam and begin an IA screening programme. Such a move would encourage other laboratories to become actively involved. The implementation of a fully fledged screening strategy for some key analytes in a routine setting is a crucial test that may well tip the balance in favour of immuno-detection screening tests. Eventually immuno-techniques shall benefit not only the analyst, but also the regulator, and the environmental scientist whose programmes are currently impeded by high analytical costs.

Literature Cited

1. Ercegovich, C.D. In *Pesticide Identification at the Residue Level*; Gould, R.F., Ed.; Advances in Chemistry Series 104, ACS: Washington, DC, 1971; 162-177.
2. Van Emon, J.; Seiber, J.N.; Hammock, B.D. In *Analytical Methods for Pesticides and Plant Growth Regulators*; J. Sherma, Ed.; Advanced Analytical Techniques, Acad. Press Inc.: New York, N.Y., 1989, Vol. XV11; 217-263.
3. Vanderlaan, M.; Watkins, B.E.; Stanker, L. *Environ. Sci. Technol.* **1988**, *22*, 247.
4. Newsome, W.H. *J. Assoc. Off. Anal. Chem.* **1986**, *69*: 919.
5. Sherry, J.P. *CRC Crit. Rev. Anal. Chem.* **1992**, *23*: 217.
6. Yalow, R.S.; Berson, S.A. *Nature* **1959**, *184*, 1648.
7. Ferguson, B.S.; Kelsey, D.E.; Fan, T.S.; Bushway, R.J. *Sci. Total Environ.* **1993**, *132*, 415.
8. *Canada's Green Plan*, Minister of Supply and Services Canada, Ottawa, 1990.
9. Newsome, W.H.; Shields, J.B. *Int. J. Environ. Anal. Chem.* **1981**, *10*, 295.
10. Afghan, B.K.; Carron, J.; Goulden, P.D.; Lawrence, J.; Leger, D.; Onuska, F.; Sherry, J.; Wilkinson, R.J. *Can. J. Chem.* **1987**, *65*, 1086.
11. Albro, P.W.; Luster, M.I.; Chae, K.; Chaudhary, S.K.; Clark, G.; Lawson, L.D.; Corbett, J.T.; McKinney, J.D. *Toxicol. Appl. Pharmacol.* **1979**, *50*, 137.
12. Sherry, J.P.; ApSimon, J.W.; Collier, T.L.; Albro, P.W. *Chemosphere* **1990**, *20*, 1409.
13. Sherry J.P.; ApSimon, J.; Collier, T.; Afghan, B.; Albro, P. *Chemosphere* **1989**, *19*, 255.
14. Sherry, J.P.; Albro, P.W. Radioimmunoassay of Chlorinated Dioxins: use of [^{3}H]-Labelled 2,3,7,8-Tetrachlorodibenzo-p-dioxin as Radioligand. Presented at *Dioxin '90* Bayreuth, Germany. September 1990.

15. Sherry, J.P.; Stanker, L.; Watkins, B.; Vanderlaan, M.; Albro, P. Monoclonal Antibody Based Radioimmunoassay for the Detection of Polychlorinated Dibenzo-p-dioxins. Presented at *Dioxin '91*, Poster #171, Raleigh Durham, N. C., September 1991.
16. Reimer, G. CanTest, Vancouver, British Columbia, personal communication, 1994.
17. Hall, J.C.; Deschamps, R.J.A.; Kreig, K.K. *J. Agric. Food Chem.* **1989**, *37*, 981.
18. Deschamps, R.J.A.; Hall, J.C.; McDermott, M.R. *J. Agric. Food Chem.* **1990**, *38*, 1881.
19. Hall, J.C.; Wilson, L.K.; Chapman, R.A. *J. Environ. Sci. Health* **1992**, *27*, 523.
20. Newsome, W.H.; Collins, P.G. *Food & Agric. Immunol.* **1990**, *2*, 75.
21. Newsome, W.H.; Collins, P.G. *Bull. Environ. Contam. Toxicol.* **1991**, *47*, 211.
22. Newsome, W.H. *Bull. Environ. Contam. Toxicol.* **1986**, *36*, 9.
23. Newsome, W.H.; Shields, J.B. *J. Agric. Food Chem.* **1981**, *29*, 220.
24. Newsome, W.H. *J. Agric. Food Chem.* **1985**, *33*, 528.
25. Newsome, W.H. In *Pesticide Science and Biotechnology*; Greenhalgh, R.; Roberts, T.R., Eds., Blackwell Scientific: Ottawa, 1987; 349-352.
26. Bissonnette, M. Environment Canada, Emergencies Science Division, Environmental Technology Centre, Ottawa; personal communication, 1994.
27. Sherry, J.P.; Borgmann, A. *Chemosphere* **1993**, *26*, 2173.
28. Hall, J.C.; Vandeynze, T.D.; Struger, J.; Chan, C.H. *J. Environ. Sci. & Health* **1993**, *28*, 577.
29. Reimer, G.J.; Gee, S.J.; Hammock, B.D. Analysis of Atrazine in Fortified Environmental Waters by a Time-resolved Fluorescence Immunoassay (TRFIA). Presented at the 8th International Congress of Pesticide Chemistry, Washington DC, July 1994.
30. Battat, A. Gouvernment du Quebec, Ministere de l'Environment, Direction des Laboratoires, Laval, Quebec, personal communication, 1994.
31. Gainer, J. Agriculture Canada, Research Station, Harrow, Ontario, personal communication, 1994.
32. Clegg, S. Ontario Ministry of Agriculture and Food, Guelph, Ontario, personal communication, 1994.
33. Wigfield, Y.Y.; Grant, R. *Bull. Environ. Contam. Toxicol.* **1993**, *51*, 171.
34. Wigfield, Y.Y.; Grant, R. *Bull. Environ. Contam. Toxicol.* **1992**, *49*, 342.
35. Ahmad, I.; Berg, B. Ontario Ministry of Environment & Energy, Laboratory Services Branch, Etobicoke, Ontario, personal communication, 1994.
36. Chaput, D. Agriculture Canada, Laboratory Services Branch, Ottawa, personal communication, 1994.
37. Smallbone, B. Hamilton Wentworth Regional Laboratory, Hamilton, Ontario, personal communication, 1994.
38. Van Emon, J. M.; Lopez-Avila, V. *Anal. Chem.*, **1992**, *64*, 79A.
39. Crozier, P. Ontario Ministry of Environment & Energy, Laboratory Services Branch, Etobicoke, Ontario, personal communication, 1994.
40. Hock, B. *Anal. Letters* **1991**, *24*, 529.
41. Anon. *J. Assoc. Off. Anal. Chem.* **1989**, *72*, 694.
42. Mastrorocco, D.; Brodsky, M. *J. Assoc. Off. Anal. Chem.* **1990**, *73*, 331.

43. Aspila, K.I.; Alkema, H.; Stokker, Y.D. *Quality Assurance Guidelines for Drinking Water Programs*, National Water Research Institute, Report prepared for the Health Canada Drinking Water Safety Program, NWRI, Burlington, Ontario; 1994.

44. Berg, B. Ontario Ministry of Environment and Energy, Laboratory Services Branch, Etobicoke, Ontario, personal communication, 1994.

45. Gubitz, G.; Shellum, C. *Anal. Chim. Acta* **1993**, *283*, 421.

46. Kim.; B.B.; Vlasov, E.V.; Miethe, P.; Egorov, A.M. *Anal. Chim. Acta* **1993**, *280*, 191.

RECEIVED October 25, 1994

Chapter 25

Panel Discussion

The speakers on regulatory and acceptance issues, P. Schuda, J. Brady, J. Rittenburg, S. Berberich, S. Coates, and M. Trucksess were panelists for a discussion which was moderated by R. B. Wong. The major points of discussion centered upon the data analysis and interpretation issues. Other questions were related to the AOCA Research Institute validation process, EPA's acceptance of immunoassay data, and problems with cross-reactivity as exemplified by the validation of aflatoxin immunoassay kits. The following is a summary of the panel discussion.

According to Dr. Brady, an immunoassay method is an analytical method and so it should satisfy the same criteria as conventional analytical methods must meet. As such, the analyst must provide evidence the limit of detection (LOD) satisfies some statistical criteria to distinguish its response from the response of the zero dose standard in addition to the LOD merely being the smallest standard (one approach to this statistical problem is to use a modified form of Rodbard's work that employs Student's t distribution for a one-sided test at 95% probability). The limit of quantitation (LOQ), by contrast, is determined by achieving satisfactory recoveries in fortified samples of each matrix the assay is intended to be used for. The lowest level at which those recoveries are achieved becomes the LOQ. The LOD thus establishes the lower limit of the standard curve but the LOQ sets the concentration at which sample residues may be quantified. The LOQ is consequently a practical, utilitarian level whereas the LOD is merely the lower end of the standard curve. Depending upon the complexity of the matrix, these levels may not necessarily be the same. A method with an LOD of 0.05 ppb, for example, may achieve satisfactory recoveries at 0.05 ppb in water but fail to do so in soil. Recoveries in soil may be that an LOQ is set at a higher level such as 0.10 ppb.

There were differences in opinion about the logit-log transformation of immunoassay data. Brady took a conservative view that the only useful portion of the sigmodial dose-response curve is the central linear dose-response region. A response corresponding to a sigmodial tail can be extrapolated back to multiple doses on the x-axis, in contrast to

0097–6156/95/0586–0354$12.00/0
© 1995 American Chemical Society

the unique one-to-one dose-response relationships found in the linear central portion. To make the tail appear to have a one-to-one dose response relationships is merely a mathematical manipulation that is unsupported by physical reality. Others in the audience consider logit-log transformation as valid, citing a long history of clinical application and also referenced Baud et al., (1) showing logit-log as the best and most accurate among the methods compared for eight different immunoassays. However, it was also acknowledged that any method of transformation or curve fitting presupposes that the data conform to a particular mathematical model. While data for microplate ELISAs and some other formats are well approximated by the four-parameter logistic equation or the logit-log transformation, data from other formats may be better fitted by other models. The most conservative view is that with any assay format it is necessary to obtain enough data to justify the use of a particular mathematical model and accurately establish the variance and confidence limits at different analyte concentrations.

The AOAC Research Institute (AOAC RI) validation procedure was clarified by Dr. Coates. The validation protocol requires that the kit manufacturer provide all the data in support of the claims. After reviewing the package, the AOAC RI will test the claims and provide a certification if validation is successful. The cost of the procedure is $7,500 per application plus billing to the sponsor for independent testing (cost range from $5,000 to $10,000). Each test kit can include any number of matrices as long as the data for such matrices are supplied at the time of application. Additional fee will be charged if data for additional matrices are provided at a later date. The AOACRI routinely check the performance of the certified kits. Adverse Advisory Statements are sent to the sponsor for unsatisfactory performance and the manufacturer has thirty days for corrective action. The manufacturer is required to send a copy of the package insert annually to the Institute and revise any changes in format, reagents or formulation. The annual maintenance fee for certification is $1,000.

The EPA-Office of Pesticide Programs views immunoassay as a positive step. This office is different in that its methods are performance based and not prescriptive. Registrants using immunoassay methods should approach the agency on an individual basis since different reviewers and their supervisors have different perspectives about the technology and no defined mechanism for immunoassay implementation is available at this time. With the work of the Analytical Environmental Immunochemical Consortium (AEIC) and the Summit Meetings sponsored by the Environmental Monitoring Systems Laboratory in Las Vegas (EMSL-LV), some of the key issues required to accept immunoassay are being resolved. Another important point Dr. Schuda brought up was a training program. While it is useful for reviewers in the agency to attend meetings such as the ACS, it is extremely difficult because of limited travel budget. Some training program will have to be devised which can accomplish the goal of introducing this technology. In response to the question of assay kit certification, the agency views certification of assay kits by recognized groups as an added measure of credibility. The idea of environmental laboratory certification adds a measure of confidence and helps to provide the agency with a "level of comfort" in accepting immunoassay methods and data.

A concern was raised about cross-reactivity in immunoassays, observation by Dr. Trucksess that four aflatoxins cross-reacted in tests that and her colleagues evaluated. Dr. Trucksess pointed out that first of all, the user must make an effort to know or determine the cross-reactivity of the kit. This information is usually provided by the kit manufacturer. Also, there are different populations of aflatoxins in various commodities. For example, in corn, B1 and B2 are most commonly found and B1 is often much greater than B2. In peanuts, all four aflatoxins can be found and in some rare instances, the amount of G 1 level can be greater than that of B1. Therefore, as a rule, when immunoassay shows a positive for total aflatoxin, a second method of analysis will be needed to quantify individual aflatoxins.

Dr. Rubio elaborated on Lawruk et al.'s (2) work which was cited in Dr. Brady's presentation. In addition to the information provided by the publication that the graph (figure 1, page 1427) represented the mean of 68 determinations, Dr. Rubio stated that the error bars represented an average of 68 runs by different analysts using different lots of reagents in a three months period.

1. Baud,M., Mercier, M., and Chatelain, F. J. Clin Lab Invest .51, (Suppl 205), 1991,120-130.
2. Lawruk, T. S.; Lachman, C. E.; Jourdan, S. W.; Fleeker, J. R.; Herzog, D. P.; Rubio, F. M. J. Agric. Food Chem. 1993, 41, 1426-1431.

RECEIVED December 7, 1994

Author Index

Affiliation Index

Subject Index

Production: Susan Antigone
Indexing: Deborah H. Steiner
Acquisition: Anne Wilson
Cover design: Alexander E. Karu & Alan Kahan

Printed and bound by Maple Press, York, PA

Highlights from ACS Books

Good Laboratory Practice Standards: Applications for Field and Laboratory Studies
Edited by Willa Y. Garner, Maureen S. Barge, and James P. Ussary
ACS Professional Reference Book; 572 pp; clothbound ISBN 0–8412–2192–8

Silent Spring Revisited
Edited by Gino J. Marco, Robert M. Hollingworth, and William Durham
214 pp; clothbound ISBN 0–8412–0980–4; paperback ISBN 0–8412–0981–2

The Microkinetics of Heterogeneous Catalysis
By James A. Dumesic, Dale F. Rudd, Luis M. Aparicio, James E. Rekoske,
and Andrés A. Treviño
ACS Professional Reference Book; 316 pp; clothbound ISBN 0–8412–2214–2

Helping Your Child Learn Science
By Nancy Paulu with Margery Martin; Illustrated by Margaret Scott
58 pp; paperback ISBN 0–8412–2626–1

Handbook of Chemical Property Estimation Methods
By Warren J. Lyman, William F. Reehl, and David H. Rosenblatt
960 pp; clothbound ISBN 0–8412–1761–0

Understanding Chemical Patents: A Guide for the Inventor
By John T. Maynard and Howard M. Peters
184 pp; clothbound ISBN 0–8412–1997–4; paperback ISBN 0–8412–1998–2

Spectroscopy of Polymers
By Jack L. Koenig
ACS Professional Reference Book; 328 pp;
clothbound ISBN 0–8412–1904–4; paperback ISBN 0–8412–1924–9

Harnessing Biotechnology for the 21st Century
Edited by Michael R. Ladisch and Arindam Bose
Conference Proceedings Series; 612 pp;
clothbound ISBN 0–8412–2477–3

From Caveman to Chemist: Circumstances and Achievements
By Hugh W. Salzberg
300 pp; clothbound ISBN 0–8412–1786–6; paperback ISBN 0–8412–1787–4

The Green Flame: Surviving Government Secrecy
By Andrew Dequasie
300 pp; clothbound ISBN 0–8412–1857–9

For further information and a free catalog of ACS books, contact:
American Chemical Society
Distribution Office, Department 225
1155 16th Street, NW, Washington, DC 20036
Telephone 800–227–5558